Distributed Energy Storage in Urban Smart Grids

Edited by
Paulo F. Ribeiro and Rafael S. Salles

The Institution of Engineering and Technology

Published by The Institution of Engineering and Technology, London, United Kingdom

The Institution of Engineering and Technology is registered as a Charity in England & Wales (no. 211014) and Scotland (no. SC038698).

The Institution of Engineering and Technology
Futures Place
Kings Way, Stevenage
Hertfordshire SG1 2UA, United Kingdom

www.theiet.org

British Library Cataloguing in Publication Data
A catalogue record for this product is available from the British Library

ISBN 978-1-83953-549-9 (hardback)
ISBN 978-1-83953-550-5 (PDF)

Typeset in India by MPS

Cover image: Sir Francis Canker Photography/Moment via Getty Images

Contents

Biography

Paulo F. Ribeiro received his B.S.E.E. from the Federal University of Pernambuco, Brazil, and the Ph.D. degree from the University of Manchester, Manchester, England, in 1985. He was a Research Fellow with the NASA Glenn Research Centers, Cleveland, OH, Electric Power Research Institute (EPRI), USA, and Erskine Fellow with the University of Canterbury, New Zealand, and with the Brazilian Institute of Electric Energy (INERGE), Brazil. He taught full-time and worked in the industry in the USA for over 24 years. He also taught at the Technological University of Eindhoven, The Netherlands. He spent 8 years as a transmission system planning engineer in Brazil. He is currently a full professor of electrical engineering with the Federal University of Itajubá. He has authored or co-authored more than 350 papers, book chapters, and 5 books. His current research interests include power electronics, and power quality, transmission and distribution systems, renewable and distributed generation, energy storage, smart grids, engineering education, and philosophy of technology. Dr Ribeiro is a fellow of IEEE and IET.

Rafael S. Salles received the M.Sc. degree in electrical engineering from the Federal University of Itajubá, Itajubá, Brazil, in 2021. He is pursuing a Ph.D. degree in electric power engineering with the Electric Power Engineering Group, Luleå University of Technology, Skellefteå, Sweden. His research interests are power quality, energy storage, and smart grids. His primary focus on research has been on power system harmonics, waveform distortion assessment, railway electrification, and power quality in general. He is an active member of IEEE and CIGRE. Currently, he is a CIGRE Next Generation Network (NGN) representative for the Brazilian C4 (Committee of Power System technical performance). He has dedicated time and efforts to promoting good research, also performing publications of book chapters, journal papers, conference papers, and technical reports.
Book Foreword by Arshad Mansoor, EPRI President and CEO.

Foreword

Energy systems around the world are undergoing a fundamental transition to meet economy-wide clean energy targets. As my good friend and former EPRI colleague Paulo Fernando Ribeiro describes it, energy systems are undergoing a "re-enchantment." As international experts in smart grid technology, Dr Ribeiro and his colleague Rafael Salles provide unique insights while examining the impacts on urban distribution energy systems.

Meeting energy grid decarbonization goals will require integrating an increasing number of distributed energy resources (DERs), such as solar and wind. This will require thorough planning, coordination, and most importantly energy storage. Distributed energy storage serves as a flexible and resilient tool for grid planners, operators, and ultimately customers as part of the urban distribution system as well as the bulk energy system.

In *Distributed Energy Storage in Urban Areas*, Ribeiro and Salles make clear that renewable energy storage is one of the most important components to urban smart grids during this re-enchantment period. They have selected, compiled, and edited contributions from expert international peers on the critical aspects of urban energy storage that must be addressed for society to integrate and benefit from this technology.

As president and CEO of EPRI, I lead a world renowned, independent, non-profit energy research and development institute. EPRI's trusted experts collaborate with more than 450 companies in 45 countries, driving innovation to ensure the public has clean, safe, reliable, affordable, and equitable access to electricity across the world. The global energy sector is going through a significant change and as the authors note, distributed energy storage has a crucial role in the future of urban smart grids.

Through current research projects and initiatives at EPRI, we are examining how distribution systems around the world, many of which were built nearly a century ago, can become "smarter" as they are used to efficiently manage the increasing volume of DERs on local grids. Distributed energy storage is a critical part of any comprehensive DER management solution.

In the past few years, EPRI has led the development of specifications for DER management system software (DERMS) that can be used to manage DERs, allowing a greater number of assets to be integrated without negatively impacting power quality or requiring expensive grid upgrades. In a recent white paper, EPRI also examined how the role of utilities' distribution control centers must evolve from mostly dispatch to actively forecasting bi-directional electricity flows, analyzing more complex data streams, and serving net loads.

Meanwhile, in Europe, EPRI launched the Advanced Grid Interface for Innovative Storage Integration – a consortium of 14 university, non-profit and private sector members who are developing methods to help energy companies rapidly deploy renewables through the advanced integration of innovative energy storage technologies.

EPRI's work dovetails with the findings and analyses by Ribeiro and Salles, outlining how numerous aspects of distributed energy storage need to be considered, including the variability of wind and solar, standards and grid codes, the impacts of increasing electrification, rising load demand, monitoring, and market impacts.

Distributed Energy Storage in Urban Areas offers detailed case studies and practical examples of successful urban energy storage use through a scientific examination of both challenges and opportunities. These observations can benefit stakeholders, including government, policymakers, energy researchers, energy providers, municipalities, the business community, and ultimately society.

This book is a compelling read for those interested or involved in the clean energy transition and distributed energy storage in urban areas should be part of a comprehensive review to assist communities and society with an affordable, reliable, and equitable clean energy future

Arshad Mansoor
EPRI President and CEO

Preface

Re-Enchanted by Electric Grids:

> *"Whether there is, or whether there is not, in this world or in any other, the kind of happiness which one's first experiences of cycling seemed to promise, still, on any view, it is something to have had the idea of it. The value of the thing promised remains even if that particular promise was false — even if all possible promises of it are false."*
>
> *"... I think there are these four ages about nearly everything. Let's give them names. They are the Unenchanted Age, the Enchanted Age, the Disenchanted Age, and the Re-enchanted Age. As a little child I was Unenchanted about bicycles. Then, when I first learned to ride, I was Enchanted. By sixteen I was Disenchanted and now I am Re-enchanted."* C.S. Lewis (1946).

In the last few decades, electrical power systems have undergone a re-enchantment via paradigm shifts associated with renewable energy sources, energy storage systems, power electronics, and modernization of communication and automation of infrastructures. As the world becomes increasingly interconnected and energy consumption continues to rise, the need for reliable, secure energy sources becomes ever more urgent. Urban grids are particularly vulnerable to disruptions in the energy supply, as they are the lifeblood of large cities and the modern economy. Energy storage technology systems are of vital importance in the integration of renewable energy sources to ensure higher power systems performance and other benefits that reach various stakeholders of the electric grid. The combination of renewables in a smart grid context makes distributed energy an asset for the entire power system. This book addresses the new infrastructure in smart urban networks, from a distributed energy storage perspective, as this emerging technology is becoming increasingly important in addressing the challenges of urban grids.

This book explores the latest developments in distributed energy storage technology and its application to urban grids, examining the potential and limitations. It outlines the main technologies involved, and examines the advantages and drawbacks of different approaches. It also looks at the challenges of integrating distributed energy storage systems. Through detailed case studies and practical examples, this book provides the reader with a comprehensive understanding of the current state of play for distributed energy storage in urban grids

The chapters deal with changes in urban grids, system architecture, standardization, technical challenges, and multidisciplinary aspects, including case studies. With an emphasis on distributed storage and urban networks, the book aims to contribute to a more integrative approach to systems planning and studies. It is an essential resource for energy professionals, academics, policymakers, and anyone with an interest in the future of urban energy systems, especially in the field of energy storage, smart grids, power electronics, and power systems.

Chapter 1: This chapter introduces concepts regarding energy transition, urban smart grids, and energy storage. The electrical energy infrastructure is one of the key life-sustaining technologies of contemporary world. This infrastructure is extremely complex due to its size, its multifarious technologies, and its interweaving with societal structures. Urban distributed energy storage in the context of urban smart grids is an important component for the future infrastructure. The transformations on paradigms regarding more sustainable ways for generating energy and more reliable systems have created several challenges and opportunities of technology deployment, and distributed energy storage has a crucial role on the future of urban smart grids.

Chapter 2: In this chapter, a multi-dimensional view of DG in the existing and future power system is explored. The main drivers that motivate DG penetration are also investigated in this chapter. Particular attention is given to the benefits that they bring about to the grid, as well as the technical challenges that they incur. To assist with the understanding of the challenges of DG integration, ES technologies are investigated emphasizing their role in the future DN, particularly in terms of services required by the grid to ensure a safe and reliable operation. Finally, the opportunities of DG and ES, and their coordinated control strategies are investigated to identify the great many benefits that they will bestow upon the future electric power grid.

Chapter 3: It has become clear that energy storage will be a critical component in the future electric power grid. As society moves to carbon free electric power generation, the intermittent solar and wind energy sources will need to be complemented with energy storage. This upcoming presence of significant levels of storage and inverter-based resources will provide both opportunities and challenges to power grid operation. This chapter discusses a number of the issues that will be involved in this transition

Chapter 4: This chapter is dedicated to analyzing energy storage experiences, bringing information about countries' electrical matrix, how storage services are reimbursed, and the regulatory practices that support storage systems. It covers how implementation and operation are structured since the regulatory framework (comprising laws, regulations, and operating procedures) can interfere with the storage industry's growth, creating opportunities or inhibiting them. Based on those experiences, it is possible to identify issues and challenges in the existing systems, highlight detachable experiences, and suggest paths for other markets.

Chapter 5: This chapter introduces control and optimization techniques for distributed energy storage systems, in the context of modern power systems. The optimization and control strategies mainly address issues of power quality and provision of ancillary services, aiming at the operation of the system with the integration of renewable sources, power converters, electric vehicles, and even microgrids. In this way, storage systems are applied for the optimal operation of the electrical network from an operation perspective, considering the restrictions of the system, such as operational and physical limits of the equipment. At the end of the chapter, the internal control loops of the converters applied to energy storage are presented.

Chapter 6: This chapter extends the discussion on the charging infrastructure to the electrical installation requirements derived from the functions incorporated in the electric vehicle supply equipment (EVSE) and presents an overview of the electrical installation requirements for safe charging infrastructure. EVs represent a nonlinear load on the grid. As with all new technologies, there are challenges and opportunities. The challenges associated with the increase in the adoption of EVs and their impact on the grid are briefly discussed.

Chapter 7: This chapter seeks to provide an overall perspective of the main efforts toward the establishment of standards and grid codes for distributed energy storage systems employment. As outcomes, readers should be able to identify and put into context the key standards, grid codes, and opportunities for energy storage systems employment, as well as the expected challenges based on the prior lessons learned.

Chapter 8: Monitoring urban grid systems, including distributed energy storage, is of paramount importance for their control and protection, and for understanding their behavior when interacting with other system components. This interaction may degrade the Power Quality (PQ), so monitoring PQ in both DC and AC sides are necessary. In this way, this chapter will show the definition of some PQ parameters in the DC side and the concept of Novelty Detection for waveform recording, which can be used in both DC and AC signals. Some new aspects of PQ in AC systems will be presented, such as the increase of supraharmonic distortion and concerning with the time varying harmonic phasors. The description of a monitoring system based on the Substation Edge Device (SED) will be discussed as the way to unify the monitoring of both DC and AC side. Some results of a SED implemented in a real transmission systems will be presented.

Chapter 9: This chapter presents applications developed for battery energy storage systems of different sizes, which are: small, deployed mostly in residential and commercial customers; medium, deployed mostly in industrial customers and low voltage (LV) distribution systems; and large, deployed mostly in medium voltage (MV) distribution systems and distribution substations. The results presented in this chapter are based on field experiences from a Brazilian distribution utility.

Chapter 10: This chapter proposes an evolved concept of "hosting capacity" using the term of "feasible region" for installing additional loads or generations. Through converting the grid model into a more compact one, "hosting capacity region" not only is promising to further exploit the grid potential for power delivery but also benefits grid operation feasibility investigation with concise formulas. Facing the derived hosting capacity, originally complicated energy storage optimization problems can be represented algebraically, which is more efficient and friendly for computer processing. The case study based on a 10.5 kV Dutch grid has been implemented, eventually demonstrating the validity of relevant assessment and optimization methods.

Chapter 11: The increasing power demand, due to the population and economic growth, is pushing the operation of urban power grids to its capacity limits. The installation of distributed energy resources (DER) like photovoltaic (PV), wind power, and energy storage (ESs) with proper control and coordination mechanisms can offer a possibility to improve grid resilience. In this chapter, emerging coordination utilizes dispatchable sources to enhance the restoration capability under different disruptive events such as grid malfunctioning, severe weather, malicious attacks, and operation missteps. These coordinated control methods are based on a multiple-time-scale hierarchical framework that can effectively manage a complex, multi-target requirement of grid resilience.

Chapter 12: This chapter seeks to provide details of the impacts of distributed energy storage in intelligent electricity markets. Business models and ancillary services for energy storage systems are described, as well as the changes on electricity market agents due energy storage system deployment taking in consideration regulatory issues. The chapter also approaches the context of energy storage policies and programs like demand response in the context of high renewable resources penetration and energy storage. Lastly, policies and flexibility is briefly explored.

Chapter 13: This chapter aims to stress the value added by energy storage applications for residential, commercial, and industrial customers, as well as the seamless integration of electric vehicles as mobile sources of energy both in the forms of privately owned resources and in public transportation. Finally, as different market models arise thanks to the proliferation of such resources, some approaches and practical outcomes related to the community and (locally or virtually) aggregated use of energy storage assets are presented with the intent to integrate all aspects previously discussed and further evidence the significance of energy storage for the present and future of the power sector.

Once, Prof. Paulo F. Ribeiro (one of the editors) asked his Ph.D. advisor if he knew a book about a subtopic of my thesis. He replied, "No, it is time for you to write one, after studying this subject for so long." That seems to go along what Prof. Lewis said about writing books: "I wrote the books I should have liked to read if I could have got them. That's always been my reason for writing." The authors and editors of this book are delighted to prepare this manuscript because they like to read and learn more about distributed energy storage and its ramifications applied to urban smart grids.

Now we are also aware that a new book is still on its trial and the layperson is not in a position to judge it. It has to be tested against the great body of electrical engineering knowledge and the wisdom of experts down the time, and all its hidden implications (often unpredicted by the authors themselves) that have to be brought to light.

We are also mindful that every age has its own viewpoint. It is especially good at seeing certain truths and especially predisposed to make certain errors. We all, therefore, need the books that will correct the characteristic mistakes of our own period. And that means the classical engineering old books. All present authors share to some extent the contemporary view—even those who seem most opposed to it. To be sure, the books of the future would be just as good and correct as the books of the past, but unfortunately, we cannot get at them yet. But we are open to corrections and criticisms.

Finally, the goal of this book is to ensure that energy storage technologies are utilized effectively and efficiently to support the transition towards a more sustainable, decarbonized and resilient energy system. At the heart of sustainability of renewable electric energy generation, storage for transmission and distribution there must be an acknowledgment of certain boundaries and limitations in the nature of all stages of production, operation, and disposal, and the willingness of stakeholders to submit to sustainable normative practices. The authors and editors sincerely hope this book will contribute to these objectives.

Acknowledgements

If this book is useful, most of the credits go for those friends, colleagues and students who inspired us with their questions and encouragement.

Rafael S. Salles would like to thank his closest loved ones, family, better half, friends, and colleagues for their continuous and unconditional support. He also would like to thank his supervisors from his academic life who taught him more than excellent electric power engineering research but also how to go beyond as a correct and good person. He also would like to thank the Electric Power Engineering Group from Luleå University of Technology for all support for his career enhancement.

Paulo F. Ribeiro would like to dedicate this publication to his wife (Adriana), daughters (Ana, Priscila, Adriana and Ruth) and grandchildren (Bella, Benny, Lilly, Mia, Alexa, Valeska, Ezekiel, Eliel and Naomi) the greatest inspirations of my life.

Thanks also to CAPES, CNPq and FAPEMIG for their support.

Chapter 1

Introduction: energy transition, urban grids, and energy storage

Rafael S. Salles[1] and Paulo F. Ribeiro[2]

This chapter introduces concepts regarding energy transition, urban smart grids, and energy storage. The electrical energy infrastructure is one of the key life-sustaining technologies of the contemporary world. This infrastructure is extremely complex due to its size, its multifarious technologies, and its interweaving with societal structures. Urban distributed energy storage in the context of urban smart grids is an important component of future infrastructure. The transformations in paradigms regarding more sustainable ways of generating energy and more reliable systems have created several challenges and opportunities for technology deployment, and distributed energy storage has a crucial role in the future of urban smart grids.

1.1 Introduction

In the last few decades, electrical power systems have undergone paradigm shifts associated with renewable energy sources, energy storage systems, power electronics, modernization of communication, and automation of infrastructures. This book describes and discusses this transition and then explores the technologies and frameworks particularly associated with distributed energy storage.

Energy storage technology systems are vital for integrating renewable energy sources to ensure higher power systems performance and other benefits that reach various stakeholders of the electric grid. Moreover, the combination of renewables in a smart grid context makes distributed energy as an asset for the entire power system. This book addresses the new infrastructure in smart urban networks from distributed energy storage perspective. The book chapters deal with changes in urban grids, system architecture, standardization, technical challenges, and multi-disciplinary aspects, including case studies. With an emphasis on distributed storage and urban networks, the book aims to contribute to a more integrative approach to systems planning and studies.

[1]Luleå University of Technology, Sweden
[2]Federal University of Itajubá, Brazil

Urban power grids play a crucial participation in society and the development of modern civilization. Supporting this world's transition toward green energy is to provide sustainable technology deployment and improve human life, nature, and climate. The disruption of renewable generation leads to a paradigm that includes other transformations in the traditional power systems infrastructure. In this scenario, decentralization is characterized by the intermittent generation closer to the load side, a data-based system with heavy deployment of sensing, automation, and innovative technologies. Looking into this context, energy storage provides solutions for power systems that fit with this new concept of urban grids requiring flexibility, smart energy usage, and reliability. Furthermore, distributed energy storage provides the opportunity to strengthen the non-carbon revolution through cost-effect solutions in the setting explored in this book, modern urban grids.

Further, in this chapter, the frameworks for urban smart grids and energy storage technology will be explored to provide a strong basis for the following chapters in this book.

1.2   Urban smart grids

Urban smart electric grids use advanced technologies and systems to optimize the distribution and use of electrical power in cities. They use digital technologies to monitor and control the flow of electricity in real time, ensuring reliable and efficient delivery of power to homes and businesses. Smart grids also enable two-way communication between the power provider and the customer, allowing for better energy management and integration of renewable energy sources. Urban smart grids aim to improve energy efficiency, reliability, and sustainability in urban areas.

This new context requires a major change in the operational paradigm and key technologies to create a better performance of power systems. The urge for this transformation comes from changing the only central generation for the distributed one. Here, it is more characterized as being closer to the load, in the distribution grid. Also, the load in urban centers increases more and more with time, presenting different and non-linear behavior if compared with the past, more dynamic and complex. New paradigms for the consumer's role, market dynamics, and technical challenges to maintain the reliability of energy supply are emerging.

> Soon, our energy systems will change further. It is believed that large-scale power plants will be complemented by a large number of small-scale energy generation units. Among others, individual households will generate solar or wind energy. It is also believed that intelligent systems will be used to communicate comprehensively, control, protect, and balance supply and energy demand. The whole structure of central and local energy generation, transmission and distribution, and enabling intelligent control and information systems is called a smart grid. Smart grids will be integrating microgrids (local systems) and super grids (high voltage transmission and bulk generation systems) [1].

The modern grids will spare no effort to guarantee a more reliable, sustainable, automated system proposal characterized by the two-way flow of energy and information. It also has, at its core, a broad application of measurement and integrated sensing. Furthermore, grids must be able to have a quick response to undesired events, robust control strategies, and self-diagnostic capacity to overcome the barriers already mentioned. Also, the future infrastructure organization, from the traditional centralized generation for decentralized flexible grid, is fertile soil for the development of microgrids and remote systems, which can bring several benefits for the power systems, utilities, and end-consumers but are challenging as beneficial. Below are some features associated with the urban smart grid concept [2–5]:

- Increased use of digital information and controls technology to improve reliability, security, and efficiency of the electric grid.
- Better facilitate the connection and operation of the traditional and distributed generation.
- High penetration and integration of renewable energy sources.
- Capable of meeting increased consumer demand without adding infrastructure.
- Usage of key technologies and applications to improve grid performance.
- Development and incorporation of demand response, demand-side resources, and energy-efficiency resources.
- Integration of smart appliances and consumer devices.
- Increasingly resistant to attack and natural disasters.
- Capable of delivering power quality and improved hosting capacity.
- Consider social and multidisciplinary aspects.

In this way, the new grid architectures appear to meet all these points. It is possible to divide the complexities of this new context into dimensional, technological, and stakeholders [1]. Figure 1.1 illustrates this scenario of complexity.

Transportation electrification is also part of this vast transformation in urban environments that is firmly attached to seeking carbon emission neutrality. The use

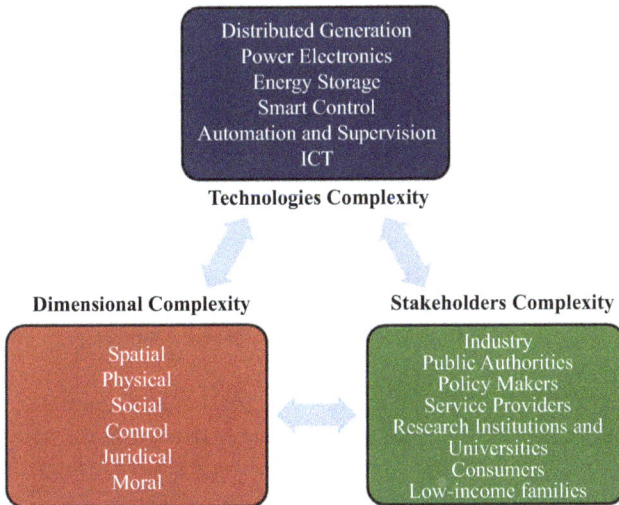

Figure 1.1   Integrated perspective of urban smart grids complexity (adapted from [1])

of electricity from external sources of electrical power is referred to as transportation electrification. This change allows utilities and society to support sustainable development goals and promote better assets to urban communities. Electrified transportation refers to the transition from conventional internal combustion engine-powered vehicles to more efficient and cleaner electrified vehicles. It regards of different types of transportation: electric vehicles (EVs), railway-based transports, ships, etc. Electric mobility is the vanguard of the energy transition in the transportation sector [6]. Energy storage technologies and devices continue to be developed as the foundation of a fully electrified transportation system integrated into a clean energy network. Lowering battery costs while increasing energy density and lifetime will hasten the electrification of road transportation [7].

In view of these changes, the new infrastructure and future urban grids have challenges regarding the level of complexity and impacts associated. In addition, the intermittency of renewable sources makes the operation more unpredictable and requires advanced management solutions to maintain a reliable power delivery. In the last decades, power quality has become a significant concern to the power sector due to deregulation, the widespread use of sensitive loads, power electronic devices, and the industrial processes transformation [8]. Also, the wide spreading of power electronic interfaces associated with distributed power generation, renewables, and transport electrification infrastructure can be a source of power quality issues and challenging aspects of coordination. Issues like voltage fluctuation, overvoltage, overloading, line losses, protection failure, waveform distortion, and other power quality issues can be more aggravated without good coordination of the new assets based on dynamic operations and power electronic conversion. On these challenges, energy storage plays on both sides of the issues, as a support and service to smooth the critical operation of renewable generation and transportation a system and as a possible cause as an interface for power quality performance compensation provided by the power electronic conversion associated to the technology for end-use applications (i.e., EVs and battery storage systems).

For example, EV charging infrastructure employs power electronics within the charge controllers that interface the vehicle's electric power system with the grid. EV charging performs as a non-linear load that needs AC–DC conversion and charging control for these electronic converters interface. Thus, inherent waveform distortion is caused by harmonics, interharmonics, and supraharmonics injection on the grid into the process. Waveform distortion injected by EV chargers into the power grid will negatively affect electric power system components designed to be supplied by pure sinusoidal waveform and could increase system losses. With a large-scale increase in EV chargers' penetration, the negative effects associated with power quality will impact the distribution network. And that is only one aspect to consider among others for this apparatus.

In this context, the concept of hosting capacity for distributed renewable generation and EV charging infrastructure arises to determine the penetration level limit for those assets to guarantee coordination between the penetration and allowable deterioration according to the reference standards [9]. Considering the many worst scenarios and uncertainties, it can be explained as the maximum generation or loading that can be integrated into the distribution grid without causing an excessive negative effect on the power quality.

Bellow, there are other challenges associated with urban smart grids.

1. *Cybersecurity*: Smart grids and distributed resources are networked systems, which makes them susceptible to cyberattacks. Secure communication protocols must be in place to protect the grid from malicious actors.
2. *Interoperability*: Smart grids and distributed resources must be able to communicate with each other to ensure efficient energy management. It requires developing standards and protocols that allow different devices to interact with each other.
3. *Cost*: Installing and maintaining a smart grid can be expensive, making it difficult for some utilities or customers to afford.
4. *Grid reliability*: Smart grids and distributed resources can be unreliable, as they are vulnerable to outages due to equipment failure or natural disasters.
5. *Regulation*: Utility companies must adhere to numerous regulations regarding installing and operating smart urban grids.

1.3   Energy storage disruption

The field of energy storage has been further explored and advanced in the last decades, both in research and project development for utilities, industry, and consumers. Looking at the documentation published until 2022 through the Scopus publication database, using the keywords "energy storage OR distributed energy storage OR battery energy storage systems," 165,834 documents in this field. Figure 1.2 shows the variation of this number per year. It is essential to note the massive increase since the beginning of the 2000s. From this total amount, the majority of documents regarding articles (journal papers) and conference papers, as illustrated in Figure 1.3. Figure 1.4 shows the distribution of research fields, and it is clear that most documents are related to engineering (28.5%) and energy (20.7%) aspects. In addition, it is highlighted that other subjects approach the field of energy storage, showing a multidisciplinary characteristic of technology that affects society and different systems.

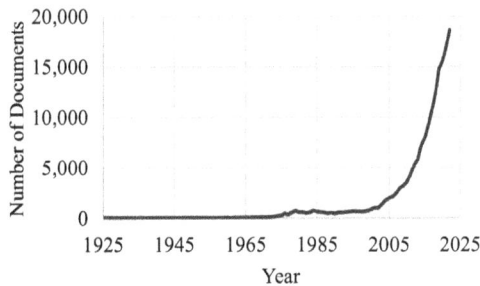

Figure 1.2   Number of publications on the field of energy storage per year

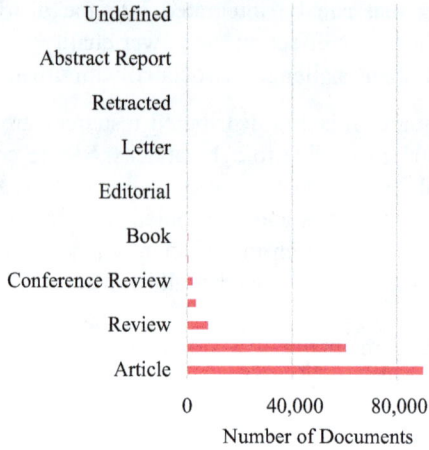

Figure 1.3   *Number of publications on the field of energy storage per type of publication*

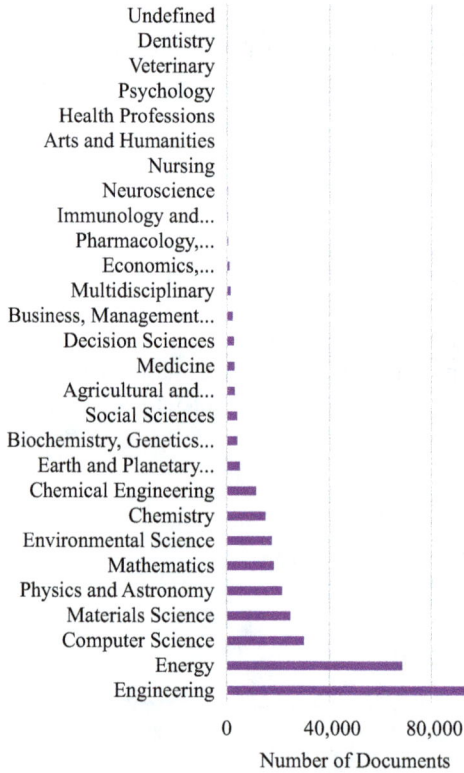

Figure 1.4   *Number of publications on the field of energy storage per subject area*

The definition of a storage system is essential since the way we characterize this system interferes with the types of services that can be provided to the electrical system. From an operational point of view, a storage system can absorb and contain a certain amount of electrical energy in a controlled manner (charging) via the interface with the electrical system or directly via the primary energy source and release it in a controlled manner during a time (discharge), through the interface with an electrical system. This process is repetitive according to the needs of the network, the storage capacity itself, and the operational limitations of the storage technology used. From a functional point of view, the storage system has been considered a subsystem of the generation system and, therefore, subject to selective dispatch rules, regulation, and tariff and tax collection.

The storage system may also provide ancillary normally distributed services unavailable in traditional generation systems. However, a regulatory framework must exist so that storage systems can produce the full potential that accompanies supporting the improvement of the electrical system's performance.

The basic principle for energy storage technology is that conversion from electricity is necessary, electromagnetically, electrochemically, kinetically, or as potential energy. The energy storage application mainly depends on two features, the total energy storage capacity and the power rating those systems can deliver [10]. The first is highly dependent on the nature of technology, and the second is related to the power conversion unit integrated into the system [10]. In addition, there are a set of characteristics of energy storage that plays a role and determine the application, duration, and sustainability aspects for those systems [11–14]: power rating, discharge time, power and energy density, response time, self-discharge losses, life and cycle time, etc. There are different types of energy storage technologies, and they are based on their energy conversion and storage principle. Below are a few examples for each physical principle [15,16].

- **Mechanical:** Pumped hydro energy storage, gravity energy storage, compressed air energy storage, flywheel energy storage.
- **Chemical:** Hydrogen energy storage, solar fuel, synthetic natural gas storage.
- **Electrochemical:** Battery energy storage, flow battery energy storage, etc.
- **Electrical:** Super magnetic energy storage, electrostatic energy storage (capacitor and supercapacitor).
- **Thermal:** Sensible heating storage, latent heating storage, thermochemical energy storage, and pumped thermal energy storage.

Those technologies with their particular features can be applied for different grid support services, like peak shaving, energy arbitrage, integration of renewables, voltage and frequency regulation, harmonics compensation, spinning and non-spinning reserves, black start, and deferral of distribution/transmission infrastructure upgrade [14,15]. It is possible to differentiate applications by their level of implementation and by stakeholders involved. Those aspects and applications will be explored in the following chapters of this book. But it is really important to highlight at the beginning the main opportunities, roles, gaps, and aspects involved in those systems in order to qualify the disruption factor of those technologies for the urban power grids. The benefits associated with each service can be associated in Table 1.1, per key aspect of

Table 1.1 *Summary of application and benefits [15,17]*

Application/benefits	Cost reduction	Power quality	Customer service	System enhancement
Loading shift	• Increased load factor • Reduced generation costs • Deferral investment and transmission and distribution assets	• Increased base load and continuity	• Reduced power chargers • improved security of power supply	• Elimination of generation and thermal constraints
Integration of renewables	• More forms of energy resources	• Improved dispatchability and availability of renewable energy	• Access to customer generation plants	• Capacity firming • Curtailment reduction or elimination
Frequency regulation	• Longer mean time between failures	• Increased grid stability	• Avoidance of load shedding	• Increased stability
Spinning reserve	• Increased base load efficiency	• Fast compensation	• Improved security of power supply	• Increased system stability
Black start	—	—	• Improved security of power supply	• Increased system stability
Harmonic compensation	• Reduction on investment in infrastructure	• Harmonic filtering	• Better power delivery	—
Energy arbitrage	• Profits from arbitrage	—	• Reduction in the electricity cost	—
Deferral of infrastructure investments	• Reduction on investment in transmission and distribution assets	• Increased base load and continuity	• Improved security of power supply	• Reduction on transmission congestion • Elimination of generation or thermal constraints

Source: W. Luo, S. Stynski, A. Chub, L.G. Franquelo, M. Malinowski, and D. Vinnikov, "Utility-scale energy storage systems: a comprehensive review of their applications, challenges, and future directions," *IEEE Industrial Electronics Magazine*, vol. 15, no. 4, pp. 17–27, 2021, doi: 10.1109/MIE.2020.3026169.

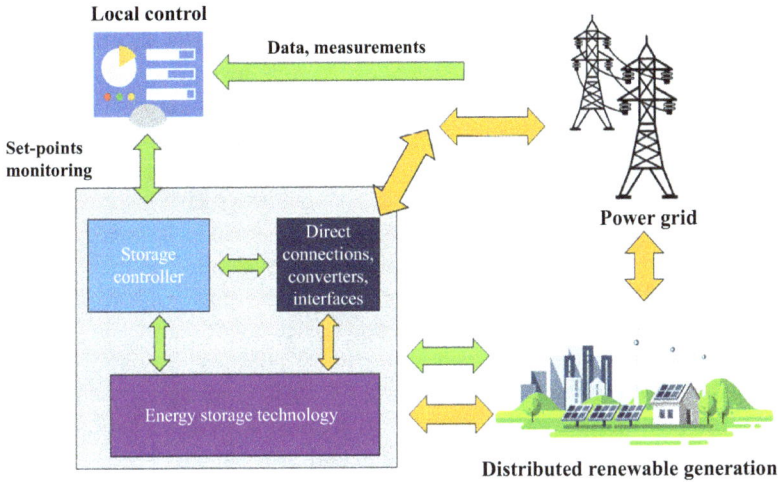

Figure 1.5   Basic scheme for energy storage systems

the power grid. Furthermore, the energy storage system, especially distributed energy storage systems, can provide applications and support for off-grid systems, microgrids, and transportation.

Given a broad view, energy storage systems comprise various aspects and associated systems. In addition to the technology with physical properties and control to reserve the surplus, the storage systems also have an electrical–electronic interface to connect with the electrical grid. In order to achieve an expected performance, which brings flexibility to power systems, the storage system must have robust control loops at different network control levels. Thus, the purpose of these systems must meet connection requirements, regulatory and tariff issues, network reliability, and services related to the consumer/concessionaire. Figure 1.5 shows a basic scheme of energy storage systems in which the storage technology, the interface with the electrical network, and the control and monitoring subsystems stand out.

Significant research and development effort has been made to seek flexible storage technologies with greater mobility and energy density to counteract fossil fuels and with reasonable costs. If successful, this effort will represent the breaking of a paradigm in the electricity sector that still places it as a unique industry since carbon-free electricity needs to be consumed at the same time it is produced. This break could even change the whole way we know the power grid. In this way, distributed energy storage systems can play a great role in energy transition and in society's infrastructure.

1.4   Holistic view for distributed energy storage

The electrical energy infrastructure is one of the key life-sustaining technologies of the contemporary world. This infrastructure is extremely complex due to its size, multifarious technologies, and interweaving with societal structures. Urban distributed energy storage in the context of urban smart grids is an important component of the future infrastructure. This book presents several elements of this new context and offers insights into understanding this intricate complexity more fundamentally. Key concepts are the ideas of holistic intertwinement of practices.

Energy storage infrastructure refers to the physical components and systems used to store energy generated from various sources, such as solar, wind, and hydropower, for later use. This infrastructure can include batteries, flywheels, pumped hydro storage, and compressed air storage systems. The main purpose of energy storage infrastructure is to balance the supply and demand of energy in order to improve the stability, reliability, and efficiency of the power grid.

Energy storage framework refers to a set of policies, regulations, and systems that support integrating and managing energy storage systems into the power grid. The framework is designed to provide a stable and reliable energy supply to consumers, manage energy demand and support renewable energy generation. It typically includes aspects such as market rules for integrating energy storage system (ESS) into the energy market, technical standards for energy storage system equipment and operation, and incentives for deploying energy storage technologies. An energy storage framework aims to ensure that energy storage technologies are utilized effectively and efficiently to support the transition toward a more sustainable and resilient energy system.

Urban distributed energy storage applications deal with the use of energy storage systems in cities to manage the distribution of energy generated principally from renewable sources such as solar and wind. These systems help to store excess energy during periods of low demand, and release it during periods of high demand, thus improving the stability and reliability of the power grid. The usage of battery systems is prominent in those applications because of the efficiency associated with other features of technology, including flexibility and size. Usual applications are microgrids, building-level energy storage, community energy storage, EV charging stations, etc. As highlighted in this chapter, all those deployments of distributed energy storage help to increase the use of renewable energy, reduce greenhouse gas emissions and improve energy efficiency in cities.

Challenges for urban distributed smart energy storage include:

1. *Integration with existing infrastructure*: One of the biggest challenges is integrating smart energy storage systems with existing power grids and energy infrastructure in urban areas.
2. *Cost-effectiveness*: Smart energy storage systems can be expensive to install and maintain, making it difficult for cities to implement on a large scale.
3. *Regulation and policies*: The energy storage market is still largely unregulated, making it difficult for cities to plan and implement these systems.
4. *Technical expertise*: The deployment and maintenance of smart energy storage systems require specialized technical expertise, which can be difficult to find in many urban areas.

5. *Energy management*: Effective energy management systems are critical for the success of smart energy storage projects, as they allow cities to monitor, control and optimize the use of stored energy.
6. *Public acceptance*: There may be resistance from the public to the installation of new energy storage systems in urban areas due to concerns about safety and aesthetics.

The ongoing transformations of the network of the future, which are becoming increasingly intelligent, raise a number of technical, social, and economic issues that cannot be addressed separately. Technological project developments must be aligned with all the faces and impact that the final deployment is influencing. The changing scenario has promoted the development of new concepts in which urban smart grids have become the new design approach for the development of future electric networks, allowing integrated and enhanced performance and diagnosis. An integrated view can allow addressing the challenges and developments of new technologies and applications in power systems [18]. It is also valid for distributed energy storage.

The view of engineering systems is essential to play this context in the design process blending engineering with perspectives from management, economics, and social science to address the design and development of complex, large-scale, sociotechnical systems. One also needs engineering systems for a complete holistic view of distributed energy storage, which is interested in addressing large-scale complex systems that involve uncertainty. This approach presents social and natural interactions with its technology and integration. The four underlying subfields for engineering systems are systems engineering (including systems architecting and product development), operations research and systems analysis (including system dynamics), engineering management, technology, and policy. System engineering is a holistic, integrative discipline wherein the contributions of structural engineers, electrical engineers, power engineers, human factors engineers, and many more disciplines are evaluated and balanced, one against another, to produce a coherent whole that is not dominated by the perspective of a single discipline [19]. In other words, those tools support planning and technology integration toward an integrated approach and broad view of the problems. Table 1.2 illustrates the views for engineering systems and system engineering for distributed energy storage. From that, it is possible to address several aspects at different levels of systems.

Following this integrated approach, other aspects beyond technical ones should be addressed in integrating distributed energy storage. Social, environmental, economic, political, and grid transformations/impacts aspects will play major importance in the society transition and the role of distributed energy storage in urban grids. The successful design should not neglect the distributed energy storage dependency of scares materials and minerals, the social inclusion and fair technology promotion, social and energetic policies, etc. Further in this book, the understanding of chapters will consolidate in a holistic visualization of the applications, technical challenges, regulatory issues, economics, and innovations.

*Table 1.2   Engineering system and system engineering aspects regarding
distributed energy storage (adapted from [18])*

Aspect	Engineering system	System engineering
Scope	Urban grids, smart energy systems	Energy storage, renewable power plan, household, isolated microgrid system, substation
Focus	Philosophy of technology, design criteria, impact on human life	Focus on storage and distribution technology, equipment, controllers, applications
Policy	Use of solutions to adapt the advances of urban smart grids in society	Use of standards, methodology, and requirements for better integration of distributed energy storage in urban grids
Sociotechnical	Crucial for distributed energy storage projects to present a holistic and empowering vision for society as a whole	Important for distributed energy storage projects to work properly, matching quality and reliability
Stakeholders	Focus on the environment, under-privileged by modern technologies, researchers	Focus on end-consumers, utilities, professionals
Roles	Grids architecture, energy markets, projects design, social aspects from urban grids, politics, biology, and others	Construction of new environments, smart grid performance, electrical engineering, information, and communication

1.5   Contents overview

The book provides an introduction to a number of different aspects and issues
regarding urban smart grids and distributed energy storage. Chapter 1 details
and describes distributed energy resources and energy storage systems. The
presentation of energy storage as a pillar for a resilient architecture of power
grids is stated and explored in Chapter 3. Aspects regarding the regulatory frameworks or international experience are covered in Chapter 4, and standardization is in Chapter 7.

The control and optimization structure for distributed energy storage is
presented in a comprehensive manner, considering several architectures and
control loops in Chapter 5. Also, the view of electric vehicle infrastructure and
its impacts is fully discussed, especially for power quality aspects in Chapter 6.
Furthermore, challenges for monitoring new infrastructures and distributed
energy storage are presented in Chapter 8. Lastly, the application and study
cases regarding distributed energy storage finish the book in the following
topics: application of battery energy systems to distribution systems (Chapter 9),
hosting capacity approach considering distributed energy storage (Chapter 10),
energy neutrality (Chapter 11), electricity market (Chapter 12), and urban
microgrids (Chapter 13).

References

[1] P. F. Ribeiro, H. Polinder, and M. J. Verkerk, "Planning and designing smart grids: philosophical considerations," *IEEE Technol. Soc. Mag.*, vol. 31, no. 3, pp. 34–43, 2012.

[2] T. Vijayapriya and D. P. Kothari, "Smart grid: an overview," *Smart Grid Renew. Energy*, vol. 02, no. 04, pp. 305–311, 2011, doi: 10.4236/sgre.2011.24035.

[3] U.S. Department of Energy, "The Smart Grid: An Introduction," 2008. https://www.energy.gov/oe/downloads/smart-grid-introduction-0.

[4] IEEE SA, "IEEE smart grid vision for computing: 2030 and beyond," *IEEE Smart Grid Vision for Computing: 2030 and Beyond*. pp. 1–133, 2013, doi: 10.1109/IEEESTD.2013.6577594.

[5] K. Al Khuffash, "Smart grids—overview and background information," in *Application of Smart Grid Technologies*. New York, NY: Elsevier, 2018, pp. 1–10.

[6] S. Rivera, S. Kouro, S. Vazquez, S. M. Goetz, R. Lizana, and E. Romero-Cadaval, "Electric vehicle charging infrastructure: from grid to battery," *IEEE Ind. Electron. Mag.*, vol. 15, no. 2, pp. 37–51, 2021, doi: 10.1109/MIE.2020.3039039.

[7] European Commission, "Transport electrification (ELT)," *TRIMIS*, 2023.

[8] S. Rönnberg and M. Bollen, "Power quality issues in the electric power system of the future," *Electr. J.*, vol. 29, no. 10, pp. 49–61, 2016, doi: 10.1016/j.tej.2016.11.006.

[9] A. F. Zobaa, S. H. E. Abdel Aleem, S. M. Ismael, and P. F. Ribeiro, Eds., *Hosting Capacity for Smart Power Grids*. Cham: Springer International Publishing, 2020.

[10] P. F. Ribeiro, B. K. Johnson, M. L. Crow, A. Arsoy, and Y. Liu, "Energy storage systems for advanced power applications," *Proc. IEEE*, vol. 89, no. 12, pp. 1744–1756, 2001, doi: 10.1109/5.975900.

[11] D. O. Akinyele and R. K. Rayudu, "Review of energy storage technologies for sustainable power networks," *Sustain. Energy Technol. Assessments*, vol. 8, pp. 74–91, 2014, doi: https://doi.org/10.1016/j.seta.2014.07.004.

[12] E. Barbour, I. A. G. Wilson, J. Radcliffe, Y. Ding, and Y. Li, "A review of pumped hydro energy storage development in significant international electricity markets," *Renew. Sustain. Energy Rev.*, vol. 61, pp. 421–432, 2016, doi: https://doi.org/10.1016/j.rser.2016.04.019.

[13] W. Choi, Y. Wu, D. Han, *et al.*, "Reviews on grid-connected inverter, utility-scaled battery energy storage system, and vehicle-to-grid application–challenges and opportunities," in *2017 IEEE Transportation Electrification Conference and Expo (ITEC)*, 2017, pp. 203–210, doi: 10.1109/ITEC.2017.7993272.

[14] L. Chang, W. Zhang, S. Xu, and K. Spence, "Review on distributed energy storage systems for utility applications," *CPSS Trans. Power Electron. Appl.*, vol. 2, no. 4, pp. 267–276, 2017, doi: 10.24295/CPSSTPEA.2017.00025.

[15] A. A. Akhil, G. Huff, A. B. Currier, *et al. DOE/EPRI Electricity Storage Handbook in Collaboration with NRECA*. United States. doi: https://doi.org/10.2172/1431469.

[16] J. Mitali, S. Dhinakaran, and A. A. Mohamad, "Energy storage systems: a review," *Energy Storage Sav.*, vol. 1, no. 3, pp. 166–216, 2022, doi: https://doi.org/10.1016/j.enss.2022.07.002.

[17] W. Luo, S. Stynski, A. Chub, L. G. Franquelo, M. Malinowski, and D. Vinnikov, "Utility-scale energy storage systems: a comprehensive review of their applications, challenges, and future directions," *IEEE Ind. Electron. Mag.*, vol. 15, no. 4, pp. 17–27, 2021, doi: 10.1109/MIE.2020.3026169.

[18] P. F. Ribeiro, M. J. Verkerk, and R. S. Salles, "Toward a holistic normative design," in *Interdisciplinary and Social Nature of Engineering Practices*. Cham: Springer, 2022, pp. 57–77.

[19] S. R. Hirshorn, L. D. Voss, and L. K. Bromley, *NASA Systems Engineering Handbook*. Washington, DC: NASA Headquarters, 2017.

Chapter 2

Distributed energy generation and storage

Ujjwal Datta[1] and Enrique Acha[2]

This chapter explores a multi-dimensional view of distributed generation (DG) in the existing and future power systems. The main drivers that motivate DG penetration are also investigated in this chapter. Particular attention is given to the benefits they bring to the grid and the technical challenges they incur. To understand of the challenges of DG integration, energy storage (ES) technologies are investigated, emphasizing their role in the future distribution network, particularly in terms of services required by the grid to ensure a safe and reliable operation. Finally, the opportunities of DG and ES, and their coordinated control strategies are investigated to identify the many benefits they will bestow upon the future electric power grid.

2.1 Introduction

To a greater or lesser extent, all electrical power networks around the world are undergoing a metamorphosis, particularly in respect of their power generation sectors where the long-enduring, conventional power generation sources are being relentlessly displaced by the so-called renewable energy sources (RES) of various kinds but mostly, solar photovoltaic and wind power generators, which exhibit a certain degree of randomness. Consequently, the power grid is experiencing new challenges in generating, distributing, and consuming electrical energy. The penetration of RES is pervasive, it involves not only large-capacity generation connected directly to the high-voltage transmission network but also the decentralized paradigm of energy generation at the distribution level and behind the meter at the residential and industrial customer ends. The rapid pace of adoption of these disrupting power generation technologies has strained every fiber of the power industry, which is regrouping to reflect and to take stoke of the new operating reality, embarking on a direction to retrain their workforce and employ the most advanced power system analysis tools in order to ensure the reliable and secure operation of today's electricity system.

[1]Australian Energy Market Operator, Australia
[2]Independent Energy Consultant, Mexico

With the ever-growing energy demand for electricity resulting from a global increase in population and its associated industrial activity, climate change and the ensuing needs for heating and cooling in extreme weather conditions, as well as the widespread adoption of awareness measures, such as electric vehicles, the distribution network (DN) is more prone to experience the overloading of transformers and distribution feeders. Over the last two decades, distributed generation (DG) has emerged in DN as a viable alternative to conventional, large generation systems, owing to its many technical advantages [1]. However, the absolute impact on the grid will vary depending on the type of DG, its rating, space availability, financial resources, and other ancillary distributed energy resources already installed in the vicinity. Such a transition would bring about substantial economic and environmental advantages but at the expense of exacerbating many of the operational challenges already faced by the power network and even introducing new ones, such as over-voltages, reverse power flows, protection failures, harmonic resonances, sags and spikes [2–4]. To reign in such undesirable effects, energy storage systems (ESSs) may play a crucial role in co-residing and managing efficiently the power regulation of DG units. Furthermore, the control and design of the power electronics converters of DG units can provide real-time control of active and reactive power regulation against frequency and voltage variations.

In the last decade, the control and efficiency improvements of ESSs have been explored quite extensively to support operational flexibility and to maximize their benefit for grid integration. With the maturity of ESS technology and decreasing cost per kWh of ESSs, the DN and end-users are in a pole position to maximize the techno-economic benefit of ESSs. However, such benefits can only be accrued with appropriate control strategies [5], such as storing power at an off-peak time and transferring it during peak time and with regulatory subsidies, that is, installation cost reduction, higher peak tariff, etc. [6]. However, if tariff-based incentives diminish, innovative solutions are imperative to attain economic benefits, such as participating in the ancillary services market, virtual power plants, etc.

This chapter explores a multi-dimensional view of DG in the existing and future power systems. The main drivers that motivate DG penetration are also investigated in this chapter. Particular attention is given to the benefits they bring about to the grid and the technical challenges. To assist with the understanding of the challenges of DG integration, ES technologies are investigated emphasizing their role in the future DN, particularly in terms of services required by the grid to ensure a safe and reliable operation. Finally, the opportunities of DG and ES, and their coordinated control strategies are investigated to identify many benefits they will bestow upon the future electric power grid.

2.2   Distributed generation phenomenon

From a conventional fashion point of view, centralized large-scale generation was driven by high efficiencies, economic operation, long-distance transmission feasibility, country-wide and continental-size systems interconnections, reliability of the grid, etc. However, the growing global demand for electricity, environmental concerns and fossil fuel depletion have motivated the shift toward DG. Although there may not be a conclusive definition of DG, the generic term of DG refers to small-scale generation resources at the distribution voltage level at or in proximity to utilities and end-users. However, a given definition may vary depending on the resolution, the installation location, rating of the DG unit, type of DG technologies, environmental impact, etc.

2.2.1 Distributed generation: classification and technologies

DG can be classified into several categories depending on their active power ratings, such as micro (<5 kW), small (>5 and <5 Megawatt (MW)), medium (>5 MW and <50 MW), and large (>60 MW). DG can be further classified based on their active and reactive power regulation as follows [7]:

- DG unit provides active power at unity power factor (pf), such as photovoltaic cells.
- Active and reactive power support provided by the DG unit at 0.8–0.99 leading pf, such as wind power, combined heat and power.
- Only reactive power support is provided by the DG unit, such as synchronous condensers.
- Active power injection and reactive power injection/absorption operating at 0.80–0.99 leading/lagging pf, such as doubly-fed induction generation and PMSG-regulated wind turbines.

Major DG technologies can be divided into renewable and non-renewable sources, as shown in Figure 2.1. The non-renewable energy power output is regulated by primary "actual" fuel; hence, the power is firm. On the contrary, the power output of renewable energy output depends on environmental conditions; hence, the power output is not firm.

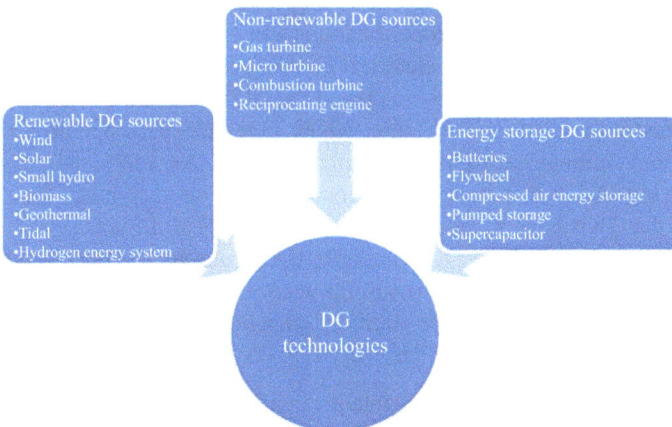

Non-renewable DG sources
•Gas turbine
•Micro turbine
•Combustion turbine
•Reciprocating engine

Renewable DG sources
•Wind
•Solar
•Small hydro
•Biomass
•Geothermal
•Tidal
•Hydrogen energy system

Energy storage DG sources
•Batteries
•Flywheel
•Compressed air energy storage
•Pumped storage
•Supercapacitor

DG technologies

Figure 2.1   Classification of DG technologies

It should be noted that renewable power generating sources are classed as zero-emissions, most notably wind and solar-driven power sources, but also geothermal and ocean waves. There is debate in some quarters as to whether or not the hydro plants ought to be classified within the renewable power category – this includes both the macro and micro scales, even though they are classed as zero-emissions. The *raison d'etre*, it is argued, is that global warming is working against the certitude counting with this, up to quite recently, plentiful resource. There are other power sources, which are also zero-emissions but are not classed as renewables, such as nuclear.

2.2.2 Driver and barriers in connecting distributed generation

Several factors are driving the growth of DG. The main ones are summarized below [8]:

- Global environmental awareness is putting increased pressure on governments and the electricity supply industry, to increase its production of electricity generation from cleaner sources. From a practical perspective, 100% clean energy may be hard to achieve in most countries; hence, an energy mix with a high renewable energy content is most desirable. For instance, some countries in South Asia, that is, Bangladesh, would face difficulty in reaching 100% penetration of clean energy sources as they have limited land for solar or wind power installations. However, in other regions of the world, many countries are in a pole position to achieve such an undertaking since they have a vast hydro resource in addition to plentiful solar and wind resources. Cases in point are the Central and South American countries, for example, Costa Rica and Brazil. Canada and the Nordic countries are also good examples of having huge hydro resources and excellent wind regimes. Most DG technologies are based on renewable resources, such as wind, solar, small-scale hydro, etc. All of this lowers the impact of carbon footprint.
- Many countries rely heavily on fossil fuel imports, so their electricity prices often fluctuate to reflect the international prices of gas, oil, and coal. In addition, fossil fuels are becoming increasingly difficult and expensive to extract and, above all, subject to geopolitical ups and downs. Hence, RES may be the ultimate viable solution for electrical energy generation.
- Since DGs are connected to the DN, there is little requirement for transmission-related infrastructure, which gives DG a competitive advantage over conventional, centralized power generation. An additional advantage of deploying DG is that transmission losses would be very much reduced, provided that DGs have sufficient capacity to feed local demands.
- From the vantage of the DG paradigm, the electricity is consumed mostly at the point of generation, and the surplus energy is fed back to the grid, resulting in lower electricity bills. In addition, generators with higher capacity can participate in delivering ancillary services, resulting in additional returns.

Although there are multiple advantages in adopting DGs' solutions, care must be exercised when implementing DG options since they still possess numerous economic, technical, and regulatory barriers, as shown in Figure 2.2.

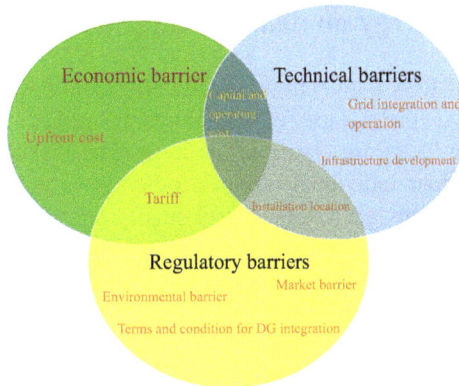

Figure 2.2   Barriers to integrating DG in the power system

- Technical barriers focus on uniform standards for DG integration and regulation.
- Non-renewable DGs might not be as efficient as conventional, centralized power plants; hence, non-renewable DGs might be producing more carbon than centralized ones for comparable amounts of electricity production.
- Regulatory barriers contain tariffs, lay out terms and conditions to connect and regulate the utility market.
- The implementation of DGs may impact the existing business model of centralized power stations, which may decrease their revenue and thwart the expansion/infrastructure development of the power systems' planning departments. These circumstances may become more complex in deregulated markets where DG investors aim to maximize DG penetration, while distribution/transmission system operators would be rightly concerned about their adverse impact on the grid.
- Several technical issues arise when the level of DG penetration surpasses the system's hosting capacity threshold. Uncertainty in technical constraints, such as power output variations, reverse power flows, generation outages, etc. Moreover, uncertainty in economic constraints, such as large fluctuations in energy pricing (e.g., price/MWh can fall negative in midday), results in a complex decision-making process. Hence, it is recommended that an in-depth techno-economic investigation must be carried out to assess whether or not the power grid supports additional DG integration at specific points, bearing in mind the secure and reliable operation of the power grid.
- An optimal DG planning, that is, correct size, best possible location, and a well-coordinated control scheme, are essential to extract the maximum benefits of the new DG installation. However, ensuring that all these issues are correctly accounted for, is a complex task that requires extensive analysis of various kinds and scenarios.

2.3   Renewable energy on distributed generation

The global growth of RES is relentless, specifically wind power and photovoltaic (PV) solar power. Since the majority of wind power resources are located in remote onshore and offshore locations, owing to the large size of today's windfarms and the colossal size of their individual turbines, other types of renewable resources such as rooftop/community PV, storage, and small-scale wind plants are directly connected to the DN and end-user terminal. Thus, the integration of DG has turned the conventional passive DN into an active power network, bringing in-built technical and economic advantages [9].

2.3.1   Technical benefit of DG installation

The technical benefits of DG installation in the DN vary according to the size, location, and type of DG installation. An overview of the DG's benefits is shown in Figures 2.3 and 2.4.

- Power loss reduction
  The power losses in the network are dependent on several factors, chiefly among them the loading conditions, that is, the power losses are higher in a heavily loaded system. The DG installation takes away the loading from the network and thus reduces the active power flows within the network. This, in turn, cuts the power losses in the AC network. The authors in [10] have concluded that a 50% penetration of distributed PV (DPV) may yield up to 40% of active power loss reduction compared to a non-DPV scenario of comparable size. Moreover, close-proximity DG installation to a significant loading point can significantly lessen the ensuing electric line losses. Nevertheless, an

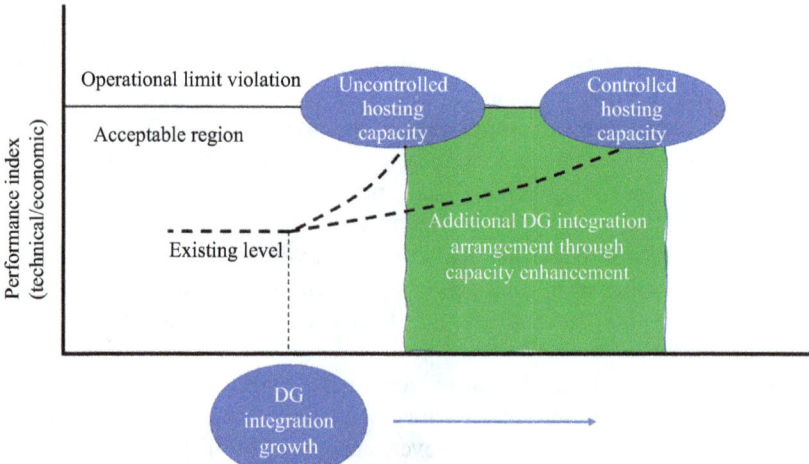

Figure 2.3   The impact of DG integration on the performance index

Figure 2.4   Technical benefits of DG integration in the power system

optimal DG size and strategic placement are key to achieving maximum power loss reduction. On the contrary, incorrect location and sizing will lead to increased power losses [11]. These twin problems are considered the primary objectives of an optimization problem aiming at minimizing the power losses in distribution systems and are solved through various optimization methods, such as Particle Swarm Optimization, Genetic Algorithms, Artificial Intelligence, and War Optimization [12,13]. Several other factors affect the outcome of the optimization, such as uncertainty of wind speed and direction, uncertainty of solar irradiances, temperatures and clouds, and the time-varying nature of the loads. All these issues must be taken into consideration in a large-scale optimization algorithm. Such a comprehensive solution approach would yield the optimal location and sizing of DG installations to ensure the highest decrease in power losses.

• Network overloading

The network elements, specifically transmission and distribution lines, generators, and transformers may experience overloading during peak demand periods or planned/unplanned network outages events. DG is ideally placed to ease this problem by generating a part of the local demand, avoiding overloading of the network elements, and increasing their lifetime. The study in [10] has shown that a moderate reduction in the overloading of the large synchronous generators is possible to achieve by DG installations.

• Improved voltage profile

The widespread inclusion of DG may function as a possible alternative to keep major blackouts at bay. Since DG provides active and reactive power to the load, it then minimizes the current flow throughout the DN, given that the active power does not have to travel far to meet the energy demand. This will boost the voltage magnitude at the load point. This would be particularly helpful to reinforce weak systems or weak busbars in a network [10]. The study in [14] suggests that beyond a certain level of DG penetration, further commissioning will result in the violation of the network's specified voltage limits during steady state and transient periods. Nevertheless, with careful planning, DG would perform as expected, even under severe unforeseen events [15].

- Power quality improvement
  Power electronics devices are an integral part of DG sources. They fulfill an essential role in connecting and controlling DC power sources and storage into the AC power grid, but they may cause some undesirable side effects on power quality. This is particularly the case if they are not functioning as intended – as in the design stage – due to aging. Some of these side effects are: voltage and current harmonics, voltage sags and swells, spikes, voltage flicker, and voltage unbalances. However, multifunctional DGs may be used to carry out corrective voltage actions and counteract such power quality issues. This involves the use of switching methods or harmonic compensation schemes integrated into the DG unit and their control loops. This may be particularly useful in weak points of the DN. The incorporation of multifunctional capabilities will increase the cost–benefit of DGs [16]. Hybrid DG units have been shown to improve power quality performance [17].

- Enhanced reliability and security
  The reliability and security of a power system are vital issues which must be kept under continuous watch, since a weakening of the indices associated with such vital issues may lead to blackouts with serious consequences and massive costs involved. A considerable portion of the reliability concerns arises from supply unavailability in the distribution system, which in turn may affect the transmission and generation systems. Reliability in a DN is quantified with multiple indexes, that is, SAIDI, CAIDI, and MAIFI, and efficient optimization is imperative to ensure the reliability of the system under contingencies [18]. Generally speaking, DG inclusion in the distribution system fulfills two purposes, acting as a main local supply or as a backup supply. An optimally placed DG can significantly improve the reliability of the network [19]. However, reliability improvement in highly loaded systems is a more tangible result of DG [19].

- Enhanced stability
  Stability performance is a key concern with DG inclusion, since the conventional DN is becoming progressively complex in terms of stability and control. A distributed, cooperative control is quite effective for enhancing power system stability [20]. However, DG grid-tied topologies play a major role in determining stability performance improvements [21]. Furthermore, a small signal stability analysis indicates that the total active power outputs of DG units, inverter's control parameters, and strength of the AC network are the critical aspects of a robust DG-integrated network [22].

- Transmission and distribution systems expansion
  Power system expansion is an essential part of a power system that, in practice, never ceases to change and that often needs repairing and upgrading to

accommodate a growing electricity generation and load demand. DG inclusion is one of the new solutions for emerging capacity augmentation; particularly, a deregulated market structure and limited transmission capacity have led to a growing reliance on DG sources. An in-depth analysis shows that DG integration reduces the requirement for additional substations and transmission lines constructions [23], and their ensuing associated costs. Moreover, coordinated planning of DG integration is essential to reducing the planned and unplanned outage costs of subtransmission lines.

2.3.2 Economic advantages of DG

The economic benefits of DG span a wide range of cost-saving factors, including fuel cost, transmission and distribution costs and electricity prices; a brief summary of the economic advantage of DG is given in Figure 2.5.

DG may defer the requirement of transmission and distribution systems' infrastructure upgrades, enabling an attractive amount of annual savings [24]. Some DG technologies are amenable to operation and maintenance cost reductions. For instance, a part of operating cost–benefit may come from peak shavings. Fuel cost reduces because in renewable DGs the "fuel" carries no cost, and improves overall efficiencies. The dependency on the wholesale spot market decreases; thus, purchasing electricity from customer-owned DGs in peak periods is a more economic option. Since DGs are distributed in nature, the need for reserve and associated costs reduce. Besides, there is a cut in customer interruption costs. DGs can participate in providing ancillary services such as frequency/voltage regulation, capacity market participation, energy arbitrage, etc., all of which incur profits for DG owners.

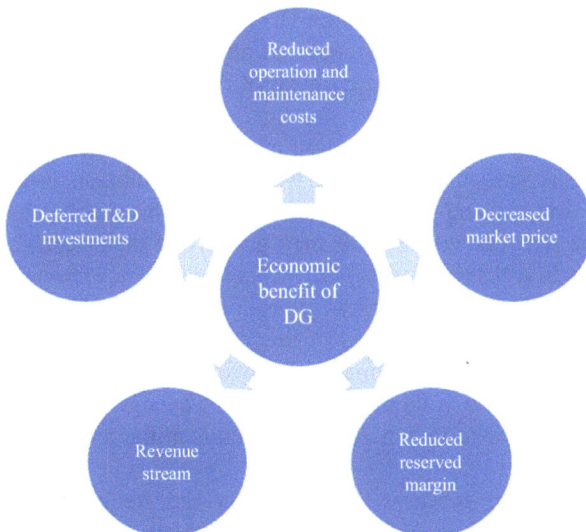

Figure 2.5   Economic benefits of DG integration in the power system

2.3.3 *Issues with large-scale DG penetration*

In general, the low-capacity factor and non-dispatchable nature of RES pose a significant challenge to grid operators. Large-scale DG penetration poses the following challenges to the grid.

- Power output volatility
  Forecasting DG's output is a hard task, and this makes the power balancing in the system difficult to achieve, both technically and economically.
- Reverse power flows
  Conventionally, the power flow direction is from the central generation center to the end-users through the distribution system. However, as there are more and more DG installations, reverse power flows may occur, that is, flowing upstream toward the network. For a small-size DG unit, the impact can be negligible, but as the percentage of DG penetration increases and becomes substantial with respect to the actual demand in the local network, the potential impact is no longer insignificant and may affect the network performance [25]. The answer lies in upgrading the network and in implementing additional security measures to ensure a safe and reliable operation.
- Reactive power
  Induction generation-based DG technologies consume reactive power from the power grid if the DG is not fitted with power electronics converters capable of injecting/absorbing reactive power. This is the case with type 1 wind turbines, which employ squirrel-cage induction generators. These are fitted with fixed capacitor banks, which enable operation at the unity power factor; hence, they do not contribute to reactive power regulation at the point of common couple (PCC). This deteriorates the grid performance at certain operating conditions, such as low-load demand with high DG generation. To resolve this, a power electronics-based converter capable of injecting/absorbing reactive power requires to be connected at the terminal of the wind farm.
- Voltage levels
  A high percentage of DG penetration may lead to changes in voltage profile. This effect is particularly marked when DG production is high, and load demand is low, causing voltage rises at the PCC. This scenario leads to voltage rises in the network voltage that may go beyond their acceptable operating limits [26]. In a distribution system, voltage rises are expected at peak generation of photovoltaic (PV) installations during times of high irradiation since this mostly coincides with times of low demand. Such voltage rises may be complementary in cases of congested networks, as congestion usually comes accompanied by low voltages. In the case of a large PV, higher amount of reverse power flow results in voltage reduction at the receiving end [27].
- System frequency
  Large penetration increases of DG affect the frequency of the system. Traditionally, frequency support is provided by large centralized synchronous generators (SGs). However, as more DG units are in operation, displacing more and more large SGs, the number of units capable of providing inertial

response, aka, frequency support, decreases, with the power system being less able to respond appropriately when there is a generation-demand imbalance of active power. Moreover, as a result of having a smaller number of SGs, the rate of change of frequency increases, resulting in a higher frequency nadir.

• Stability issues

With RES-based DG technologies, maintaining a stable grid is often cumbersome due to their stochastic generation pattern. For instance, a large voltage drop at PCC may disconnect a wind farm momentarily as mandated by grid codes. Subsequently, ancillary services must be obtained to compensate for the power generation loss. High penetration of wind turbines increases the post-fault oscillations, resulting in longer settling times and prolonged responses, making the system prone to instabilities. Concerning PV systems with no dynamic reactive power capabilities, they generate oscillations, causing system-wide stability concerns, such as transient overvoltage phenomena, overloading of transmission lines, generation tripping, etc. The risk of power systems instabilities may be substantial if several MWs of renewable distributed generation (RDG) power sources are connected into a weak area of the power grid.

• Increase in power loss

There is a certain maximum limit beyond which DG penetration would increase both the active power loss and the reactive power loss in the network [28]. To prevent this from happening, suitable network reinforcements are essential for the massive insertion of DG.

• Harmonic injection

The power electronics interface in the DG unit may produce harmonics if the filtering system does not work as intended at the design stage. This can distort the otherwise sinusoidal waveform and deteriorate the voltage and current of the network producing voltage fluctuations and flicker. When several DGs are placed in close-proximity, oscillations may occur in the range of 0.7–2 Hz, whereas DG inter-area oscillations may be observed in the range 0.1–0.7 Hz.

• Protection schemes

In conventional radial networks, the implementation of protection schemes is straightforward. However, in DN with DGs the existence of bi-directional power flows becomes a real possibility, rendering inapplicable the existing protection schemes. This requires re-evaluation and re-design of such protection schemes. The fault current contribution of inverter-type DG units is generally less than two times the rated current. In a DN with minimum or no DGs, the power grid contributes the fault current, and the circuit breaker will detect the high fault current. However, in DN with significant DG units, these may provide a degree of voltage support during a fault, such that the fault current from the grid would be below the breaker's pickup level or that the circuit breaker will detect the fault current with a delayed response. DG units may experience sympathetic tripping if the fault current cannot be detected by the network overcurrent protection and isolate the faulty part of the network.

- Economic impact

  Although DG has already been widely adopted in the DN industry worldwide, the high capital cost of energy/kWh continues to be the main concern of its return on investment. However, the actual cost of energy/kWh is not the same for the various DG technologies, that is, combustion turbines have a lower cost/kWh than fuel cell-based DGs. Technological advancement in DG technologies is the main driver for decreasing the cost of energy/kWh.

2.4 Energy storage technologies and its role in power systems transition

The interest in energy storage (ES) in power systems is growing due to breakthroughs in Li-ion battery technology and its decline in cost. Further progress in these directions is expected over the following years, which would benefit current and future developments in power grids with high RES penetration. ES provides various forms of support to the power grid since they respond quickly and responsively in overcoming the gap between generation and demand: they provide support against the variability of RESs, enhanced grid stability, reliability, and security. With the broader availability and applicability of storage devices and their increased efficiency, the cost-effectiveness of installing and maintaining a storage device becomes more and more attractive, thus paving the way for decarbonization in power systems.

2.4.1 Energy storage technologies

The various types of ES technologies are at different maturity levels. Some of them have been in the market for decades, and some others are still in their prototype stage. The ES equipment may be classified according to the duration of the service (long and short-term) and storage media (electrical, chemical, thermal, etc.). Electrical storage encompasses capacitors, supercapacitors, and superconducting magnetic energy storage (SMES). Capacitors are mainly low-energy, high-power applications technology. Although capacitors have fast-charging capability, self-discharge, and high energy dissipation are some of their drawbacks. Supercapacitors have a very high energy capacity compared to normal capacitors. Their energy efficiency lies in the range of 85–98%, with power levels up to the range of MWs, longer life cycles, high tolerance to deep discharges, and response times of less than 5 ms [29]. The main drawbacks are their very short charging/discharging times, low-energy densities and high self-discharging rates, up to 14% per month. SMES is quite a mature piece of technology, with high-power density, high energy efficiency (90–95%), fast response times, and low degradation. The main shortcomings of this technology are its high cost and self-discharge rates.

Electrochemical technologies store chemical energy in their electrochemical cells, which are then transformed into electrical energy. Batteries and redox flow batteries are the two main groups of batteries. Lead-acid batteries are low cost, have relatively high energy efficiencies, fast response times ($<$5 ms), and low self-discharge (0.1–0.3%/day) [30]. On the downside, they have a very limited lifespan and slow charging capability. However, an advanced lead-acid battery design yields higher capacities and elongated lifespans. Advanced lithium-ion batteries possess high power and energy density and are suitable for the transportation industry and, at the utility level, power system applications. Their energy efficiency is superior to the lead-acid batteries, observing similar lifespans and self-discharge rates. Sodium sulfur and sodium nickel chloride batteries have 80–90% round trip efficiency and a service life of up to 15 years – round trip being the percentage of electricity put into storage that can be retrieved later for actual usage. Nickel metal hybrid is one of the most popular batteries employed in the transportation industry, with an efficiency of 70% and a service life of 5–10 years. The nickel–cadmium battery has a higher power density and a longer lifespan than other battery technologies, say up to 20 years, but with a higher self-discharge rate of 0.03–0.6% per day. Flow batteries are one of the most promising battery technologies for large-scale applications due to their long-term storage capacity, almost zero self-discharge, and long lifespan. However, it requires high investment costs. Vanadium redox flow batteries, zinc–bromine batteries, and polysulfide bromide batteries are among the most popular types of flow batteries used in stationary storage applications.

The hydrogen fuel cell converts hydrogen into electricity. The process is reversible; hence, during off-peak hours, the fuel cell, turned electrolyzer, produces hydrogen. During times of high electricity demand, it is used to generate electricity. This technology has a high energy density and requires low maintenance, both attributes that put it into the realm of a long-term storage device. However, they have a low to medium efficiency (20–50%), high storage costs, and a typical lifespan of 15 years [29].

Mechanical type storage may be broadly classified into two groups: response time, power rating, and energy capacity, with slow and fast responses. Pumped Hydro Storage (PHS) and Compressed Air Energy Storage (CAES) are usually of large capacity and have slow responses. PHS operates by pumping water into an upper reservoir during off-peak periods and later on, when the water is needed, discharging it into a lower reservoir to spin a turbine-generator set to export electricity using the power grid. PHSs have high efficiencies, long lifetimes (30–60 years), long-duration storage capacities, and large power capacities in the range of 100–1,000 MW. However, these technologies have several drawbacks: long construction periods, large building area requirements, and high capital costs. CAES stores energy in the form of compressed air, which is then released into a turbine generator set to produce electricity. CAES is considered a long-term energy storage solution that can reserve energy for up to several days with a typical power capacity of 100–300 MW and lifetimes of up to 40 years. Nevertheless, it has a low-energy density of about 12 kWh/m^3 [29]. Flywheels store energy in the form of rotational energy and are considered fast-responsive units. The energy is discharged through a rotating shaft directly attached to a generator. Flywheel devices have power ratings ranging from 100s of KW to a few MWs.

In thermal energy storage (TES) systems, the energy can be stored at either low or high temperatures. TES discharges the stored energy in a thermal reservoir into a heat engine to generate electricity. Low-temperature TES technology comprises aquiferous low-temperature TES and cryogenic ES TES. High-temperature TES comprises concrete storage, phase change materials, molten salt storage, and room temperature ionic liquids. More details of the ES technologies can be found in Refs. [29,30].

2.4.2   *Energy storage role on power systems transition*

One of the main strategies to achieve total decarbonization is the ordered transition toward renewable energy. The transition plan requires various aspects to be considered, including economic feasibility, the kind of technology to be used, planning and social impacts, etc. Among various industry sectors, such as transport, agriculture, and buildings, the power sector contributes the largest part of the total greenhouse gas emissions. Harnessing the so-called renewable energy (e.g., hydro, geothermal, wind, solar, and ocean) yields electricity and, indirectly, it helps to decarbonize other sectors, such as transport, energy-efficient buildings, etc. Certain types of dispatchable renewable energy, such as hydro energy, bioenergy, etc., are a limited resource in many countries. Thus, wind and solar resources are the main drivers of the renewable energy transition. Due to the variable generation patterns of these resources, flexible technologies are often amalgamated to ensure robust and reliable operation [31].

There is a wide range of technological options to facilitate RES integration: ES is one of them, as exemplified in Figure 2.6. The storage systems operate in three different modes: charge, discharge, and idle (with a certain amount of self-discharge, which varies from one storage technology to another). When imbalances arise in the power system in terms of active power, ES is injected/absorbed at the PCC to maintain the power balance. Hence, a balance must be maintained to ensure that the ES device has sufficient power and energy capacity to participate when requested. Diverse storage technologies may coexist owing to their operational flexibility, to support grid stability and reliability. Here-to-fore, the most popular

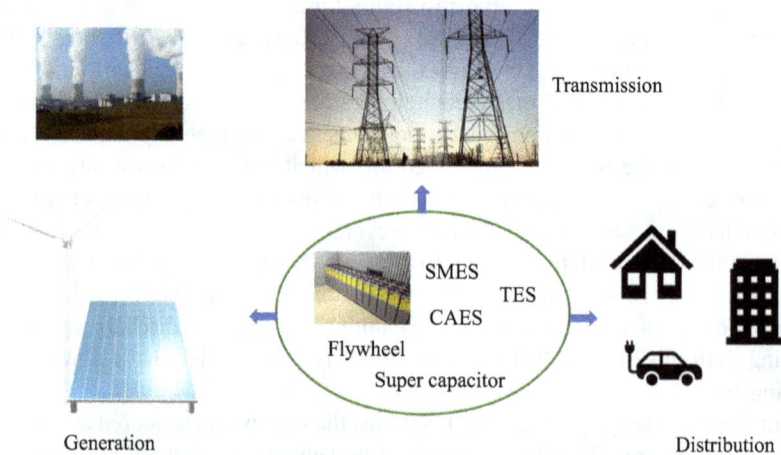

Figure 2.6   ES applications in the power grid

grid-scale application of ES is energy arbitrage, which allows generators with low variable costs to operate in high-demand periods instead of during low-demand periods, and store electricity during the low pricing periods and sell back to the grid during peak hours [32,33]. With the ever-increasing penetration of the so-called RESs, it becomes mandatory to maximize the technical requirements and economic benefits. At the utility level, time-shifted use of ES may be short-term, as it is the case in energy arbitrage, and long-term, involving days and even months, which then must have very little self-discharge losses [33]. The same service can be provided at the end-user level to reduce grid consumption and, eventually, the cost of electricity.

As the penetration of variable renewable energy resources increases, the conventional, large SGs are becoming displaced in direct proportion to the load demand. In this circumstance, ES will play a key role in maintaining the balance between generation and demand, smoothing out the imbalance. The capacity to fulfill this requirement is potentially quite large, ranging in the tens of MWs in a typical installation. In such a case, the ES must be quickly responsive to ramp up and down to follow swiftly the power generation variations of RES. Large-scale RES farms are usually located far from the end-user points, which puts additional stress on the transmission and distribution systems, which may need to be suitably upgraded to ensure adequate power transfer capacity. Optimally placed ES can improve dispatching operations, reduce the grid's stress, and in many cases, defer the infrastructure investment.

The grid regulation is being challenged by the large share of RES and the ensuing displacement of the large SGs. ES maintains the demand–supply balance during steady-state and transient conditions. The latter ensure grid security following a contingency. Short-term energy storages compete with other flexible options, such as the trip of large loads following an under-frequency event, and the active power reduction of REs during an over-frequency event. Furthermore, a fast frequency response yields fast responses to arrest the frequency nadir. When fitted with an adaptive control technique, ES can inject/absorb reactive power to enable voltage regulation, but this service is not incentive-driven/mandatory, and, currently, the energy generator companies/users may not be willing to provide it.

Grid-scale ES will be the main driver for supporting RES uptake in the power sector. Several studies have found that the need for storage is low with little RES penetration and that this grows rapidly with significant increases in RES integration. The storage is required to store the surplus of RES generation and to provide transient stability support, which becomes a necessity when the share of SGs decreases. At the distribution level, an array of electric vehicle batteries can be designed and controlled to assist in grid support and balance the PV variability. TES carries low investment costs and may be used as seasonal ES to balance wind power variations in winter-dominated countries. At the end-user point, battery storage has gained popularity over other storage technologies; it can be used for demand peak shaving, thus reducing electricity costs.

2.5   Opportunities for distributed energy generation and storage

As the penetration level of DG increases, the transformation of the conventional, centralized power system into a decentralized one, becomes apparent; at 100% DG penetration, the grid will be completely decentralized. Such a shift may happen with the assistance of present and future enabling technologies. This projected trajectory revolves around the traditional centralized electric grid migrating into a fully decentralized power grid, as illustrated in Figure 2.6. DG has a key role to play in this inexorable change. The pace of such a change is largely dominated by the level of uncertainty of each factor related to DG, such as the kind of "fuel" available and their reliance at PCC, the system strength at PCC, etc.

2.5.1   *The main factors that may influence the future role of DG in electricity systems*

The geographical location and climatic conditions would have a major influence on what role DG will play and the type of technology to be deployed [1]. Solar PV would be an ideal option for an area where solar insolation is a plentiful resource, whereas in areas with strong winds, wind power would be the ideal alternative. There may be areas that enjoy both kinds of resources, and then economics and efficiencies may dictate which kind of renewable energy resource is preferred. There may be other areas where neither resource is an option, but micro-hydro may be plentiful there. In the future, when the full potential of hydrogen technology is unleashed, most households may have a microgenerator.

Existing infrastructure will have an impact on DG integration. If a certain DG type is geographically suitable for a maximum generation but connected to a weaker part of the grid, maintaining network stability becomes challenging. With electric vehicle and heat pumps adoption, electricity demand in DN will change. Combined heat and power technology can avail the maximum benefit of DG within a district heating in winter-dominated countries. However, cost-effectiveness is driven by the population density in the area where DG is installed. Therefore, the arrangement of ES is another factor that needs to be considered when evaluating DG's role in the future power grid.

Demand response in conjunction with DG generation can facilitate a positive impact on the grid, that is, charging an electric vehicle, using heavy electric appliances during the DG peak generation. Regulatory requirements and economic incentives will significantly influence the deployment and role of DG. It can provide better grid stability and security services as the high capital investment cost diminishes. A partially or fully decentralized electric grid requires support from DG units to ensure the secure and stable operation of the grid. In the near future, when the need arises, DG units must be able to operate in grid-forming mode instead of the conventional grid-following mode.

2.5.2   Opportunities for DG and ES

Here-to-fore, reverse power flows are perhaps the most common technical issue that the transmission and distribution power grids are experiencing as the DG increases. With integrated battery energy storage systems (BESS), reverse power flows can be avoided in the transmission-DN [34]. The absolute benefit depends on the size and location of the BESS installation, which can, in turn, contribute to reducing energy losses. As generation from DG sources increases, the voltage at the generation connection point also increases. The situation worsens when DG generation is at maximum output, and the network demand is low. Hence, a means for voltage regulation needs to be in place to solve the overvoltage issue. However, existing voltage control methods, and step voltage regulators are not designed to regulate network voltage in the presence of DG penetration. It is observed that BESS charging/discharging and reactive power support from the BESS inverter mitigate the overvoltage issue. However, the coordinated control of on-load tap changers, DG and step voltage regulators to control BESS charging/discharging is essential for optimum results [35]. Also, DG can contribute to voltage regulation by controlling the DG converter. With high DG penetration, active power may increase in the grid, which requires reconfiguration of the existing electric grid. A coordinated dispatch of DG and ES minimizes the cost of power purchase by reducing DG curtailment, lowering active power loss, and reducing the penalty cost from voltage deviation [36,37]. Nevertheless, an economic dispatch strategy is imperative to ensure sufficient financial gain from ES, by decreasing sufficiently the curtailment of DG generation. This is achieved by exploiting the active/reactive power injection from ES to balance uncertainties of DG generation.

Provided the DG/ES resources are homogeneously distributed in the power grid, the network performance would be reasonably stable and reliable. However, due to geographical location, climatic conditions, one area might have a significant

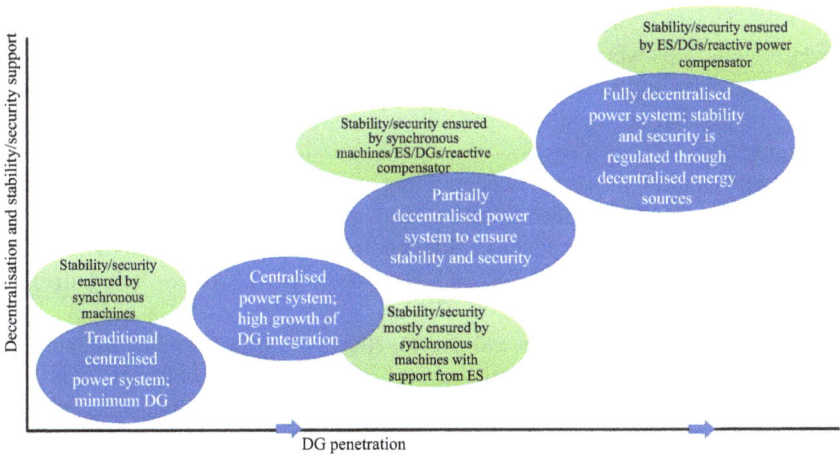

Figure 2.7   Projected electric power grid reformation in the future electricity grid

number of a specific type of DG, while another area may have a substantial storage capacity. In this situation, it would be of maximum interest for both areas to enhance coordination to balance the DG variations, ensuring better reliability and security of the grid [38], as observable in Figure 2.7. In addition, the combined DG and ES would be an economically viable solution when the initial cost of storage units drops sharply. Peer-to-peer trading can also be seen as an alternative for energy trading and gaining a sustained income [39].

2.6   Conclusion

The conventional power generation production, based on large SGs, is being disrupted by the pervasive penetration of distributed renewable/non-RES. DG may not impact much on the power grid much when it is only a small percentage of the total generation capacity. However, as the share of DG participation in the existing power grid increases, technological advancements and alternative technology adoption are a must to ensure a reliable and secure future power system. DG inherently brings about a host of new challenges when replacing the existing conventional, large SGs. DG may reverse the direction of the power flows, affecting the existing protection systems and creating a potential threat to personal and equipment. The intermittent nature of some kinds of DGs induces voltage and frequency risks, which become more severe following a contingency in DG-dominated power grids. ES is the cornerstone technology for the future energy power system, owing to its operational flexibility and fast-acting nature. The short-term and long-term ES technologies afford great flexibility to select the best option according to the power grid requirements, that is, long-term storage facility from hydro/CAES or fast response from BESS/SMES. Furthermore, ES technologies are capable of relieving stress from the transmission and distribution system when used in a planned way. Combining the best attributes of DG along with ES technologies, that is, coordinated control, plays a crucial role in ensuring a reliable and secure power system. This would enable extracting the maximum technical benefit of the DG and ES technologies with optimal costs. It is imperative that both technologies operate in parallel to maximize their benefits with enhanced resiliency of the power grid.

Notwithstanding these most encouraging developments, several challenges are yet to be overcome in the near future, to draw the most benefits from DG and ES technologies. Currently, the cost of ES technologies is considered the main barrier to the wide-scale adoption of ES, and a great many efforts are being made to reduce this.

Coordinated control of DG and ES, together with other parallel technologies operating in the power system, will become more complex to operate. It is envisaged that the use of artificial intelligence to enable more sophisticated controls of equipment/systems will gather pace. The future power grid will continue to experience further upgrading in terms of DG, ES, and other parallel technologies; hence, the way these technologies interact would require in-depth and accurate modeling and analysis of the electric power grid, taking due account of its future reality. Without a comprehensive understanding of these enabling technologies and new arrangements of the power grid, it would make it difficult to operate the power grid in a secure and the most economic manner.

References

[1] Mehigan L., Deane J.P., Gallachóir B.P.Ó., and Bertsch V. 'A review of the role of distributed generation (DG) in future electricity systems'. *Energy*, 2018;163:822–836.

[2] Bangash K.N., Farrag M.E.A., and Osman A.H. 'Manage reverse power flow and fault current level in LV network with high penetration of small scale solar and wind power generation'. In *2018 53rd International Universities Power Engineering Conference (UPEC)*. 2018, pp. 1–6.

[3] Wang Y., Silva V., and Lopez-Botet-Zulueta M. 'Impact of high penetration of variable renewable generation on frequency dynamics in the continental Europe interconnected system'. *IET Renewable Power Generation*, 2016;10:10–16.

[4] Mohammadi P. and Mehraeen S. 'Challenges of PV integration in low-voltage secondary networks'. *IEEE Transactions on Power Delivery*, 2017;32(1):525–535.

[5] Solano J.C., Brito M.C., and Caamaño-Martín E. 'Impact of fixed charges on the viability of self-consumption photovoltaics'. *Energy Policy*, 2018;122:322–331

[6] Koskela J., Rautiainen A., and Järventausta P. 'Using electrical energy storage in residential buildings – sizing of battery and photovoltaic panels based on electricity cost optimization'. *Applied Energy*, 2019;239:1175–1189.

[7] Singh B. and Sharma J. 'A review on distributed generation planning'. *Renewable and Sustainable Energy Reviews*, 2017;76:529–544.

[8] Gupta N., Joshal K.S., and Tomar A. 'Environmental and technoeconomic aspects of distributed generation'. In Tomar A. and Kandari R. (eds.), *Advances in Smart Grid Power System Network, Control and Security*. London: Academic Press, 2021, pp. 237–263 (Chapter 9).

[9] Chofreh A.G., Goni F.A., Klemeš J.J., Malik M.N.N, and Khan H.H., 'Development of guidelines for the implementation of sustainable enterprise resource planning systems'. *Journal of Cleaner Production*, 2020;244:118655.

[10] Suresh M.C.V. and Edward J. B. 'A hybrid algorithm based optimal placement of DG units for loss reduction in the distribution system'. *Applied Soft Computing*, 2020;91:106191.

[11] Georgilakis P.S. and Hatziargyriou N.D. 'Optimal distributed generation placement in power distribution networks: models methods and future research'. *IEEE Transactions on Power Systems*, 2013;28(3):3420–3428.

[12] Mahmoud K., Yorino N., and Ahmed A. 'Optimal distributed generation allocation in distribution systems for loss minimization'. *IEEE Transactions on Power Systems*, 2016;31(2):960–969.

[13]   Quadri I.A., Bhowmick S., and Joshi D. 'A comprehensive technique for optimal allocation of distributed energy resources in radial distribution systems'. *Applied Energy. Electric Power Systems Research*, 2018;211:1245–1260.

[14]   Karatepe E., Ugranli F., and Hiyama T. 'Comparison of single- and multiple-distributed generation concepts in terms of power loss, voltage profile, and line flows under uncertain scenarios'. *Renewable & Sustainable Energy Reviews*, 2015;48:317–327.

[15]   Öner A. and Abur A. 'Voltage stability based placement of distributed generation against extreme events'. *Electric Power Systems Research*, 2020;189:106713.

[16]   Yahya N., Hosseini S.H., Zadeh S.G., Mohammadi-Ivatloo B., Vasquez, J.C., and Guerrerov J.M. 'An overview of power quality enhancement techniques applied to distributed generation in electrical distribution networks'. *Renewable and Sustainable Energy Reviews*, 2018;93:201–214.

[17]   Elbasuony G.S., Shady H.E.A.A., Ahmed M.I., and Adel M.S. 'A unified index for power quality evaluation in distributed generation systems'. *Energy,* 2018;149:607–622.

[18]   Mitra J., Vallem M.R., and Singh C. 'Optimal deployment of distributed generation using a reliability criterion'. *IEEE Transactions on Industry Applications*, 2016;52(3):1989–1997.

[19]   Xu Z., Liu H., Sun H., Ge S., and Wang C., 'Power supply capability evaluation of distribution systems with distributed generations under differentiated reliability constraints'. *International Journal of Electrical Power & Energy Systems*, 2022;134:107344.

[20]   Wu X. and Shen C. 'Distributed optimal control for stability enhancement of microgrids with multiple distributed generators'. *IEEE Transactions on Power Systems*, 2017;32(5):4045–4059.

[21]   Zhou H., Shi C., Jingang L., *et al*. 'Modeling and synchronization stability of low-voltage active distribution networks with large-scale distributed generations'. *IEEE Access*, 2018;6:70989–71002.

[22]   Nasri S., Zamanifar M., Naderipour A., Nowdeh S.A., Kamyab H., and Abdul-Malek Z. 'Stability and dynamic analysis of a grid-connected environmentally friendly photovoltaic energy system'. *Environmental Science and Pollution Research*, 2021:1–13

[23]   Rad H.K. and Moravej Z. 'Coordinated transmission substations and sub-transmission networks expansion planning incorporating distributed generation'. *Energy,* 2017;120:996–1011.

[24]   Dixit M., Kundu P., and Jariwala H.R. 'Incorporation of distributed generation and shunt capacitor in radial distribution system for techno-economic benefits'. *Engineering Science and Technology, an International Journal*, 2017;20:482–493.

[25]   Demazy A., Alpcan T., and Mareels I. 'A probabilistic reverse power flows scenario analysis framework'. *IEEE Open Access Journal of Power and Energy*, 2020;7:524–532.

[26] Carne G.D., Buticchi G., Zou Z., and Liserre M. 'Reverse power flow control in a ST-fed distribution grid'. *IEEE Transactions on Smart Grid*, 2018;9 (4):3811–3819.

[27] Iioka D., Fujii T., Orihara D., *et al*. 'Voltage reduction due to reverse power flow in distribution feeder with photovoltaic system'. *International Journal of Electrical Power & Energy Systems*, 2019;113:411–418.

[28] Ogunjuyigbe A.S.O., Ayodele T.R., and Akinola O.O. 'Impact of distributed generators on the power loss and voltage profile of sub-transmission network'. *Journal of Electrical Systems and Information Technology*, 2016;3 (1):94–107.

[29] Argyrou M.C., Christodoulides P., and Kalogirou S.A. 'Energy storage for electricity generation and related processes: technologies appraisal and grid scale applications'. *Renewable and Sustainable Energy Reviews*, 2018;94:804–821.

[30] Kebede A.A., Kalogiannis T., Mierlo J.V., and Berecibar M. 'A comprehensive review of stationary energy storage devices for large scale renewable energy sources grid integration'. *Renewable and Sustainable Energy Reviews*, 2022;159:12213.

[31] Stephan A., Battke B., Beuse M.D., Clausdeinken J.H., and Schmidt T.S. 'Limiting the public cost of stationary battery deployment by combining applications'. *Nature Energy*, 2016;1:16079.

[32] Schill W.P. 'Electricity storage and the renewable energy transition'. *Joule*, 2020;4(10):2059–2064.

[33] Gallo A.B., Moreira J.R.S., Costa H.K.M., Santos M.M., and Moutinho E.d. S. 'Energy storage in the energy transition context: a technology review'. *Renewable and Sustainable Energy Reviews*, 2016;65:800–822.

[34] Unahalekhaka P. and Sripakarach P. 'Reduction of reverse power flow using the appropriate size and installation position of a BESS for a PV power plant'. *IEEE Access*, 2020;8:102897–102906.

[35] Tshivhase N., Hasan A.N., and Shongwe T. 'An average voltage approach to control energy storage device and tap changing transformers under high distributed generation'. *IEEE Access*, 2021;9:108731–108753.

[36] Kalantar N.M. and Cherkaoui R. 'Coordinating distributed energy resources and utility-scale battery energy storage system for power flexibility provision under uncertainty'. *IEEE Transactions on Sustainable Energy*, 2021;12 (4):1853–1863.

[37] Li X., Wang L., Yan N., and Ma R. 'Cooperative dispatch of distributed energy storage in distribution network with PV generation systems'. *IEEE Transactions on Applied Superconductivity*, 2021;31(8):1–4.

[38] Baker K., Guo J., Hug G., and Li X. 'Distributed MPC for efficient coordination of storage and renewable energy sources across control areas'. *IEEE Transactions on Smart Grid*, 2016;7(2):992–1001.

[39] Álvaro D.J.do.N., and Ricardo R. 'Evaluating distributed photovoltaic (PV) generation to foster the adoption of energy storage systems (ESS) in time-of-use frameworks'. *Solar Energy*, 2020;208:917–929.

Chapter 3

Energy storage as a pillar of the architecture of a resilient electric grid

Thomas H. Ortmeyer[1], Tuyen Vu[1], Jianhua Zhang[1] and Leo Yazhou Jiang[1]

It has become clear that energy storage (ES) will be a critical component in the future electric power grid. As society moves to carbon-free electric power generation, the intermittent solar and wind energy sources will need to be complemented with ES. This upcoming presence of significant levels of storage and inverter-based resources will provide both opportunities and challenges to power grid operation. This chapter discusses a number of the issues that will be involved in this transition.

3.1 Recent technological developments

There is a global need for ES systems with increased capability and reduced cost. This need is widely recognized, and there are significant investments being made into research, development, design, and field demonstrations of storage technologies. In 2020, the World Energy Council published their assessment of a number of the competing technologies [1]. A portion of this assessment is shown in Table 3.1. Table 3.1 shows the wide range of readiness, efficiency, response time, and discharge time between technologies. There are also significant differences in the feasible sizes of the different storage technologies. Of the options shown in Table 3.1, supercapacitors and flywheels would be small in ES capability, pumped storage hydro would be large, hydrogen would be medium to large, and the electrochemical solutions are scalable. In rough numbers, small would be less than 1 MW, and large would be greater than 100 MW. It is clear that a decarbonized power grid will require ES assets with a wide range of capabilities.

The New York Independent System Operator's (NYISO's) report "The State of Storage" [2] lists the following potential services for providing revenue to ES installations:

- Energy, day ahead market
- Energy, real-time market

[1]Electrical Engineering, Clarkson University, Potsdam, NY, USA

Table 3.1 Sample overview of several storage technologies (used by permission of the World Energy Council)

	Electrical		Mechanical		Electrochemical			Chemical
Technology	Supercapacitor	PSH	Flywheels	Sodium sulfur	Lithium ion	Redox flow		Hydrogen
Maturity	Dev'mt	Matur	Early Comm	Commercial	Commercial	Early Comm		Demonstration
Efficiency	90–95%	75–85%	93–95%	80–90%	85–95%	60–85%		35–55%
Response time	ms	sec–mins	ms–sec	ms	ms–sec	ms		sec
Discharge time	ms–60 min	1–24 h+	ms–15 min	s–h	min–h	s–h		1–24 h+

- Capacity
- Voltage support
- Regulation
- Operating reserves

While black start is also a potential service, New York currently has sufficient black start capacity that meets their requirements.

3.1.1 ES systems for urban environments

The term "Distributed ES" is typically defined to include all storage systems connected to the electric power distribution system. This includes systems connected on the primary distribution lines owned by the electric power companies as well as those connected to customer owned systems connected at the service voltages. Installations on customer premises are often referred to as "behind the meter" systems. Residential behind the meter storage systems are typically connected at 120 V/240 V single phase, with size ratings up to about 25 kW. Commercial and industrial systems would be connected to three phase systems of 480 V and above, and range in size from 100's of kilowatts to many megawatts in power rating.

Residential systems: Historically, residential electricity users were similar in nature—using electricity for heating/cooling of space, water and food, lighting, and entertainment. Residential rates were based on energy usage, with a small monthly fixed charge. The meter required for this service was relatively cheap, and the electricity cost across the sector was considered fair.

New uses have recently emerged:

- Usage of computers/electronics for work at home
- Solar photovoltaic (PV) installations in the home
- Electric vehicles (EV) charged at home
- Residential ES
- Advanced technology electric heating

Each of these uses changes the power consumption profile of a given residence. In particular, residential PV systems reduce the amount of energy a residence draws from the grid, without impacting a significant portion of the costs of serving that customer. In extreme cases, residences generate more energy than they consume over a year. Net metering rates can allow these residences to be compensated for the excess energy, typically at the wholesale rate. These customers then pay only the monthly fixed fee, which historically has been low. While net metering has provided a useful incentive for installing residential PV, it is now generally recognized that net metering can distort the market, with non-PV customers providing a subsidy for customers with PV. As a result, amendments to net metering tariffs and alternate pricing structures are under active consideration. Alternate pricing schemes include time of day energy rates, market based rates, and employing a demand charge to recover a portion of the monthly costs. Increases in the monthly flat rate are also occurring. Separate tariff structures for customers with these new characteristics are being considered, and in some cases are in place.

Most of those currently in place are put in place through opt-in actions. Some may become mandatory in the future.

The use of constant energy rates provides no financial incentive for the application of behind the meter residential ES. The changes in rate structure mentioned above may begin to provide incentives for residential behind the meter storage. ES systems with the capability to provide load ride through for momentary sags/interruptions or to serve critical loads during extended interruptions will provide an additional benefit for those residential customers valuing those services.

Commercial/industrial systems: Commercial and industrial power grid customers in urban areas have ES opportunities that are also based on cost and reliability issues, but the scale and nature of these customers leads to significantly different outcomes in their benefits and costs. These include:

- Commercial and industrial customers currently have a demand component in their billing structure, and many have market based energy rate structures.
- Commercial and industrial customers can suffer significant losses during momentary events and/or sustained interruptions that can be mitigated by storage assets.
- The size of the storage installation at commercial and industrial customers will generally be larger than at residences. This impacts the following:
  o The cost per kilowatt-hour of the installation
  o The potential of the storage owner to bid directly into ancillary service markets of the independent system operator (ISO) or regional transmission organization (RTO).

These factors will vary significantly from case to case, and a careful benefit–cost analysis is necessary before any investment decision can be made.

Urban grid level storage needs: Urban areas typically import electric power from generation sites that can be significant distances from the urban area. The bulk power transmission network has evolved to move the power from the generation source to the load centers, both urban and rural. In order to maintain reliability and quality of service, the power generation assets and the bulk power transmission network must be planned, designed, and operated to:

- Maintain the system voltage within accepted bounds of magnitude and frequency
- Avoid equipment overloads
- Maintain the system stability

Many urban load centers currently have natural gas fired peaking units that help them to meet these three reliability goals. These peaking units are being phased out, in order to reduce carbon emissions and improve the urban air quality. ES is expected to play the most significant role in providing these reliability services to the urban grid as these peaking units are decommissioned. While renewable energy sources (RESs) located in an urban area will also play a role, population density and zoning restrictions will limit the level of the renewable resource capability within a given urban area.

ES is currently more expensive than gas fired peaking units, and the power grid must adapt to this difference in order to ensure that the overall cost of electricity remains at a reasonable value. Just as importantly, storage assets have significantly different operating characteristics than gas fired generation. Many of the storage technologies connect to the grid through inverters rather than synchronous generators. Storage assets by nature have a finite amount of energy available at any given moment, and when depleted must be recharged from the grid before providing additional energy. Gas fired units, on the other hand, can provide electric power generation whenever natural gas is available from the pipeline.

The remaining sections of this chapter describe several challenges that must be addressed to optimize ES's role in the upcoming transition from today's grid to the future low carbon grid.

3.2   Integration of ES with renewable DG

Energy transition: Transition to distributed renewable variable energy resources (VER) such as wind and solar in power distribution systems can benefit the end-users in clean, reliable, and affordable energy. However, the increasing penetration levels of VER may challenge the grid operation in both deployment and the operation due to their intermittencies. In the short term, the power quality with voltage flicker, harmonics, bidirectional power flows, and changing of voltage profiles are the concerns. In the medium term, the mismatch between load and renewable energy generation may worsen the peak load demand and challenge the operation of power systems. In the long term, the energy adequacy (for off-grid applications such as microgrids) is an issue due to their intermittencies, weather dependencies, and seasonal dependencies. The involvement of ES can help address most of these issues.

Advantages: The primary advantages of combining VER and ES include:

1. *Balancing loads*: Without ES, VER are non-dispatchable and may need to be curtailed if generation exceed the loads. On the other hand, ES can support the loads under intermittent power output, especially true to distributed VER.
2. *Firming VER generation*: ES helps smooth the output of VER, and can reduce the flicker and voltage variation caused by VER.
3. *Increased resilience*: Collocation of VER and ES can serve loads during extended power grid disruptions (with microgrids or nanogrids).

ES can be standalone, collocated with VER (hybrid) and loads, or play a role in microgrids. It is worth noting that EVs can be considered as ES if being managed efficiently. Distributed ES can be applied to both front-of-the-meter (FTM) and behind-the-meter (BTM) applications.

3.2.1   Challenges and solutions for ES

There are both technical and economic challenges to adopting ES in power distribution systems with high renewable energy generations.

Technologies and cost: Technologies for ES includes batteries (e.g., lithium-ion, sodium sulfur, lead acid, flow, etc.), pumped storage hydro, compressed air, thermal (e.g., molten salts), hydrogen, and flywheels [3]. Each type of storage has their own application niches based on the power and energy requirements of the application. For distributed applications, lithium-ion batteries have been the main trend as their cost has declined by significantly (nearly an order of magnitude) over the last decade mainly driven by the high demand of EVs. The battery pack price is expected to drop below $100/kWh by 2024 [4]. Although the average cost of battery packs has dropped significantly, the system cost remains high. As shown in Table 3.2, the 2021-cost for battery ES systems (BESS)/kWh for residential, commercial, and utility application stands at $1,354, $481, and $379, respectively [5]. This BESS cost consists of the individual storage units, integration into a pack, power conversion system (PCS), energy management system, and other components, as shown in Table 3.3.

As the cost of BESS remains high, to date their adoption to distributed VER generation is limited to be effective in some special applications such as balancing loads, peak shaving or some primary frequency response, and other demand response programs from local regions/utilities. For example, the cost of a 1 MW-PV system costs

Table 3.2   Projected costs by NREL of typical sizes of battery installations by use class [5]

	Residential	**Commercial**	**Utility**
2020 cost, $/rated kWh	$1,499	$538	$437
2021 cost, $/rated kWh	$1,354	$481	$379
% reduction, 2020–2021	9.7%	10.7%	13.1%
Typical size	5 kW/12.5 kWh	600 kW/240 kWh	60 MW/240 MWh

Table 3.3   Battery energy system components

Storage units	**Integration**	**Power conversion system**	**Energy management system**	**Soft costs**
Storage cells	Container housing	Bi-directional inverter	Charge/disch. control	Overhead, net profit
Battery management and protection	Wiring	Switchgear	Load management	Interconnect uncertainty, Delay, permitting
Racking	Climate control	Transformer	Ramp rate control	Labor
		Interconnect equipment and controls	Grid stability	Supply chain cost

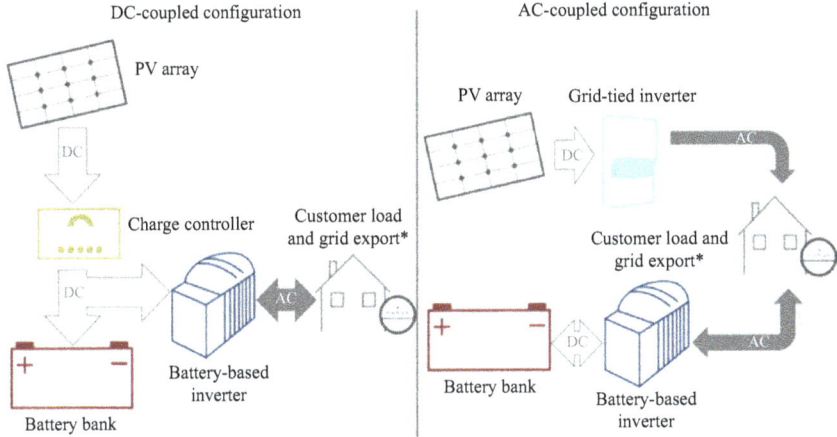

Figure 3.1 Simplified diagram of coupled PV-storage options. From US DOE publication NREL/TP-7A40-80694 [15].

about ~$1.0 million while the cost of a comparable 600 kW/2,400 kWh would cost ~$1.2 million. If the BESS is distributed, the cost can reach to $1.5 million. Therefore, the total cost of the PV-BESS can easily double the original PV-system alone cost. Integration of BESS with VER can be shown in Figure 3.1, which is for solar but also applicable for distributed wind. The BESS can be coupled either in DC or AC sides of generation resources. Analyzing cost/benefits for ES units will therefore depend on the local subsidies/incentives. On the other hand, system service resilience against extreme weather events is also a consideration. For example, Tesla Powerwall owners in Texas avoided extended outages during a winter storm event in February 2021 [6].

3.2.2 Controls and management for BESS

To realize the benefits distributed ES can bring when being integrated with the VER, advanced controls must be in place. BESS can play different roles in supporting renewable-based generation, especially when there is a high level of penetration. Some of primary controls of ES in conjunction with VERs can be detailed as follows:

- *Voltage and frequency support*: Controls of stand-alone BESS or hybrid BESS-VER can be considered to follow IEEE 1547-2018 [7] to provide active and reactive power control, and voltage and frequency support.
- *Inertia support*: Physically, BESS has stored energy that can mimic the inertia, which can provide the primary frequency support to low-inertia systems having high VER penetration. Alternatively, in microgrid operation, the BESS inverter can be controlled in isochronous mode with the BESS energy capability supporting frequency stability.
- *Ride-though controls*: Controls of inverters for BESS systems also need to follow IEEE 1547-2018 to provide essential voltage and frequency ride-through

capabilities in different modes from momentary cessation to mandatory operation, and continuous operations.

- *Energy management system*: Managing BESS can come from local level (home/building or standalone system) to microgrid/feeder level and distribution levels. The management scheme would be depending on certain applications of BESS. For example, a high level distributed energy resource management system (DERMS) could manage a local distribution system, and schedule the operation distributed generation and distributed storage systems for optimal power flow (OPF) within that portion of the grid. BESS can also play other roles in virtual power plants to provide the secondary frequency response, energy trading, and demand management. In the microgrid scale, managing BESS is especially important for the system voltage/frequency supports, OPF, and resilience. In hybrid system, smoothing generation profile also needs to consider managing BESS carefully as their power/capacity is limited and the exhaustive operation of BESS would decrease the lifetime of batteries significantly.

- *Microgrid operation*: In grids with high penetration of distributed VER, it may be possible to form microgrids to provide local service during extended power interruptions. In these cases, BESS must be included to complement the variable generation. The microgrid must have black start capability. There will be challenges with black starting in the microgrid, due to the limited overload rating ($\sim$1.1 pu) of inverter connected resources. Therefore, if relying solely on the BESS to black start any grid, the power rating of storage inverter must be oversized. Advanced controls need to be deployed to combine the black start capabilities of BESS with other VERs to ensure the overall power grid resilience.

- *Communication*: IEEE 1547 requires any DER larger than 250 kVA to be monitored PCC voltage, active power, reactive power, and connection status. The issue could be that there are many small DER to be deployed at the lower power level; which causes the challenges the operations of power grid (e.g., anti-islanding capability to increase the safety for operations and maintenance).

In summary, practical ES systems are a key factor in transitioning to the low carbon economy. In order to fully utilize their capability, they would need to be integrated into smart controller that would manage the generation, power system, and loads of a given system.

3.3   ES as part of the architecture of the future grid

Driven by high penetration of RESs, electrification of other energy sectors (e.g., transportation and heating), liberalization of electricity market and high-impact low-probability (HILP) natural disasters and physical intentional attacks (PIAs), the ES system is widely recognized as the next major disruption to the current electricity grid architecture, and it is expected to improve the flexibility, reliability, and resiliency of the future grid [8]. As discussed in Section 3.1, the ES system is the enabler of multiple benefits along the entire value chain of the grid, including [9–11]

- voltage regulation, power loss reduction, peak shaving, capacity support, black start for the isolated mode, network congestion relief, emission reduction, improvement of resiliency to natural disasters/attacks, and deferral of infrastructure investment for the distribution system operator (DSO),
- congestion reduction, and deferral of infrastructure investment for the transmission system operator (TSO),
- frequency regulation, fast frequency regulation, spinning reserve, non-spinning reserve, black start, and price arbitrage to support energy market for the ISO,
- high power quality, reliability and resiliency improvement, reduction of usage time, and demand charges for customers.

To achieve these benefits, a challenge of the ES system adoption into the grid is storage expansion planning which is expected to address these questions:

(i) how to optimize location and sizing of power rating and energy capacity of the ES systems to support grid services in multiple subsystems of the grid;
(ii) how to mix different storage technologies to mitigate multi-scale issues that arise in RES applications, such as energy/price arbitrage and load balancing;
(iii) how to provide multiple functional services at the different stages for the whole lifetime to maximize the overall revenue in the energy market;
(iv) how to coordinate ES expansion planning with other components, such as wind and/or solar RESs, power networks, EV charging stations, and demand response-based load.

Thus, the traditional deterministic generation and network planning methods have become unsuitable. Aiming to develop appropriate models and methodologies for the ES planning and its variant joint planning, the key features of ES planning problem have been identified from different perspectives, such as the allocation of RESs, ESs and network components, high-level uncertainties, and coupling of operation and planning. Within this context, this subsection focuses on overview of the envisioned grid architecture with high penetration of ESs and the optimal ES planning in both transmission and distribution grids.

3.3.1 Architecture of the future grid with ES

Future demand for storage generally is categorized into three groups in terms of installation location:

(a) *Grid-scale ES*: Large ES systems such as hydrogen storage system, pumped storage plant, and lithium-ion/flow battery system deployed in the transmission grid, including some long-term storage systems.
(b) *Distributed ES*: Storage systems such as lithium-ion batteries or flywheels directly connected on primary distribution systems. Typically have several hours of storage intended for daily cycling.

(c) *Household ES*, also called Behind-The-Meter (BTM) storage units such as Tesla batteries, and thermal ES units of an electric water tank.

According to the transportability, the storage units also can be divided into two groups: stationary and mobile storage systems. Mobile ES system refers to the transportable modular battery ES modules located on train cars or trucks. As an innovative solution, they can provide the same utility services as stationary ES does, but with more flexibility.

To achieve this envisioned architecture, different ES planning techniques in both transmission and distribution grids will be introduced and described from different perspectives including objectives, constraints, network modelling, uncertainty management, and solution methods, as below.

3.3.2   ES planning in transmission grids

Grid-scale ES consists of a number of technologies that have widely different specifications in terms of cost, spatial footprints, capacity, and response timescales that are shown in Table 3.1. Identifying the appropriate technology, size, and location is critical for realizing multiple or specific potential benefits (e.g., fast frequency regulation, congestion reduction, black-start, and resilience to natural disasters and PIAs) of ES systems in the transmission grid. Inspired by the inherent coordinated generation and transmission network planning framework, these ES benefits/applications can be achieved, quantified, and calculated by steady-state balanced studies, and they are considered as a goal to plan ES in the transmission network by network operation frameworks namely OPF, unit commitment (UC), and ES. The grid-scale ES applications and its associated planning objective functions and constraints are discussed below.

Grid-scale ES applications and planning objectives and constraints: The main purpose of the grid-scale ES integration into the generation and transmission systems is to improve the RES utilization, which usually refers to the minimization of renewable energy curtailment in the planning objective function. Assuming that the future generation will be based primarily on RES, the system operator needs to optimally control flexible assets (such as ES and series capacitors) at the transmission level as well to maximize utilization of non-controllable RES generation assets. As a promising tool, the battery ES enables shifting energy in time, from periods of RES overproduction to periods of insufficient RES output. Similarly, for the application *Network Expansion Deferral*, it is embedded in the minimization of investment cost of equipment. *Resiliency Improvement* considers the grid resilient to natural disasters and PIAs, and supplies the black-start service. The customized resilient index is modeled as one objective function. The rest of applications include *frequency/voltage regulation, congestion management, and spinning reserve service*, which usually are formulated as constraints in the optimization planning framework. It is worth to note that ESS operation costs include the maintenance and lifetime degradation cost. Table 3.4 shows the representative ESS applications and objective function terms, optimization techniques and considerations in the coordinated ES planning problem.

Table 3.4 *Grid-scale ESS application and objective functions*

Ref #	ES application	Objective function	Optimization type	Joint planning with	Network modeling
[12]	Resiliency improvement—against PIA Arbitrage	Weighted average energy not served Weighted costs of load shed Investment costs of new transmission lines and ESSs Operation cost of ESSs and generators	Single objective, MNILP and big-M method Single stage Scenario-based stochastic optimization	Transmission network	DC PF model PIA and load uncertainties
[13]	Congestion management Arbitrage Spinning reserve service	Investment costs of new transmission lines and ES systems Operation costs of generators, load shedding penalty, BESS transportation	Single objective Single stage Scenario-based optimization	Transmission network Mobile battery ES system (BESS),	DC PF model Renewable uncertainty in spinning reserve
[14]	Resiliency improvement—black start Arbitrage	Investment costs of ESSs Operation costs of generators, load loss	Multi objectives Two-stage robust optimization	Transmission network	DC PF model Transmission line random outage
[15]	Arbitrage Voltage regulation	Investment costs of new transmission lines, series compensators, and ESSs Operation costs of generators	MILP Bender decomposition	Transmission network, series compensators	Linearized AC OPF

Uncertainty modeling of grid-scale renewable generation and aggregated load: To address the uncertainties of grid-scale renewable generation and aggregated load, the optimal ES planning in the conventional coordinated expansion planning of distributed RES generation and transmission/distribution networks using the following techniques:

1. scenario-based stochastic programming (SBSP) [16–18];
2. chance-constrained stochastic programming to handle multi-scale and multi-type uncertainties [19];
3. robust optimization [20]; and
4. distributionally robust optimization [21].

The uncertainty modeling techniques have been widely used in the ES planning problem [22] include

1. Markov transition probability matrix
2. Point estimate method
3. Monte Carlo simulation
4. Beta probability density function
5. Weibull probability density function
6. Heuristic moment machine (HMM)
7. Clustering techniques such as k-means

The planning results of both robust optimization and stochastic programming rely strongly on the model deployed for the uncertain parameters. Robust optimization-based planning must construct a proper uncertainty set, targeting a balance between robustness and economic efficiency. As for the stochastic programming-based methods, the selection of probabilistic scenarios is nontrivial, with a trade-off between computational complexity and representativeness of the scenario set. To address the complexity issues arising from a large number of scenarios, scenario reduction and decomposition techniques, Monte Carlo simulation, and heuristic algorithms are applied to make the stochastic programming model computationally tractable. However, for practically large systems, computational burden remains a challenge [20].

Transmission network model: In a network setting, both capacity requirements and the location of storage resources strongly depend on the network properties [23]. The most common way to constrain the physical transmission network operational principle is through OPF framework, which optimizes a cost function subject to physical and operational constraints on both the network and at the individual network nodes or buses. OPF with storage formulation has been proposed for operational and planning problems. However, the general OPF with storage problem (as well as the basic OPF) is difficult to solve because it is a nonconvex, NP hard problem. Thus, for the ES planning problem, the DC OPF considering storage components, which is a linear approximation, has been extensively used to investigate optimal storage siting and sizing.

3.3.3 ES planning in distribution networks

The distribution power system structure is evolving dramatically to fit the increasing penetration of the RES, EV charging stations, demand response programs, and home energy management system (HEMS). ES systems can provide various grid services due to its fast time-response and diverse power and energy density in different technologies. In particular, the radial distribution network has a different risk structure compared to the transmission system. As a result, optimal placement, and optimal sizing of distributed energy resources (DERs) including RES and ES is becoming the emerging significant aspect for the DSO in planning the smart distribution network, especially along with these specific issues, such as power exchange between grid and DER owners, high initial capital investment cost associated with DERs, economics consideration of emission cost, and bi-directional power flow led by the ES integration. In the early stage, the simplified ES planning models in distribution network have two general drawbacks: (a) the production cost simulation does not account for the detailed operation strategy; and (b) probabilistic modeling approaches cannot effectively address the uncertainties of multiple DERs. To address these two issues, recent studies focus on the advanced ES planning models, which are enabled by the development of two-stage stochastic optimization techniques and accurate uncertainty modeling techniques.

ES planning on the distribution system is usually formulated as an optimization problem and that is approached by using one or more of these techniques:

1. analytical approach,
2. conventional optimization techniques, and
3. meta-heuristic approach [24].

ES applications and planning objectives at smart distribution grid: In the past, the distribution system planning process primarily considered investment, replacement, maintenance, and other operational costs. Recently, multiple additional and diverse objectives have emerged. These can be broadly classified into three categories: technical, economic, and environmental aspects. The objectives on technical aspects mainly constitute power loss reduction, voltage profile enhancement, equipment overloading, systems reliability, and sensitivity [22]. Among of them, equipment overloads, power loss reduction, and voltage regulation dominate most of the existing research work. The economic objectives mainly comprise annual energy savings, reduction of power import from the grid, minimizing renewable generation curtailment, and minimizing system total cost including investment and O&M cost. For the mobile ES planning, the economic objectives also account for the transition delay cost. While considering the environmental criteria, the objectives deal with primarily enhanced social welfare and reduced carbon emissions. There are three aspects expected to be duly considered and properly formulated into the coordinated DER (including distributed RES, charging station, and ES) planning model featuring with multi-level multi-objective. As comparison with conventional distribution network planning problems, the distributed RES, charging station, and ES sizing and siting problems are generally different, because the main idea is that this new planning problem is to determine the optimal capacity and location of these smart grid components.

The coordinated expansion planning framework for various combinations of DERs that include wind turbines, PV panels, and battery ES is the same with the conventional coordinated generation and transmission network planning. A typical state-of-the-art planning tool for stationary ES units is routinely formulated as a two-stage/master-subproblem stochastic MILP and considers ES units as stationary resources. In these tools, the first stage optimizes the ES locations and sizes, while the second stage fixes the first-stage decisions and co-optimizes the operation of existing resources and newly installed ES units. While the planning tools for mobile ES units become more complex due to two facts: (a) the mobile ES units' routing paths are the decision variables, called recourse decisions; and (b) the recourse decision on each unit is binary, where "1" denotes move, and "0" is stay [15,26]. It results in the two-stage stochastic MILP with binary recourse decisions that are more computationally demanding, and usually the existing Bender's decomposition technique performs poorly. Thus, the progressive hedging (PH) algorithm is applied to address this computational complexity issue and it partitions the problem in a scenario-based fashion, called scenario-based decomposition and each stage decisions are optimized for each scenario independently.

Distribution network model: In the transmission network, the DC power flow approximation is employed to further reduce the computational complexity. It should be noted that the DC power flow approximation is less accurate on the distribution system, due to its lower X/R ratio. Modeling of RES and load is different on the distribution system than on the transmission system. Due to single phase loads, three phase load flow studies are often needed to account for the phase imbalance.

Uncertainty modeling of renewable energy generation and EV charging load: In locations close to specific loads or RES, the statistical smoothing of loads or sources will not be as pronounced as on the transmission grid, and there can be a stronger statistical correlation in geographically small areas [26]. In some cases, the use of the central limit theorem for uncertainty modeling for the distribution grid is not practical, and a more general non-parametric stochastic modeling technique can be considered for the ES planning problem [27]. Point estimate method (PEM) is employed for probabilistic OPF [27]. The application of optimization methods in distribution system ES planning is a complex but a necessary topic that is evolving to meet the new challenges of the modern distribution system.

In summary, both the grid level planning and the distribution level planning practices will need to evolve to have the capability to plan for the integration of the time varying and stochastic RESs and battery ES systems that will interact with them. This section has discussed a number of the potential methods that have potential for accomplishing these tasks.

3.4   Long-term storage—needs and requirements

The increasing penetration of variable renewable energies, e.g., wind and solar, and emphasis on grid reliability and resiliency to keep the lights on have driven the strong need of long-duration storage in recent years:

- The falling cost of renewables together with government subsidies makes it economically preferable over traditional fossil-fuel-based generation resources, e.g., nuclear, coal, and gas. The continuing buildout of renewable projects and the intermittency of renewable generation expose the grid with long stretches of surplus and/or lack of renewable generation supplies, which create the need for ES to charge during the period of surplus renewable generation to minimize renewable curtailment and discharge when renewable generation is not available. Take New York State for example. In 2020, there were 74 instances when the wind resources combined across the state supplied less than 100 MW (5% of the installed wind capacity) to the grid for periods of more than 8 consecutive hours [28]. These low wind events will lead to periods with reduced wind energy production to serve load, which makes the long-duration storage highly desired.

- Emerging policies at the wholesale level that are identifying the need for long-duration storage energy and establishing its defining characteristics. In California's resource adequacy program, an ES resource is required to have at least a 4-h duration in order to warrant the full capacity contribution with respect to its rated power capacity. However, the PJM's (PJM Interconection LLC is regional transmission organization (RTO) in USA) capacity market requires an ES to have at least a 10-h duration for its capacity contribution to match its rated power capacity.

- Individual state policies require the future procurement of significant amounts of ES technologies and renewable energies to achieve the goals of 100% carbon-free electricity systems, which in turn create the opportunity for long-duration ES for consideration. New York State mandates to procure 6GW storage and 70% of its electricity from renewable sources by 2030 [29]. A study conducted found that, in order for California to achieve the 100% clean electricity goals, California will need approximately 2–11 GW of new operational long-duration ES by 2030, and between 45 and 55 GW of long-duration ES by 2045 [30].

What is the definition of a long-duration ES?: An ES installation is measured by both its rated capacity (power) and stored energy. Its duration is typically quantified by the number of hours that a storage system can provide the continuous output at its rated capacity. However, across the literature, there has not been a consensus of the definition of long-duration ESs and the defined duration varies from a few hours (e.g., 4 h or larger) to multiple days or even months. A commonly accepted definition from the US Advanced Research Project Agency-Energy (ARPA-E) is that long-duration ES must have a duration of greater than 10 h [31].

How to monetize the service from long-duration ES?: The value of long-duration ES will be tied to the applications for the provision of grid services. In the

United States, there has been a focused effort in wholesale market design over the past decade to facilitate the entry of smaller and short-duration storage devices, due to the high initial costs of these resources. The salient feature, i.e., more energy from long-duration ES over short-duration storage systems, has not been fully valued from the current market design. Below is a discussion of the typical applications of ES in the current power market and comments are given if a particular grid service helps justify the financial investment of long-duration ES projects.

Capacity firming: With the retirement of the conventional fossil-fueled synchronous generators and the growth in electricity demand driven by the continued electrification of the transportation and heating sectors, stronger economics, and increasing population, ES is increasingly deployed for the purpose of providing firm capacity and supporting the resource adequacy of the power grid. Different from conventional generation resources, the capacity credit, or effective load-carrying capability (ELCC), of an energy limited resource, such as the ES, is highly dependent upon its duration. ELCC reflects the load serving availability of a generator during the period of highest risk of a power outage. These high-risk periods typically have corresponded to hours of peak demand. However, the peak net demand (where net demand is measured by the gross demand minus the generation from variable renewable energies) is increasingly important. Take a 1 MW, 2 MWh storage system as an example. If the installation of this storage enables the system to supply additional load of 0.6 MW while maintaining the same level of reliability, the ELCC is calculated to be 0.6 MW, which means that this 2-h duration storage has a capacity credit of 60% of its rated power capacity. If the duration of the storage increases to 6 h and the ELCC is increased to be 1 MW, the 6-h storage will then have a capacity credit of 100% of its rated capacity. The capacity credit of a storage system is highly dependent upon the load shape or net load shape, and it is location and time dependent. In many parts of the United States that are summer peaking, it has been found that 4-h duration storage devices can provide high capacity credit [32]. This desired storage duration is also consistent with the threshold value set for full capacity credit established by many market regions in the United States, such as New York Independent System Operator. The system level resource adequacy program in the United States does not currently reward a longer duration ES system and a diminishing return on investment should be expected when the duration of an ES device is increased.

It is worth noting that with the continued electrification by replacing fossil fuel space heating to electricity heating and the grid integration of EVs, the load profiles across regions in the United States are likely to shift significantly. Expected impacts include: (1) EV charging and residential space heating lead to early evening peak loading and (2) winter peaking of the residential heating load. Already, a double daily net peak load has been observed in places, with morning and evening peaks, and a valley during peak solar output. The changing load profile will impact the capacity credit from ES.

It is also worth noting that market rules for storage duration are based on the system resource adequacy analyses. For example, required storage duration could be based on the requirement that the Loss of Load Expectation (LOLE) be limited to 1 day of interruption every 10 years. Resource adequacy provides guidance to maintain generation resources to serve the load and reduce the average risk of lost load due to generation capacity deficiency in the long run. However, it does not provide an effective means to assess the risk of losing power supply to load under interdependent forced outages of generation assets, integrated energy infrastructure issues such as the gas-power network, and extreme events. When these factors are considered in resource adequacy programs, there may be some room for boosted return on investment for a longer duration of ES.

Energy arbitrage: An ES resource will shift energy by charging and discharging at different time instants when net wholesale market revenues are positive after factoring in the cycling efficiency, i.e., the charging efficiency and discharging efficiency. Wholesale energy prices vary across the day, reflecting diurnal changes in load or net load and the resulting difference of the marginal cost of energy supply. This variation means that potential energy arbitrage is not equal in value across the day and that the incremental economic benefit for each additional hour of energy capacity declines due to the suppressed price spread. This phenomenon may potentially lead to storage saturation issues for energy arbitrage. If a long duration ES has a roundtrip efficiency of 90%, the storage earns profit through energy arbitrage by charging at the low energy price and discharging at the high price if the price ratio is no larger than 0.90 before incorporating the variable operational and maintenance cost and startup/shutdown cost. It has been found that the potential storage energy value in the day-ahead market from energy arbitrage has increased significantly in the last few years due to the effect of solar energy in suppressing midday prices, which creates a large price spread [33]. However, with the increasing penetration of solar energy, the double daily valley energy price occurs at early morning and early afternoon, and the period from the valley price to peak price hours is decreased (e.g., approximately 6 h), which indicates that the incremental economic benefit from a longer ES may decrease. This is testified by the preliminary analysis results of the Southern California Edison (SCE) utility transmission system zone, where the analysis showed in 2019, the 4-h and 8-h ESs achieve 76% and 91% of the maximum potential revenue, respectively, as measured by the very-long duration result with a 20-h ES installation [33]. The analysis results indicate that even though the high penetration of solar energy will lead to increased opportunities for ES for energy arbitrage, additional duration of the ES will however add diminishing profit from energy arbitrage despite a higher if not proportionally increasing capital cost for the added storage duration.

Ancillary services: Ancillary services consist of frequency regulation, primary frequency response, contingency reserves, voltage control, and black start. Depending on the grid services, both capacity and energy commitment at different time scales may need to be fulfilled from the service providing resources. Long-duration storage, such as some pumped storage hydro, has long been proven for supplying these services for decades due to its flexibility. However, the short-time scale grid services with high prices, such as frequency regulation markets, have been gradually taken by recent installed short-term storage including lithium-ion batteries. The market size for ancillary services is generally small and can be

Table 3.5   Ancillary service procurement quantities (MW) and selected prices ($/MW) for US RTOs in 2019 [33]

RTO	ISO-NE	NYISO	PJM	SPP	ERCOT	CAISO
Freq. reg. (MW)	50–200	175–300	525–800	Variable	100–700	350–500
Avg. freg. reg. price ($/MW)	$21.96	$9.08	$16.30	RU* $9.45 RD* $7/60	RU-$23.14 RD-$9.06	RU-$13.27 RD-$11.74
Contingency reserves (MW)	1,500–1,770	1,310	2,000–4,000	~1,500	2,300–3,200	700–1,100
Avg. spin. res. price ($/MWh)	~$2	$4.39 (NYC zone)	$3.01 (RTO zone)	$5.17	$26.61	$7.39
Ramping res. (MW)	None	None	None	None	None	0–700
Primary freq. res. (MW/Hz)	38	48.8	258.3	86.9	381	197.6

*RU, regulation up; RD, regulation down.

saturated by the increasing availability of flexible resources, e.g., energy limited resources and demand responsive resources. The average hourly procurement quantifies (MW) of the key ancillary services in the US Regional Transmission Owners (RTOs) and the average prices are summarized in Table 3.5 [33]. Due to the changing load profiles, the procured ancillary services may be varying in a range.

Frequency regulation: This ancillary service can have the highest market price and may get additional financial compensation for performance as measured by the regulation mileage such as in PJM's wholesale market. While frequency regulation is often the source of the greatest potential revenue from ES, the size of the frequency regulation market is typical only a fraction, e.g., 1–3%, of the size of the energy market, making it prone to saturation with wide deployment of flexible resources. Note that the long-duration ES such as pumped storage hydro (PSH) is also playing as the service supplier when the PSH is running in the generating mode and the unit output can be flexible. The newly developed variable speed PSH and ternary PSH will enable these traditional long-duration ES assets to provide frequency regulation in both pumping and generating modes. However, it is projected that within a few years, most of the frequency regulation market in the United States will be provided by the short-term ES, such as lithium-ion batteries due to the fast response and high roundtrip efficiency, and the market will be saturated attributed to its limited market size [33]. Lower prices may be expected over the coming decade as more battery storage resources enter and displace conventional resources. As such,

long-duration ES is expected to continue to participate in the frequency regulation market while there will be rapid declines in revenue streams from this
service in the years to come.

Contingency reserve: Currently, the contingency reserve provided by ES is
primarily from the traditional long-duration ES, e.g., the existing PSH in both
pumping and generation modes. For constant speed PSHs in pumping, the total
pumping demand is provided as the contingency reserve. The primary feature for
different storage resources to provide contingency reserve is the required minimum
continuous energy. In the United States, the continuous energy supply is about
30 min in several regions, which make it suitable for both short-term and long-
duration ES to provide these services even though the market prices for contingency is low as shown in Table 3.5. The revenue stream for ES from contingency
reserve services only accounts for a fraction of the total revenue. It is also worth
noting that contingency reserves are usually procured over large load zones, e.g.,
there are 11 load zones in New York Independent System Operator's service territory. As such, there is no location limitation and no specific differentiation in
these services from ES with different durations.

Primary frequency response: This service is to arrest frequency deviations when the system experiences disturbances and/or contingencies until
generation resources are re-dispatched to restore frequency back to the predetermined range. The eligible resources to provide this service are required to
autonomously respond within milliseconds to seconds. This service for each
Regional Transmission Owner or large utility as required by NERC is mostly
met from committed synchronous generators even though some literature has
reported that inverter-based resources, e.g., wind farms and solar farms, may
potentially be able to provide this service. ES is allowed to participate in this
market per FERC Order 841. As the energy needed for primary frequency
response is small, the short-duration battery storage with advanced inverters of
autonomous frequency–watt response functions is well positioned to provide
this service. This does not exclude long-duration ES to provide this service.
However, this service is unlikely to be a main revenue stream for future long-
duration ES projects. It is worth noting that currently, only ERCOT has developed a wholesale market product for resources to provide primary frequency
response in the United States.

Ramping reserve: This service is held in system operation to address
forecast uncertainty related to intra-hourly system ramps, which are primarily
due to increasing production variability of variable renewable energy and load
demand uncertainty. This reserve service, largely provided by conventional
resources, such as thermal generators and hydropower, has been implemented
by CAISO and MISO. The dispatch intervals for this reserve ranges from 5 min
to 15 min without additional continuous energy requirement. As such, this service can be supplied by both short-duration and long-duration ES in principle.
However, due to the active charging and discharging, the management of state
of charge plays a critical role for this service, which makes long duration ES
more feasible to provide this service. The need of long duration ES for this

service is location-dependent. For New York State, the large availability of hydropower provides the flexibility to meet the ramping need. The hydropower is subject to the amount of available water and the service may be impacted by the amount of snow fall each year, i.e., dry year versus wet year, which makes the long-term revenue stream from long-duration ES for this service hard to estimate.

Voltage control: This service is provided to regulate system voltage by injection and/or absorption of reactive power at a particular location in the transmission network. The RTOs procure this service using cost of service or energy opportunity costs from resources with volt-var or volt-watt functionalities. This service can be provided by any resources that are able to provide reactive power regulation regardless the energy/power capacity. Therefore, the advanced converter for inverter-based resources, including wind, solar, and storage, is able to provide voltage control service in addition to static/dynamic var compensation devices and the conventional generators through automatic voltage regulators. This wide range of resources to provide voltage control makes it hard to be the main revenue stream for long duration ES projects in the future.

Blackstart: This service is to provide the self-starting capability to accelerate system restoration by cranking generators and energizing transmission assets after a complete blackout. In the United States, three methods of procuring blackstart resources are being implemented currently: (1) cost of service, which has been adopted by NYISO, CAISO, and PJM; (2) flat-rate payment, which is implemented by ISO New England; and (3) competitive procurement in the market as implemented by ERCOT. Regardless of the procurement mechanism, the blackstart resources are expected to sustain the response for minutes to hours to allow for the restart-up and resynchronization of generators with the power grid. Therefore, long-duration ES is favored over short-duration ES for the prolonged energy supply. In New York State, the Blenheim-Gilboa Pumped Storage Hydro is providing the blackstart service for example. It is also worth noting that short-duration ES can be coupled with other assets, such as heavy duty gas turbines, to provide the blackstart service, which has been demonstrated by the technology provider in collaboration with the utility in California [34]. This service provides an opportunity for long-duration ES. However, there is a lack of clear requirement of the minimum blackstart capability for each RTO from the NERC standards and it is hard to provide main revenue stream from this service for long-duration ES projects.

It is worth noting that ES in general can provide multiple grid services and values from different applications are stacked to maximize revenue to the asset owners/developers or to maximize the grid benefit. The products based on the current market structure in the United States are summarized in Table 3.6 and some brief discussions of short-duration versus long-duration ESs have been included. Currently, the ES is still relatively expensive grid asset compared to the conventional fossil-fueled generators. Value stacking provides an effective way to increase the potential of an ES project to be financed and approved. It is also worth noting that some applications such as transmission and distribution upgrade deferral by ES

Table 3.6   Grid supporting market products for ES technologies

Products		Short-duration ES	Long-duration ES
Capacity firming		Higher return per MWh	Continue to serve but with a diminishing return with longer duration
Energy Arbitrage		Higher return	Continue to serve but with a diminishing return with longer duration
Ancillary Serv	Frequency regulation	Preferred	Continue to provide this service
	Contingency reserve	Less preferred	More preferred
	primary frequency response	Continue to provide	Continue to provide
	Voltage control	Continue to provide	Continue to provide
	Blackstart	Less preferred but could be coupled with other technology	More preferred with longer duration

are not fully captured by the market products as discussed above. Even though transmission and distribution deferral may only need ES to charge and discharge for as little as a few hours per year for load flattening. The required energy may make longer duration storage more favorable considering the uncertainties of both loads and state of charge management of storage.

How does long-duration ES contribute to grid resilience: Resiliency is defined as "the ability to withstand and reduce the magnitude and/or duration of disruptive events, which includes the capability to anticipate, absorb, adapt to, and/or rapidly recover from such an event" [35]. Natural hazards and extreme weather conditions, equipment and network failures, and cyber/physical attacks have the potential to negatively impact grid resilience. These hazards and events, including high impact low probability events, are commonly excluded from reliability calculation. The 2021 Texas statewide rolling blackouts showed that the extreme event can lead to unserved load for multiple days with economic damages in the millions of dollars or more. These high costs and long-duration power outages make the long-duration ES attractive for some consumers. Currently, even though the community has put efforts to develop the resilience metrics, there are no widely adopted standards by the power industry yet. This makes it difficult to quantify the benefits from different ES scenarios. This service from ES, especially long-duration ES with superior energy reserve for improved grid resiliency, has not been fully monetized in current market structures.

3.5 Economic aspects of behind the meter storage

Previous sections have discussed the role of storage systems in the future power grid. While not all storage technologies are suitable for geographic location in urban areas, they can all benefit the urban consumers through their economic, reliability, and/or resilience capabilities.

Storage assets will be owned by the power utility, the electricity consumer, or a third party provider. This section will discuss "behind the meter" storage owners, both the individual consumers and the distribution companies such as municipal electricity departments. The individual consumers cover the range of residential, commercial, and industrial. While the scale of these consumers vary over many orders of magnitude, the basic economic decision making is largely similar.

The storage decision for consumers boils down to economics. There are, however, a number of aspects that go into this economic decision:

1. "Blue sky day" economics of normal day to day operation
2. Economics of improving reliability
3. Economics of improving resilience
4. Maintenance, local issues, and convenience

The following sections discuss these topics.

3.5.1 Blue Sky Day Economics

The power grid operates in its normal mode nearly all of the time—its availability is typically well over 99.9%. The individual consumer will work to recover a significant portion of the costs of the storage during these times.

This cost recovery is covered by the rate structure and the consumer's own load profile. The rate structure of most commercial and industrial consumers is based on a demand charge and an energy charge. Most residential consumers to date pay for electricity with an energy charge plus a fixed monthly fee. Energy rates can be market based, time of day, or fixed.

Demand charge: The demand charge is generally based on the peak monthly power demand of a consumer. It is based on the average power drawn over a period of time, typically between 15 min and 1 h. Behind the meter storage can be used to reduce this monthly peak demand, by supplying a portion of the load from storage during high demand periods, and by recharging the storage during periods of low demand. This practice is commonly referred to as "peak shaving."

Market-based energy rates: Market-based rates are based directly on the costs of generation at a given time. The utility operators—typically ISO's or regional transmission organizations (RTO's)—select the generation mix based on economic and reliability considerations. These rates vary with usage, weather, fuel cost, equipment outages, and other factors. In deregulated energy markets, generators bid into the day ahead and the real-time markets. Utilities then tend to offer their customers energy rates based on these markets. In New York, for example, National Grid offers hourly market-based rates based on the energy bids accepted

Hourly Rates, $ per MWH

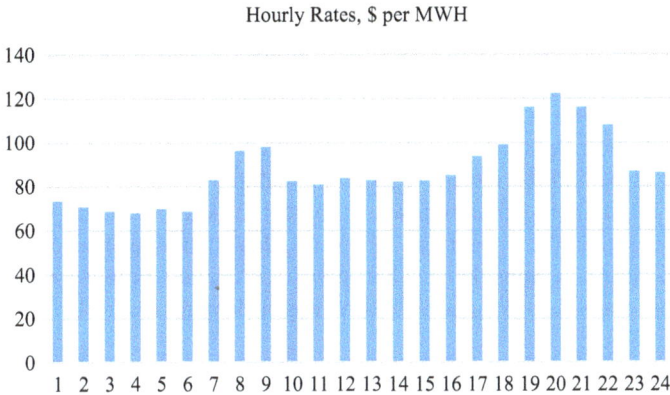

Figure 3.2   Example National Grid hourly market base rates for their New York territory

by the New York ISO. Each day, the hourly energy rates for the next day are posted by 4 pm. Their rates for their Capital Region District on April 18, 2022 for Large General Service Customers served at the primary level are shown in Figure 3.2. This data is taken from the National Grid web site https://www.nationalgridus.com/ Upstate-NY-Home/Rates/Service-Rates.

A market-based rate consumer can use this daily information to lower their energy costs through changing their usage, or through using storage to supply energy during high cost hours and store energy in low cost hours. This is often referred to as energy arbitrage. On the particular day shown in Figure 3.2, the peak power cost is nearly double the overnight cost of energy. There is a significant amount of historical data available on the web site as well. Historical data over a number of years can be used to predict future cost savings when making investment decisions. However, there is no guarantee that future energy costs will follow the same patterns.

Time of day rates: Some utilities offer time of day energy rates, particularly to residential customers. Sacramento Municipal Electric District (SMUD) offers off-peak and peak rates during non-summer months and off-peak, mid-peak and peak rates during summer months. These published rates allow potential BTM storage buyers to better estimate the cost benefits of the units they are evaluating.

Some utilities have gone to a market-based approach to their time of day rates and do not publish their future rates. In these cases, investment decisions must be made on past performance.

Flat rates: At this point in time, many residential customers are billed at a flat energy rate for their service. In some cases, this is the only rate structure offered, but in many cases, the consumers have several options and have selected the flat rate option. Flat rates without demand charges do not offer these consumers that chance to save money by installing ES systems. Flat rates also may be available to commercial or residential consumers for the energy component of their bills, either directly from their utility service provider, or through a bilateral arrangement with a generation provider. In this latter case, the consumer would pay a delivery charge to their service provider. While these consumers could save money through peak shaving, they could not save money on their energy use pattern.

The specific rate structure at a given location has a significant impact on the economics of behind the meter BESS. While adding a BESS in conjunction with renewable intermittent generation seems logical, it is the rates and rate structures that will determine if it is practical. In general, this is but one of several significant aspects that must be considered when power companies propose rate plans and public utility regulators consider in approving or denying these plans.

3.5.2   Reliability cost impacts

There are a number of things that can impact loads on the electric power distribution system. These include:

- Momentary voltage sags
- Momentary voltage swells
- Momentary interruptions
- Sustained interruptions
- Voltage waveshape

Of these, sustained interruptions are considered reliability issues, and the others are generally categorized and power quality issues. Sustained interruptions and momentary interruptions are generally the most common issues that cause load disruption. Momentary voltage sags and swells can also impact certain loads. Interruptions are considered momentary when their duration is less than 5 min. Interruptions from 5 min up to as much as a day are considered sustained. Longer term interruptions impacting many consumers are generally considered resiliency events.

Behind the meter ES systems are often used to provide ride through support for loads during momentaries and sustained interruptions. These are most generally battery systems, and are often called uninterruptible power supplies (UPS). The largest UPS's go into the MW range, and serve larger mission critical loads. Most UPS's are designed to serve one or several critical loads, such as personal computers or medical test equipment.

Each UPS system has both a power rating and a duration rating. The UPS will serve loads up to the power rating, for a period of up to the duration rating. Not all UPS's will support load for the full duration of a sustained interruption. In these cases, they do provide for an orderly shutdown, which is preferable and less costly than an unexpected sudden crash. For the most critical loads, UPS systems can be sized to provide short-term ride through while backup generators start up and come on line to serve the load.

These reliability measures are clearly cost effective and widely used. Most, however, are intended solely for this use, and do not provide for the other services that BESS systems can perform.

The cost savings due to a UPS vary widely depending on the application. These savings are generally in terms of saving per event. In a 2019 survey conducted by the firm Information Technology Intelligence Consulting, 98% of businesses estimated that a one hour outage would result in at least $100,000 in costs. Many utilities now report on the average interruption frequency (SAIFI) and duration (CAIDI) statistics for sustained interruptions on their service areas.

When facility size multi-function storage units are installed, these could be used to replace or augment existing UPS systems as capacity allows. In these cases, a portion of the BESS ES capacity must be reserved for serving the critical loads during sustained interruptions.

3.5.3 Resiliency cost impacts

Resiliency events are multi-day power grid interruptions that impact wide areas. These extreme events can be due to extreme weather, physical or cyberattacks on the grid, or unintended grid blackouts. In his March 10, 2021 testimony before the US Senate, Frank Rusco [36] reports estimates that extreme weather events have cost power grid customers $55 billion in the 2006–2019 time period. He further reports that these costs are expected to rise significantly by the end of the twenty-first century due to climate change.

Planning for resiliency events is still evolving. A recent study [37] reports on methods to estimate the cost impacts of long-term interruptions on residential, commercial, and industrial grid customers. As resiliency events generally have a low probability of happening at a single residence or commercial or industrial site, it can be difficult to justify significant expenditure for this purpose alone. In the residential sector, a potential application could be a combine PV array and ES system that include inverters that have the capability to operate independently of the power grid. This type of system could provide energy, demand, reliability, and resiliency benefits.

Finally, in addition to the issues in the previous paragraphs, behind the meter BESS must meet local building and fire codes. The power company involved will have interconnection requirements that must be met before a system can be installed. ES systems also take up space and require maintenance.

3.6 Summary

This chapter discusses the opportunities and challenges of ES systems providing services in the electric power grid. This includes the integration of these systems on to the grid, the planning process required to successfully deploy them, the need for long-term storage, and the economics of their installation behind the meter in residential, commercial or industrial settings. In this chapter, it can be seen that BESS will be critical in the success of moving the power grid to non-carbon energy sources. The chapter also discusses a number of the challenges that need to be addressed to realize this goal of an active, flexible clean power grid with significant levels of storage.

References

[1]   Five Steps to Energy Storage—Innovation Insights Brief 2020. Published by World Energy Council, London, 2020.

[2]   The State of Storage: Energy Resources in New York's Wholesale Electricity Markets. New York Independent System Operator. Rensselaer, New York. 2017.

[3]   T. Ortmeyer and T. Vu, "Energy storage peak shaving feasibility for Tupper Lake, Lake Placid, and Massena Municipal Electric Departments," NYSERDA Report Number 20-17, 2020.

[4]   A. Colthorpe, "BloombergNEF: average battery pack prices to drop below US$100/kWh by 2024 despite near-term spikes," Dec 1 2021, Energy Storage news.

[5]   Ramasamy Vignesh, David Feldman, Jal Desai, and Robert Margolis, U.S. Solar Photovoltaic System and Energy Storage Cost Benchmarks: Q1 2021. Golden, CO: National Renewable Energy Laboratory. NREL/TP-7A40-80694, 2021.

[6]   https://cleantechnica.com/2021/02/21/tesla-solar-powerwall-gives-texan-electricity-in-middle-of-blacked-out-neighborhood/

[7]   IEEE Standard 1547-2018 "IEEE Standard for Interconnection and Interoperability of Distributed Energy Resources with Associated Electric Power Systems Interfaces." IEEE, April 2018.

[8]   L. Meng, J. Zafar, S.K. Khadem, *et al.*, "Fast frequency response from energy storage systems—a review of grid standards, projects and technical issues," *IEEE Transactions on Smart Grid*, vol. 11, no. 2, pp. 1566–1581, 2020, doi:10.1109/TSG.2019.2940173.

[9]   G. Carpinelli, G. Celli, S. Mocci, F. Mottola, F. Pilo, and D. Proto, "Optimal integration of distributed energy storage devices in smart grids," *IEEE Transactions on Smart Grid*, vol. 4, no. 2, pp. 985–995, 2013, doi:10.1109/TSG.2012.2231100.

[10] V. B. Venkateswaran, D. K. Saini, and M. Sharma, "Approaches for optimal planning of energy storage units in distribution network and their impacts on system resiliency," *CSEE Journal of Power and Energy Systems*, vol. 6, no. 4, pp. 816–833, 2020, doi:10.17775/CSEEJPES.2019.01280

[11] X. Wu and Y. Jiang, "Source-network-storage joint planning considering energy storage systems and wind power integration," *IEEE Access*, vol. 7, pp. 137330–137343, 2019, doi:10.1109/ACCESS.2019.2942134.

[12] H. Nemati, M. A. Latify and G. R. Yousefi, "Optimal coordinated expansion planning of transmission and electrical energy storage systems under physical intentional attacks," *IEEE Systems Journal*, vol. 14, no. 1, pp. 793–802, 2020, doi:10.1109/JSYST.2019.2917951.

[13] G. Pulazza, N. Zhang, C. Kang, and C. A. Nucci, "Transmission planning with battery-based energy storage transportation for power systems with high penetration of renewable energy," *IEEE Transactions on Power Systems*, vol. 36, no. 6, pp. 4928–4940, 2021, doi:10.1109/TPWRS.2021.3069649.

[14] F. Yao, T. K. Chau, X. Zhang, H. H. -C. Iu, and T. Fernando, "An integrated transmission expansion and sectionalizing-based black start allocation of BESS planning strategy for enhanced power grid resilience," *IEEE Access*, vol. 8, pp. 148968–148979, 2020, doi:10.1109/ACCESS.2020.3014341.

[15] Z. Luburić, H. Pandžić, and M. Carrión, "Transmission expansion planning model considering battery energy storage, TCSC and lines using AC OPF," *IEEE Access*, vol. 8, pp. 203429–203439, 2020, doi:10.1109/ACCESS.2020.3036381.

[16] P. Xiong and C. Singh, "Optimal planning of storage in power systems integrated with wind power generation," *IEEE Transactions on Sustainable Energy*, vol. 7, no. 1, pp. 232–240, 2016, doi:10.1109/TSTE.2015.2482939.

[17] M. Asensio, P. Meneses de Quevedo, G. Muñoz-Delgado, and J. Contreras, "Joint distribution network and renewable energy expansion planning considering demand response and energy storage—Part I: stochastic programming model," *IEEE Transactions on Smart Grid*, vol. 9, no. 2, pp. 655–666, 2018, doi:10.1109/TSG.2016.2560339.

[18] M. Asensio, P. Meneses de Quevedo, G. Muñoz-Delgado, and J. Contreras, "Joint distribution network and renewable energy expansion planning considering demand response and energy storage—Part II: numerical results," *IEEE Transactions on Smart Grid*, vol. 9, no. 2, pp. 667–675, 2018, doi:10.1109/TSG.2016.2560341.

[19] B. Odetayo, M. Kazemi, J. MacCormack, W. D. Rosehart, H. Zareipour, and A. R. Seifi, "A chance constrained programming approach to the integrated planning of electric power generation, natural gas network and storage," *IEEE Transactions on Power Systems*, vol. 33, no. 6, pp. 6883–6893, 2018, doi:10.1109/TPWRS.2018.2833465.

[20] R. A. Jabr, I. Džafić, and B. C. Pal, "Robust optimization of storage investment on transmission networks," *IEEE Transactions on Power Systems*, vol. 30, no. 1, pp. 531–539, 2015, doi:10.1109/TPWRS.2014.2326557.

[21] L. Yang, R. Xei, W. Wei, C. Sun, and S. Mei, "Coordinated planning of storage unit in a remote wind farm and grid connection line: a distributionally robust optimization approach," in *2019 IEEE Innovative Smart Grid Technologies – Asia (ISGT Asia)*, 2019, pp. 424–428, doi:10.1109/ISGT-Asia.2019.8881677

[22] S. K. Wankhede, P. Paliwal, and M. K. Kirar, "Bi-level multi-objective planning model of solar PV-battery storage-based DERs in smart grid distribution system," *IEEE Access*, vol. 10, pp. 14897–14913, 2022, doi:10.1109/ACCESS.2022.3148253.

[23] S. Wogrin and D. F. Gayme, "Optimizing storage siting, sizing, and technology portfolios in transmission-constrained networks," *IEEE Transactions on Power Systems*, vol. 30, no. 6, pp. 3304–3313, 2015, doi:10.1109/TPWRS.2014.2379931.

[24] V. B. Venkateswaran, D. K. Saini, and M. Sharma, "Environmental constrained optimal hybrid energy storage system planning for an Indian distribution network," *IEEE Access*, vol. 8, pp. 97793–97808, 2020, doi:10.1109/ACCESS.2020.2997338.

[25] J. Kim and Y. Dvorkin, "Enhancing distribution system resilience with mobile energy storage and microgrids," *IEEE Transactions on Smart Grid*, vol. 10, no. 5, pp. 4996–5006, 2019, doi:10.1109/TSG.2018.2872521.

[26] X. Cao, T. Cao, F. Gao, and X. Guan, "Risk-averse storage planning for improving RES hosting capacity under uncertain siting choices," *IEEE Transactions on Sustainable Energy*, vol. 12, no. 4, pp. 1984–1995, 2021, doi:10.1109/TSTE.2021.3075615.

[27] M. Sedghi, A. Ahmadian, and M. Aliakbar-Golkar, "Optimal storage planning in active distribution network considering uncertainty of wind power distributed generation," *IEEE Transactions on Power Systems*, vol. 31, no. 1, pp. 304–316, 2016, doi: 10.1109/TPWRS.2015.2404533.

[28] NYISO, *Power Trends 2021: New York's Clean Energy Grid of the Future*. https://www.nyiso.com/documents/20142/2223020/2021-Power-Trends-Report.pdf/471a65f8-4f3a-59f9-4f8c-3d9f2754d7de

[29] New York State, *New York State of the State Book 2022*. https://www.governor.ny.gov/sites/default/files/2022-01/2022StateoftheStateBook.pdf

[30] California Energy Storage Alliance, *Long Duration Energy Storage for California's Clean, Reliable Grid*. https://static1.squarespace.com/static/5b96538250a54f9cd7751faa/t/5fcf9815caa95a391e73d053/1607440419530/LDES_CA_12.08.2020.pdf

[31] National Renewable Energy Laboratory, *Storage Future Study: The Challenge of Defining Long-Duration Energy Storage*. https://www.nrel.gov/docs/fy22osti/80583.pdf

[32] A. W. Frazier, W. Cole, P. Denholm, D. Greer, and P. Gagnon, "Assessing the potential of battery storage as a peaking capacity resource in the United States," *Applied Energy* vol. 275, p. 115385, 2020. https://doi.org/10.1016/j.apenergy.2020.115385

[33] U. Helman, B. Kaun, and J. Stekli, "Development of long-duration energy storage projects in electric power systems in the United States: a survey of factors which are shaping the market," *Frontier in Energy Research*, vol. 8, Article 539752, 2020.

[34] GE, *Hybrid Solutions: GE Completes First Battery Assisted Black Start of a GE Heavy Duty Gas Turbine*. https://www.ge.com/news/press-releases/hybrid-solutions-ge-completes-first-battery-assisted-black-start-ge-heavy-duty-gas

[35] D. Ton and W. P. Wang, "A more resilient grid: the U.S. Department of Energy Joins with Stakeholders in an R&D plan," *IEEE Power & Energy Magazine*, vol. 13, no. 3, pp. 26–34, 2015.

[36] F. Russo, Director, Natural Resources and Environment, US Government Accountability Office. "Electric Grid Resilience: Climate Change is Expected to Have Far-reaching Effects and DOE and FERC Should Take Actions." Testimony before the Committee on Environment and Public Works, United States Senate, March 10, 2021.

[37] S. Bird, C. Hotaling, A. Enayati, and T. Ortmeyer. "Resilient community microgrids: governance and operational challenges," in *The Energy Internet*, 1st ed. Sawston: Woodhead Publishing, 2019, pp. 65–95 (Chapter 4).

Chapter 4

International experience on distributed energy storage

Maise N.S. Silva[1], Rafael S. Salles[2] and Paulo F. Ribeiro[1]

Energy storage systems (ESS) can provide different types of services, which change depending on the characteristics of each technology. The electricity matrix will drive the adoption of more appropriate storage technologies. For example, pumped storage hydroelectric is a consolidated technology used in many countries to guarantee energy security, sustainability, and lower electricity bills. Therefore, the diversification and decentralization in the energy matrix raised the need for diversification and adoption of different storage systems.

This chapter is dedicated to analyzing energy storage experiences, bringing information about countries' electrical matrix, how storage services are reimbursed, and the regulatory practices that support storage systems. It covers how implementation and operation are structured since the regulatory framework (comprising laws, regulations, and operating procedures) can interfere with the storage industry's growth, creating opportunities or inhibiting them. Based on those experiences, it is possible to identify issues and challenges in the existing systems, highlight detachable experiences, and suggest paths for other markets.

4.1 World context for distributed energy storage

United Nations Climate Change Conference, COP-21, took place in Paris in 2015 and was a key event for consolidating climate change mitigation strategies. The Paris agreement, signed by the countries, comprises the electric sector as the major player in achieving the commitments. Renewable generation, energy storage systems, and energy efficiency figures as the main contributors to reducing global CO_2 emissions in the sector.

Storage systems have become the center of attention in many countries because of the impacts it causes on electric systems. Back-up power, peak shaving, ancillary services, and demand responses are services provided by ESS capable of

[1]Federal University of Itajubá, Brazil
[2]Luleå University of Technology, Sweden

interfering with climate change in the future. For this sake, governments supporting the use must consider incentives and regulations and ensure the most appropriate technologies will be applied. It is possible to optimize the operation of renewable energy sites, reduce generation costs, and minimize emissions by integrating storage systems into power systems [1,2].

Furthermore, another hazard event brought still more attention to the ESS. The COVID-19 outbreak significantly impacted the power sector globally because of the restrictions used by governments to prevent the disease from spreading. While the industrial sector suffered severe losses, the energy storage industry suffered less due to increased demand for energy storage, aiming for energy safety. Some countries developed policies, standards, and regulations to stimulate power supply through renewable energy and energy storage systems. Those initiatives should be adopted to adjust the balance between supply and demand for peak shaving when supply exceeds demand [3]. The pandemic also accelerates studies and investments in non-conventional energy storage technologies, such as fuel-cell by green hydrogen [4].

Geopolitical conflicts also interfere in the global electricity sector. Russia and Ukraine conflict started in February 2022, ignited discussions, and impacts mainly in European countries. Russia is second in natural-gas production, first in natural gas exportation, and third in petroleum production [5]. This conflict shows the urgency of an energetic policy shift, including new storage systems and renewable technologies. The race for renewable energies that was expected from this conflict favored the coal market. The imbalance between supply and demand has driven up coal prices, thus increasing investments in this polluting technology, which contrasts with the transition to cleaner energy in progress [6].

4.2   Regulatory framework

The regulatory framework for energy storage depends on the energy matrix, once the composition will guide the adoption of the best storage technologies. The initial steps to create a strong energy storage regulation are studying existing storage technologies on the matrix and the prospects for the electric system.

A specific definition for energy storage systems is needed to affect, for example, the creation of its regulation and, consequently, subsidies and market response to new electric system needs. A few countries have specific regulations, incentives, or support for energy storage systems. The availability of revenue for storage systems can help the development of new technologies and strengthen the market for consolidated ones.

Despite the participation of reversible hydroelectric power plants as a storage technology globally, it is in stationary storage systems with batteries that market expectations reside in the short, medium, and long term because of renewable energy intermittency. This type of technology has space in both small residential and large industrial systems and large-scale renewable energy facilities.

The services provided by the ESS can be negotiated in a regulated or liberalized contracting environment, varying according to the attributes of existing technologies and the needs of the electricity system in question. Germany, Australia, the United Kingdom, and the United States of America (USA – California, Massachusetts, North Carolina, and New York) are examples of countries where energy storage is contemplated on regulations created especially for or that affect it.

4.2.1 Germany

German energy matrix is quite diversified, with renewable sources and others powered by fossil fuels. An overview of the evolution of electricity generation in Germany by type of source can be seen in Figure 4.1. The installed power exceeded 200 GW, with solar and wind generation prominence among non-conventional renewables.

As can be seen, among the technologies in the matrix, pumped storage plants stand out for their storage attribute, corresponding to 6.2 GW of installed power with a production of 38.5 GWh in 2018, in addition to another 3 GW from Luxembourg, Switzerland, and Austria, plants that exchange energy with the country. The first pumped storage hydroelectric power plants in Germany were built around 1920, but the large-scale plants began operating in the 1970s and 1980s. Since then, however, the increase in capacity has stagnated. The only new plant developed in the country after 1990 was Goldisthal, with 1,100 MW of installed capacity.

The expansion of storage systems by pumped storage plants in Germany has faced geographical limitations and resistance to acceptance by the affected population. On the other hand, stationary storage systems with batteries (BSS) have shown rapid growth. Another factor that contributes to the development of the implementation of BSS is the growth of solar and wind energy projects in the German matrix. The increasing share of intermittent power plants in the matrix led the German government to start a program to install 300 MW in storage systems based on battery technology in 2013.

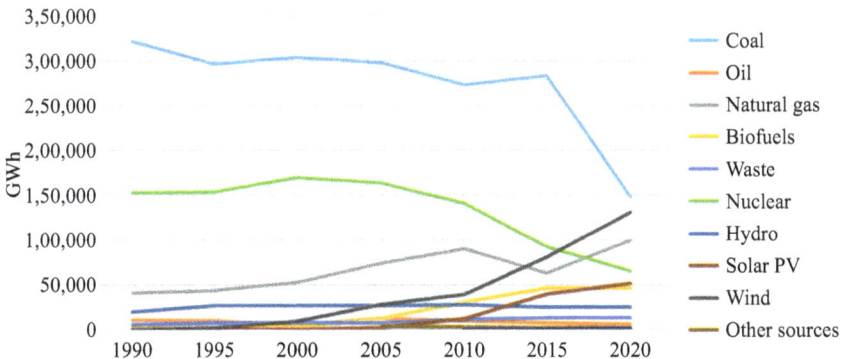

Figure 4.1 Evolution of Germany electricity matrix. Source: IEA, 2022.

There are three possible sources of revenue for pumped storage plants in Germany and, consequently, storage systems: price arbitrage, ancillary services provided to the grid, and payment for the capacity reservation. 2018 was the first year in which BSS revenues in the market exceeded those of pumped storage plants, a trend that should be maintained considering the limited growth of pumped storage plants and the growth potential of BSSs.

According to [7], small-scale storage facilities in Germany exceed 85,000 units, and distributed systems, together with collective supply points for electric cars, have been tested in pilot projects to provide services to the country's electricity grid. The insertion of storage systems behind the meter generated a reduction in payments for the use of the network, also enabling consumption profile optimization, with a decrease in costs with the acquisition of electricity.

Small-scale battery systems focus on consumer energy self-sufficiency, mainly with a capacity below 30 kWp, installed on residential rooftops. On an average, the share of own consumption of electricity generated by the photovoltaic panel increases from 35% to 70% when installed with storage systems, such as a battery.

Large-scale systems are tied to market control, playing an increasingly significant role in integrating and balancing large amounts of wind and solar energy in real-time. The German and European power regulation markets are attractive to large manufacturers and operators of battery systems. Around 1,250 MW of primary control power is sold in the German, Belgian, Austrian, Dutch, French, and Swiss coupled markets, but it can reach 3,000 MW across Europe. At the end of 2018, Germany had a cumulative capacity of around 400 MW from large-scale battery projects [8].

The "power-to-gas" technology, natural gas plants with hydrogen, will occupy a prominent place in the long-term energy storage plans. Several pilot power plants for gas are already in operation, as the country has an extensive natural gas network. On the same GTAI report, power-to-gas technology can, in addition to helping to stabilize the electricity system, replace renewable sources (solar and wind) and minimize the need for new investments in energy network expansion.

Regarding the adoption of batteries used for electric vehicles, this technology is considered the central pillar for the energy transition in the country. The transport industry and the energy sectors are the biggest generators of greenhouse gases in the European Union and Germany, accounting for a fifth of CO_2 emissions.

A relevant trend for storage systems in the German market is the trend toward the insertion of electric vehicles. At the end of 2019, around 180,000 electric vehicles were supported by an infrastructure of approximately 7,900 AC charging stations and more than 1,400 DC stations on German roads. As the German automotive industry is known worldwide for innovation, in 2019, the market had around 40 new models of electric vehicles (EV) presented.

According to [8], Germany and France lead the production of supercapacitors in Europe, although the most prominent producers are in Asia. As for superconducting magnetic energy storage (SMES), in Germany, more specifically at Karlsruhe Institute of Technology (KIT), there is a pilot project with a hybrid concept of SMES in combination with hydrogen that has been studied in detail. The first small superconducting MgB2 coil was built and tested, which combines fast SMES operation with liquid hydrogen storage for large capacities.

For compressed air storage systems, the ADELE project—Adiabatic Compressed Air Energy Storage for Electricity Supply—and another located in the city of Huntorf, which has been working with diabatic technology since 1978, stand out.

In December 2020, the German parliament passed a new version of the Renewable Energy Sources Act, known as "EEG-Novelle." According to German Energy Storage Association (BVES), the EEG Novelle needed to resolve the regulatory gap in the definition of what can be characterized as storage systems.

Germany has been a pioneer regarding energy storage technologies, but their definitions were general and related to "energy generation or charging discharging based device." There is no regulation for projects, acts, or initiatives specifically related to energy storage. Recently the energy storage definition was updated to "the final use of electrical energy is postponed to a later point in time than when it was generated." This change can provide legal security for energy storage investments, and specific regulations can be created.

Regarding the German market's legal aspects, it is essential to note that in March 2019, the European Court of Justice (CJEU) concluded that the feed-in tariff did not constitute a subsidy since it does not have state funding. This discussion took years in the European community, but it ended, allowing the maintenance of the tariff that promotes new renewable energy projects, including storage systems.

The Berlin energy policy institute Energiewende, responsible for the German energy transition program, predicts that the EEG surcharge will peak around 2023 and decrease. The reasons are that expensive projects committed at the beginning of the EEG in 2000 will expire after its 20th year, and the new projects are much cheaper, given the technological evolution, resulting in lower generation costs.

The German government at the federal and even local levels has been promoting storage systems, of which we highlight:

- exemption from paying network usage fees for "in front of the meter" storage facilities;
- guaranteed payment of the feed-in premium (Marktprämie) by the government for 20 years, depending on the technology and size of the project;
- creation of a subsidized line of credit for small-scale storage systems, managed by the state bank KfW, in partnership with the Ministry for Economic Affairs and Energy, where a low interest subsidy was established. Grant recipients are required to register with a scientific monitoring program and provide technical data for their home storage system. This program aims to provide a holistic and macroscopic picture of the storage market in Germany;
- programs that promote the construction of more energy efficient buildings, granting tax exemption to new buildings that adhere to such premises;
- Energy Research Program funding projects and companies dedicated to optimizing energy use, developing technologies (which includes energy storage systems), energy transition issues and better operation of existing electrical grids.

The German government must invest up to 80 million euros in projects that use storage as a form of energy [9]. Also, according to the report, research and development (R&D) is considered one of the most critical areas for developing the German economy. Industry and the public sector have committed to spending around 3% of the national GDP per year on R&D activities.

Germany's 7th Energy Research Program provides information on support for research and innovation measures until 2022. The plan covers many actions and can be considered complete and well-structured. It identifies five main lines of research: (i) energy transition (buildings and districts, industry, businesses and services in commerce, mobility, and transport, emphasizing "energy efficiency first"); (ii) energy generation (wind, solar, thermal generation); (iii) system integration (networks, storage, sector coupling, and hydrogen); (iv) cross-sectional research (energy system analysis, digitalization, CO_2 technologies); (v) nuclear safety, supporting the departure of nuclear energy from the matrix. However, the changes are not fully quantified in the plan. It also needs the definition of the deadline and funding targets or budgets required to achieve the desired changes.

The final plan states that hydrogen will be indispensable for the successful decarbonization of Germany's economy. Although the federal government considers only green hydrogen sustainable, the final plan indicates that other types of hydrogen can be used during the transition period due to Germany's integration with the European electricity system. Since Germany presented its final plan, the federal government has adopted a national hydrogen strategy, which includes a set of measures to increase hydrogen.

4.2.2   Australia

Australia has an electrical matrix that is heavily dependent on coal, although, when analyzing its evolution over the last 20 years, is observed a movement in favor of decarbonization, as suggested in Figure 4.2, which shows the development of the country's electricity matrix.

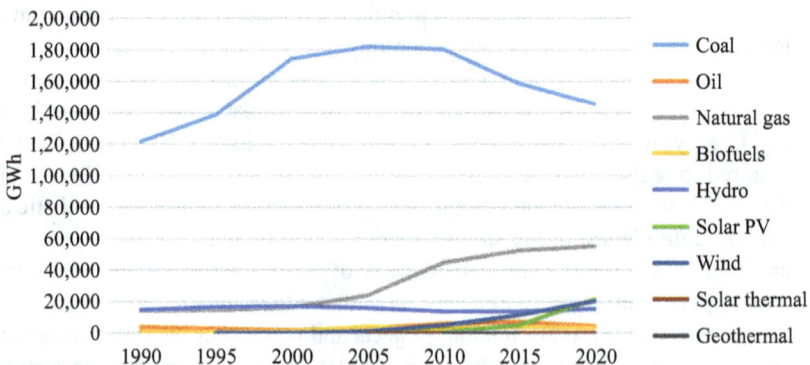

Figure 4.2   Evolution of Australia electricity matrix. Source: IEA (2022).

The energy market in Australia was built on four guiding principles contained in the electricity sector restructuring report commissioned by the Council of Australian Government (COAG) in 1991 to promote GDP growth. The restructuring of the electricity sector is based on fostering competition in generating and serving consumers, ensuring free access to the system, withdrawing public investment in assets, and improving/extending the interconnection of New South Wales (NSW) systems, Australian Capital Territory (ACT), Victoria and South Australia, and when economically feasible, to interconnect the power systems of Queensland and Tasmania.

The Regulatory Agency, the Australian Energy Regulator (AER), although responsible for regulating energy and gas services, allows the autonomy of the regions to adhere to the regulation, respecting the local reality fully.

At the federal level, the National Electricity Law (NEL) regulates the main aspects of access to the electric system, establishing the functions of the National Electricity Market (NEM). And it is responsible for developing a competitive electricity market for an electrical system less complex than the current one, where the load flows have a bidirectional energy flow, and the electrical system operator functions as the Australian Energy Market Operator (AEMO).

AEMO operates two electricity markets and power systems in Australia: the NEM, which covers Queensland, NSW, ACT, South Australia, Victoria, and Tasmania, and the Wholesale Electricity Market (WEM) in Western Australia.

The Australian Renewable Energy Agency (ARENA) is an independent agency charged with increasing the supply and improving the competitiveness of renewable energy in Australia. ARENA was created with a $2 billion federally funded fund, which aims to invest $200 million a year in renewable energy projects by 2022.

Australian regulatory framework is extensive and consists of laws, rules, and procedures that guarantee the operation of the market. Such rules and procedures are known as National Electricity Rules (NER). Both the NEL and the NER are instruments in continuous updating.

Among the initiatives of the legal framework that interferes in the evolution of storage systems in the Australian market, we highlight the following instruments:

- **Electric Industry Law** promoted the opening of the consumer market at all service levels in 2004.
- **The feed in tariff regime**, instituted in July 2010, offered $0.40 cents per kW/h injected into the grid for eligible projects, the program reached the target of 150 MW in 13 months, and the possibility of new participants ended.
- **The change in the tariff structure** in December 2014, AEMC started to require distributors to present their tariffs for approval, with the change in the tariff structure was another step in the remuneration of systems distributed in electrical systems.
- **Law on renewable energy** was passed in June 2015, which set a renewable energy penetration target (RET) at 20% by December 2020.

- **Rule 2017 No. 15** established a new methodology for the formation of short-term prices, the calculation of short-term prices started to be consolidated every 5 min (currently it is calculated for 30 min). Given the complexity and impact, the rule had a long transition period, and will be implemented from October 1, 2021, with the granularity of prices and remuneration for services that can be provided by storage systems will benefit.
- **Distributed Energy Repurchase Scheme (DEBS)** in August 2020 replaces the Renewable Energy Repurchase Scheme (REBS), where the amount paid for the supplied energy varies depending on the demand at the time of injection, while REBS the amount is fixed along the day.

Energy storage is becoming an increasingly important part of the NEM and recent predictions point to a growing role for storage in the future. ERC0280 was a public consultation, opened in 2020, to discuss changes to the rules for integrating storage systems into the NEM, which should receive contributions until February 11, 2021, in addition to a better definition of what a storage system is, the issues related to hybrid power plants are also addressed in the material under discussion.

As mentioned, although there are federal rules and procedures, some state initiatives deserve to be highlighted. Table 4.1 lists the public policies of state governments that promote the energy transition in search of a less polluting and more decentralized matrix.

Under the tax aspect, since 2015, the federal government has allowed Small Businesses to request a tax credit equivalent to 100% of the cost of acquiring energy storage systems, in the fiscal year of acquisition. These tax breaks are also observed in the United States.

Hydrogen caught the attention of politicians and industry, gained support at the federal and state levels. There are currently more than 20 hydrogen projects underway across the country; the Morrison government is betting; it will be cheaper than fossil fuel alternatives.

In March 2017, the Australian government announced the "Snowy 2.0" program to provide a substantial increase in hydraulic pumped storage capacity to the electrical system. The program proposes a single underground plant linking the Tantangara and Talbingo reservoirs. This would provide 2,000 MW of additional generating capacity and 350,000 MWh of pumped energy storage.

ARENA has estimated that the national electricity system will likely need 6–19 GW of available power by 2040 as the system is increasingly dominated by solar and wind power. According to a survey by ARENA, renewable energy can supply 90% of electricity by 2035 if backed by storage systems.

Electric car (EV) batteries can be a significant national energy asset. According to Bloomberg New Energy Finance, by 2040, 40% of Australia's vehicle fleet could be electric and act as a storage reservoir. The potential is similar to the capacity entered by the Snowy 2.0 program.

It is worth noting that there are several regulatory and technical barriers to the application of Vehicle-to-Grid (V2G) in Australia. As technical barriers to exporting electricity stand out the installation of an expensive residential metering

Table 4.1 Public policies of Australian states

State/ territory	Policy/incentive	Renewable energy objective
ACT	The $25 million next-generation battery storage scheme aims to provide subsidized systems to 5,000 homes and businesses by 2020.	100% in 2020
New South Wales	No current policy. The closing of a generous feed-in tariff spurred investment in batteries for home use.	Support the National Renewable Energy Goal
Northern Territory	No current policy. The Home Improvement Scheme previously offered vouchers of up to $4,000 for purchases including solar batteries and. Participants were required to fund at least 50%.	50% in 2030
Queensland	No interest loans and discounts will be provided in 2018 to drive battery uptake. Reverse auction of 100 MW for energy storage that is part of the 400 MW renewable energy auction. $50 incentive for homeowners who register their energy storage system in a new statewide database.	50% in 2030
South Australia	Operating 100 MW/129 MWh Li-ion battery. $100 million proposed grant program to facilitate batteries in 40,000 homes. Solar Thermal Plant in Port Augusta to supply electricity to the State Government. A $150 million renewable technology fund to support a variety of dispatchable renewable energy projects is fully allocated.	Support the National Renewable Energy Goal
Tasmania	Nation hydro-pumping feasibility study. $200,000 micronet tablet proposal.	25% in 2020
Victoria	Construction of two large-scale battery storage plants: 25 MW/50 MWh Tesla battery integrated with Edify Energy's Gannawarra Solar Farm and a 30 MW/30MWh fluence system in Ballarat. Backed by a $25 million investment from ARENA and $25 million from the government of Victoria	40% in 2025
Western Australia	No specific policy.	Support the National Renewable Energy Goal

system, the bidirectional charger function, and an intelligent inverter. Concerns about cost, driving range and access to charging stations causes a delay the Australian consumers to adopt EV's.

The Australian system compensates energy storage systems for responding to demand. When the system helps to control the amount of energy supplied in relation to the demand for energy for this type market, technologies with short response times have obtained substantial revenues [10].

The possibility to arbitrate prices in the short-term market is another possible remuneration that storage systems can take advantage of. Large increases in energy prices can produce cash flows for energy storage projects if they are optimized for discharge.

Another form of remuneration for storage systems has been revenue for ancillary services as: the frequency control service (FCAS) for both fast and slow variations; network service control (NSCAS) such as voltage control, interconnection flow control and control to improve electromechanical stability; and black start service (SRAS).

4.2.3   United Kingdom

Electricity generation in the United Kingdom has been mostly based on thermal power plants coal-based. In the 1990s, with the electrical system reconstruction, there was a shift from coal to natural gas. Following the same trend of several countries and the commitment made in the same decade to decarbonization, thermal energy sources have been replaced by renewable ones, mainly solar and wind. Figure 4.3 shows the evolution of energy sources in the generation of electricity.

The variations in energy supply by coal and natural gas in the country were due to price fluctuations, but the downward trend of coal is remarkable nowadays. The generation capacity by renewables includes storage through "pumped hydro." The increase in generation capacity between 1990 and 2019 was slight, from 73.6 GW to 103.1 GW. The replacement of coal-fired thermal plants by combined cycle plants (CCGT) and renewable generation occurs in the same period. Pumped hydro plants, as in other European countries, have existed for decades (about 2.7 GW) as they served to flatten the load curve.

The rapid growth of renewables in the United Kingdom was due to the agreement signed in March 2007 by the European community to combat climate change, known as the Renewable Energy Directive. Wind energy is the primary source besides biomass in generating electric energy. Solar energy has grown significantly since 2010 due to the incentive given by the feed-in tariff.

This transition from a reliable coal-fired system to a climate-dependent system has created significant challenges for the operation system. Due to this situation,

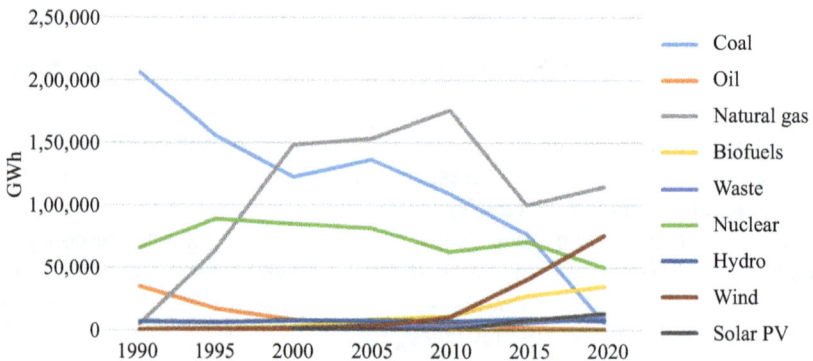

Figure 4.3   Evolution of UK electricity matrix. Source: IEA (2022).

storage systems have been suggested as an essential solution for the intermittence of renewable sources, improving the load-generation balance and the control of the electrical grid.

Nowadays, the United Kingdom has about 1 GW of operational chemical batteries and an additional 13.5 GW of projects under development, most of them with individual power below 50 MW. This action by the UK government aims to achieve zero CO_2 emissions by 2050, making storage contribute to grid regulation, as there is a forecast of total wind generation of 40 GW in 2030.

Storage systems stand out from the 2010s due to the growth of renewable energy, mainly after 2007, when IPCC confirms through climate models the trend of global warming with severe consequences. From there, several governments create mechanisms to encourage renewable energy through laws and specific regulations. In the case of England, we can mention:

- **The feed-in tariff (FIT)** was implemented in April 2010 (Energy Act 2008) in order to encourage the adoption of distributed generation (power less than 5 MW) and ended in March 2019. Only generators whose source is renewable have access to a generation tariff higher than the energy price defined by the regulator. The project receives this subsidized tariff for 20 years.
- **The Contract for Difference (CfD)** was one of the instruments created by British government to enable investments in larger renewables, taking the risks of the energy market price. In fact, a low carbon contract company (LCCC) was created, serving as an interface between the renewable generator and the consumer. The generator contract is made through a fixed price, avoiding the market risk that is allocated to LCCC. Funds for the LCCC come from positive differences in market prices and a fund imposed on energy suppliers that are licensed by the government.
- Another aspect of incentives is for domestic or non-domestic thermal heating **(Renewable Heat Incentives)**, which started in 2014 and ended in 2022, where those who join end up receiving extra payments every 4 months. Those who use biomass, solar heating, thermal energy are eligible to fall into this group.

One of the obstacles for most infrastructure projects is the permission for installation given by the central government. In July 2020, the British government published the reforms it will implement in planning concerning storage projects. Conceptually, storage is treated as another form of generation. Any generation with a capacity greater than 50 MW (350 MW in Wales) qualifies as an NSIP (National Significant Infrastructure Project) under the Planning Act 2008.

As an NSIP, the project must go through the DCO (Development Consent Order), a bureaucratic process until the Secretary of State recommends it. If the power is less than 50 MW, the local authority is responsible for authorizing it, which is more straightforward. It has led most storage projects to power below 50 MW.

The UK electricity regulator published the "Smart Systems and Flexibility Plan" in 2017, which offers information on removing barriers to technologies such as energy storage and creating a market for flexible resources [10].

On another front, the British government has been increasing its focus on rapidly introducing hybrid and fully electric cars to reduce the 35% share of transport in CO_2 emissions. Today, only a tenth of the vehicles sold fall into this category, and there is an enormous scope to improve this picture. However, as seen earlier, one-third of the electricity needed to power these cars comes from natural gas, making renewable energy more critical.

Storage is placed as a solution, including using the EVs' batteries as a large aggregate battery, minimizing the need to invest in large storage systems, also known as vehicle-to-grid (V2G). Suppose all 38 million vehicles licensed in the United Kingdom were electric, with an average of 50 kWh of power plugged into a 7 kW. The capacity of this segment would be 220 GW, which is 15 times what is planned today in electrical system storage. There is, therefore, a cheaper alternative incorporating one more function into car batteries.

4.2.4   USA—North Carolina, California, New York, Massachusetts

United States of America is perhaps one of the most complex examples, as there is no clear coordination between federal-level apparatuses and state-specific regulatory processes. Regulation 755 (2011) of the Federal Energy Regulation Commission (FERC) aimed to encourage the use of rapid response resources for injection and withdrawal of power, in addition to ensuring fair and efficient compensation for the regulatory service provided by flexible resources like batteries and flywheels. Regulation 784 (2013) had the scope to organize and make more transparent the ancillary services provided by flexible resources. Finally, regulation 841 (2018) proposes to direct regional grid operators to remove barriers to the participation of storage systems in the wholesale market, redefining market rules to recognize the unique technical characteristics of storage, that is, not classify it either as a generation or as a load.

USA has approximately 24 GW of storage installed [11,12], of which 23 GW coming from pumped storage plants installed decades ago. Nationwide, 1.2 GW of storage was installed in the year 2020 according to Wood MacKenzie [13]. This number is expected to jump dramatically over the next 5 years, rising to nearly 7.5 GW in 2025. Kelly Speakes-Backman, CEO of the US Energy Storage Association, says that the addition of battery storage has doubled in 2020 and would have tripled if was not for construction delays caused by the COVID-19 pandemic [14].

About federal incentives for storage technologies, there is no specific line targeting different technologies. However, in the case of renewable generation systems combined with storage, it is possible to qualify for investment tax credit [15,16] for renewables and thus recover part of the investment via income tax.

In summary, the Federal Energy Regulation Commission (FERC) [17,18] presents some directives that are related to storage:

- Requires from Regional Transmission Organizations (RTOs) and Independent System Operator (ISOs) to create a compensation structure for frequency regulation service provided by storage and energy resources. This compensation is based on your corresponding performances.

- Requires compensation for regulatory services to be split into payment for capacity and payment for performance separately.
- Resources that perform frequency regulation must keep part of their capacity in reserve, to be able to provide frequency regulation service at any time.
- Resources not associated with generation, such as storage and demand response systems, should be considered as network services by the wholesale market.
- Each operator must make the data more open, regarding regulatory services.
- Require that RTOs and ISOs review their tariffs and establish a participation model that recognizes the physical and operational characteristics of energy storage resources.
- Sets the minimum size requirement for a storage system to 100 kW.
- Requires some regional interconnection operators to define market and operating parameters for different types of resources.

4.2.4.1   California

California is an excellent case study in terms of renewable energy and energy storage, which is the focus of this study. The fact that the advances and initiatives started some time ago makes the state an international reference in terms of penetration and energy storage projects. The state's electrical matrix, presented in Figure 4.4, is quite varied, with a total capacity of 75.5 GW, with the following sources: natural gas (51%), hydroelectric (14%), nuclear (3%), coal (~0%), and non-hydro renewables (30%).

Among renewable energies, solar energy has an installed capacity of 12.6 GW, while wind energy corresponds to 6.1 GW of installed capacity. Finally, biomass

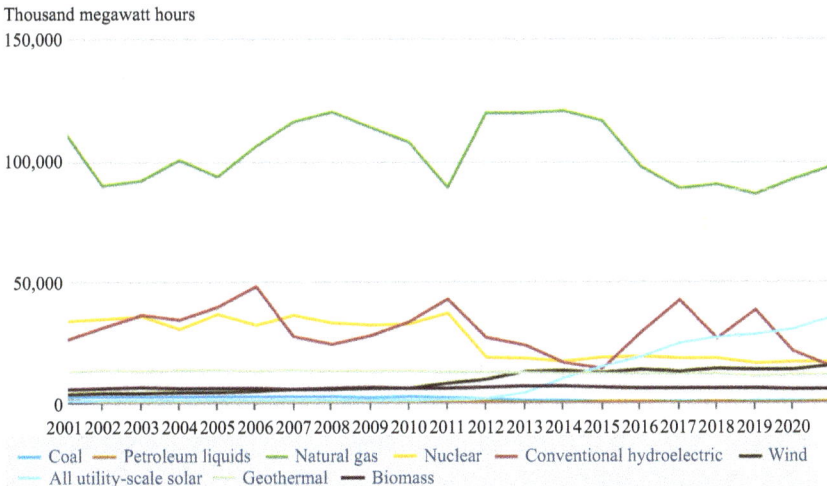

Thousand megawatt hours
150,000

100,000

50,000

0

2001 2002 2003 2004 2005 2006 2007 2008 2009 2010 2011 2012 2013 2014 2015 2016 2017 2018 2019 2020

— Coal   — Petroleum liquids   — Natural gas   — Nuclear   — Conventional hydroelectric   — Wind
— All utility-scale solar   — Geothermal   — Biomass

Figure 4.4   Evolution of California electricity matrix. Source: EIA (2022).

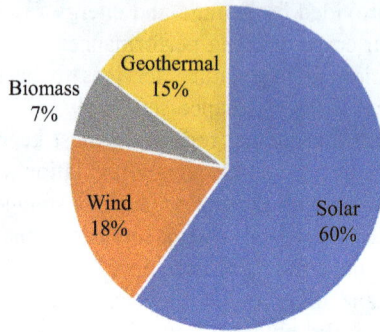

Figure 4.5   Percentage of renewable energy in California. Source: EIA (2022).

and geothermal follow with 497.7 MW and 1.8 GW, respectively. Figure 4.5 illustrates the percentage of each type of renewable in California.

California's ambitious clean energy agenda places it at the forefront of efforts nationally and globally to increase efficiency, reduce greenhouse gas emissions, and shift to a cleaner and more sustainable environment. The changes taking place to implement this vision are broad and encompass legislative and regulatory action, private sector and industry initiatives, public–private partnerships, and inter-agency coordination and collaboration. As the state progresses towards its goals, the potential for energy storage to assist in the integration of renewable resources and the maintenance of a reliable and efficient electrical grid takes on great significance.

The state aimed to meet 33% of its demand for renewable resources by 2020, and 60% by 2030. In addition, the California Senate, through the "SB100" bill, mandates that 100% of California's electricity must be supplied with carbon neutral resources by the end of 2045 (Koseff, 2018). The California Public Utilities Commission (CPUC) has imposed energy storage procurement targets, the largest in the US, for each of the California investor-owned utilities totaling 1,325 MW to be completed by the end of 2020 and implemented by 2024 (Assembly Bill No. 2514, 2010). In 2017, California Legislature passed AB 1405, which established a standard that requires a portion of peak hour demand to be supplied by clean resources. Storage systems can shift available renewable energy supply from off-peak production times to peak demand times, providing a solution to meet the "clean peak" standard.

California has 4.2 GW of installed energy storage and about 180 MW of pipeline projects. Characterizing the penetration of storage in the state, there is a list of initiatives, programs, legislation, and regulations that promote this advance, in addition to those already mentioned [19]:

- **Bill AB 2868**: Requires the three largest utilities (owned by investors): Pacific Gas and Electric (PG&E), Southern California Edison (SCE), and San Diego Gas & Electric (SDG&E) to create energy storage capacity of 1.3 GW by 2022 through the use of batteries. This also authorizes the CPUC to double public utility customer collections for energy storage and renewable generation incentives.

- **Regulation D.19-09-043**: It was determined in October 2019 that energy storage should be included in the modeling related to the Effective Load Carrying Capability (ELCC) values used by the CPUC in decision making.
- **Regulation D.17-04-017**: CPUC increased funding for the Self-Generation Incentive Program (SGIP) from $83 million to $166 million and allocated about 80% of the funding to storage incentives.
- Various new policies and rules that favor the removal of interconnection challenges, ensuring the employment of mass energy storage in the state's renewable energy landscape.
- **Regulation D.19-01-030**: Allows net metering for facilities that have energy storage, as long as the control solution prevents the storage device from loading or unloading on the network.

Pacific Gas & Electric (PG&E) is asking California regulators to approve five lithium-ion battery storage projects totaling 423 MW/1,692 MWh, including a 60 MW/240 MWh battery along with a geothermal facility. The five projects are scheduled to be in operation in August 2021 and increase the amount of storage that PG&E has contracted to over 1,000 MW. The utility plans to make another request for features that can come online in 2022 and 2023. Much of the new storage that has been contracted in California is driven by near-term system reliability needs, according to the California Energy Storage Alliance (CESA) [20,21]. On August 14 and 15, 2020, the California Independent System Operator Corporation (CAISO) was forced to institute rotating electricity outages in California because of an extreme heat wave across the west. These events make California's energy storage goals more urgent.

In this way, some projects point to California's leadership regarding the use of energy storage in the United States. One project, located in San Diego, demonstrated the integration of multiple renewable energy technologies with energy storage to support grid reliability [22]. San Diego Gas & Electric installed a 1.5 MWh solar photovoltaic system, integrated with a battery storage system at the Borrego Springs municipal substation, as part of a broader distributed energy storage plan. The 6 MW substation connected battery power storage system with nearly 5 MWh of customer-owned battery power is designed to improve the reliability of the local power grid for the city, both under normal conditions and in outage events. The demonstration was intended to prove the effectiveness of integrating multiple renewable energy technologies, energy storage, automation system technologies, interruption management systems with advanced controls, and communication systems to improve the reliability of the electrical grid.

University of California and BMW are leading a project which reuses electric vehicle batteries for stationary energy storage [23]. Located in San Diego, the project stores up to 108 kW of electricity for 2–3 h inside each battery. Although batteries used in EVs have a normal vehicle life of 8–10 years, they still have significant capacity as an alternative for use. Demonstrating this technology will help prove the feasibility of using second-life EV batteries for stationary energy storage. If the project demonstrates that used EV batteries have a viable second use, the total cost of owning an electric vehicle can be minimized since the biggest cost is precisely the battery.

4.2.4.2   Massachusetts

The state of Massachusetts also has a diversified energy matrix for electricity, but with a predominance and great dependence on natural gas (7 GW). The matrix also relies on non-hydro (1.3 GW) and hydroelectric (400 MW) renewable energy. Another important aspect is that the coal-based energy source was liquidated in 2018 and the nuclear since 2019. Figure 4.6 shows the next generation evolution of each source in the state.

In the case of renewables, Massachusetts concentrates on Solar energy its largest source at the utility level, followed by Wind energy and Biomass. Figure 4.7

Net generation, Massachusetts, all sectors, annual

Thousand megawatt hours

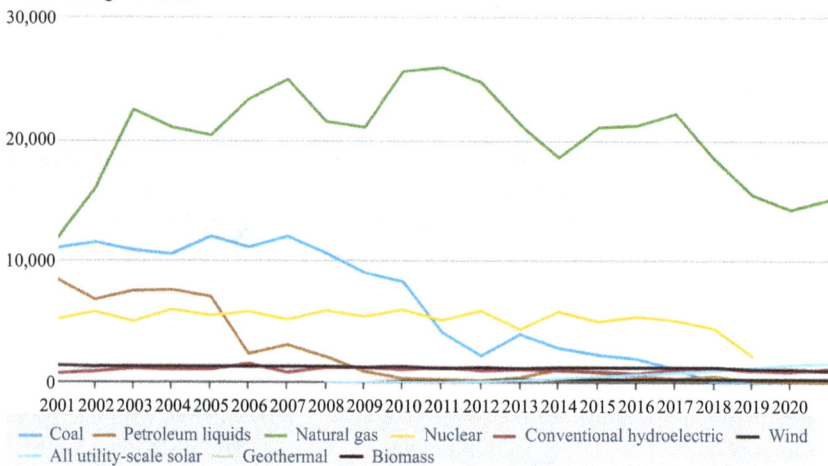

Figure 4.6   Evolution of Massachusetts electricity matrix. Source: EIA (2022).

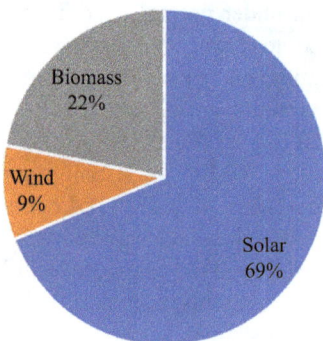

Figure 4.7   Percentage of renewable energy matrix in Massachusetts. Source: EIA (2022).

shows this percentage breakdown for renewables. Solar energy alone corresponds to 854 MW, within the total.

On September 16, 2016, the governor issued the Executive Order #569, establishing an Integrated Climate Change Strategy for the Community, which directed the Secretary of Energy and Environmental Affairs (EEA) to publish a Comprehensive Energy Plan (CEP) [24]. CEP is part of a broader strategic plan to coordinate and make consistent new and existing efforts to mitigate and reduce greenhouse gas emissions and build resilience to climate change. Specifically, the Executive Order required that the CEP should include and be based on reasonable projections of the Community's energy demands for electricity, transportation and thermal conditioning, and include strategies to meet those demands in a regional context, prioritizing meeting the demand of energy through conservation, energy efficiency, and other demand-reducing resources, including storage systems, in a way that contributes to meeting the Community's emission limits for greenhouse gases.

The Energy Storage Initiative (ESI) was launched in May 2015 with $20 million provided to fund 26 storage pilot projects across the state. In 2016, the study entitled "State-of-Charge" [25] analyzed and proposed a way forward for the Commonwealth of Massachusetts to achieve energy storage penetration within the opportunities and needs established.

The Department of Energy Resources (DOER) has set a target of 200 MWh for storage on January 1, 2020, and the Department of Utilities has approved grid modernization and storage projects in recent fee processes. To encourage consumers to buy zero-emission vehicles such as EVs, DOER has provided $20 million to offer discounts on the purchase of EVs, under the financing program called MOR-EV. Since January 2017, more than 14,640 vehicles have been financed by the MOR-EV program, including battery-powered EVs, plug-in hybrid vehicles and zero-emission motorcycles [26].

Massachusetts lawmakers continued to act through the legislation and at the end of the formal legislative process in 2018, a bill for the advancement of clean energy was passed, a bill establishing several new clean energy initiatives. The bill increases the amount of electricity needed to be supplied with renewable energy credits, increasing the Renewable Portfolio Standard (RPS), which sets an energy storage target of 1,000 MWh by December 31, 2025 and establishes a program to encourage use clean energy during times of peak demand, operating at its highest cost and emitting the most greenhouse gases. These existing policies continue to reduce the community's greenhouse gas emissions.

The state of Massachusetts has approximately 1,770 MW of installed power as energy storage and 2.2 MW of advanced projects. Storage systems are predominantly pumped storage and batteries. With programs and incentives, in 2019, 22,804 kW of power and 84,925 kWh of estimated energy storage were installed, with projects entirely based on Li-Ion batteries, with only one project with flow batteries for that year [27]. Below is a list that summarizes legislation, incentive programs, regulation and promotion, in addition to those mentioned above, for the development of energy storage in Massachusetts:

- **Legislation H.4857**: In addition to setting a target of 1,000 MWh by 2025, directs the Massachusetts Department of Energy Resources to consider policies that ensure proper assessment in planning processes associated with energy storage.
- **Legislation H.B.2496**: In 2019, added battery storage to the definition of clean energy technologies ("green energy technology"), so that in this way it is stipulated for contracts for renovation of public buildings.
- **DPU 17-146-A**: The Department of Utilities allows solar array and storage systems to participate in net metering as long as the storage cannot load from the grid or export to the grid.
- **The Advancing Commonwealth Energy Storage (ACES)**: Program that secured nearly $20 million in funding for 26 storage projects in 2017, designed to demonstrate various storage applications.
- **Third Party Financing**: To reduce the upfront cost of storage systems, many energy storage developers are offering third party rental models. These models often use shared savings agreements to split the benefits of reduced demand charges or demand response program payments between the financing entity and the consumer.
- **Tax Incentives**: Battery energy storage when paired with renewable energy technologies qualify for accelerated depreciation and a federal tax credit. The Federal Investment Tax Credit (ITC) supports combined solar + storage systems, where up to 75% of the storage system is charged from the solar system rather than the grid. Until 2019, the ITC provides a credit of 30% of the system cost; credit drops to 26% in early 2020, to 22% in early 2021, and permanently dropped to 10% in early 2022 (only commercial systems will be eligible for the 10% at this point).
- **Commercial Solar and Storage Technical Assistance and Resource Hub**: The Massachusetts Clean Energy Center (MassCEC) provides commercial services to homeowners with free resources and technical assistance on issues associated with solar energy and storage.
- **Utility-based demand management pilots**: Each utility develops pilot demand management programs that rely on automated demand response programs with energy reductions on a dollar-per-kW basis. The Department of Utilities recently approved the demand management of Eversource's pilot program, which will leverage thermal and battery applications to decrease utility peak demand. Utilities are expected to provide a wide range of demand management offerings as part of 2019–2021 efficiency plans.
- **Community Clean Energy Resiliency Initiative (CCERI)**: This was a multi-phase initiative launched by Massachusetts DOER to encourage public facilities to invest in energy resilience, with funded subsidies, feasibility assessments and implementation support. A final phase expanded the program to hospitals.
- **Community Microgrid Program**: The community microgrid program provided grants for feasibility assessments for microgrids, which may include renewable energy, CHP and energy storage. Projects had to be community-led,

but could include commercial and residential areas. The program is closed, but may accept candidate rounds in the future.

• **Commercial and Industrial Storage Feasibility Grants:** To encourage storage through development in commercial and industrial facilities, MassCEC has provided grants for feasibility assessments for combined solar and storage in manufacturing facilities.

In this way, Massachusetts has made great strides in developing energy storage policies and other states use the initiatives as a case study to address complex policy issues in their own states. Some projects may exemplify advances in storage in Massachusetts. National Grid has a project under development that will be one of the largest battery energy storage resources in the Northeastern United States [28]. Built on the island of Nantucket, Massachusetts, the facility will ensure electrical reliability for customers during the peak summer months and defer the need to build an additional underwater supply cable to the island. To meet Nantucket's growing demand, National Grid has developed an integrated plan, IslandReady, to upgrade the island's electricity infrastructure. A critical component of that plan involved the development and construction of a 6 MW/48 MWh Battery Energy Storage System (BESS), the largest of its kind in New England. In addition, a new 15 MW diesel generator and power control house were installed. The project cost approximately $81 million.

Eversource presents a 25 MW lithium-ion battery system, to be located in a building in Provincetown, designed to provide backup power to customers in Provincetown, Truro and Wellfleet during an interruption in the distribution line running along of Route 6 [29]. Provincetown residents unanimously voted in 2020 to approve a lease for Eversource to operate the battery storage project at the city's transfer station. The so-called community battery will also defer the need to install 21 km of distribution line, according to Eversource, and is designed to provide 10 h of backup power in winter and up to 3 h in summer when tourists flock to the region.

4.2.4.3 New York

In the United States, sources of electricity generation have been shifting from coal to natural gas and renewables since the mid-2000s. Changes in the composition of New York State's electricity generation portfolio have contributed to this trend. Coal's share of the state's electricity generation dropped from 14% in 2005 to less than 1% in 2019, and natural gas electricity grew from 22% to 36%. Figure 4.8 shows the evolution of electricity generation by source in the state from 2001 to 2019.

In 2019, New York's Clean Energy Standard [30] was revised to require electricity production to be 100% carbon-free by the year 2040. Until that year, 29% of generation in the state came from renewables, including large- and small-scale installations. Approximately one-third of the state's net generation comes from nuclear power plants [31]. Note that the revision of New York's Clean Energy Standard includes nuclear as a carbon-free source that can be part of the state's generation portfolio by 2040. Hydroelectric energy is the renewable source with the largest share in the energy matrix (78% of the total of renewable generation and 23% of the state's total generation), with the state being the third largest producer

Thousand megawatt hours

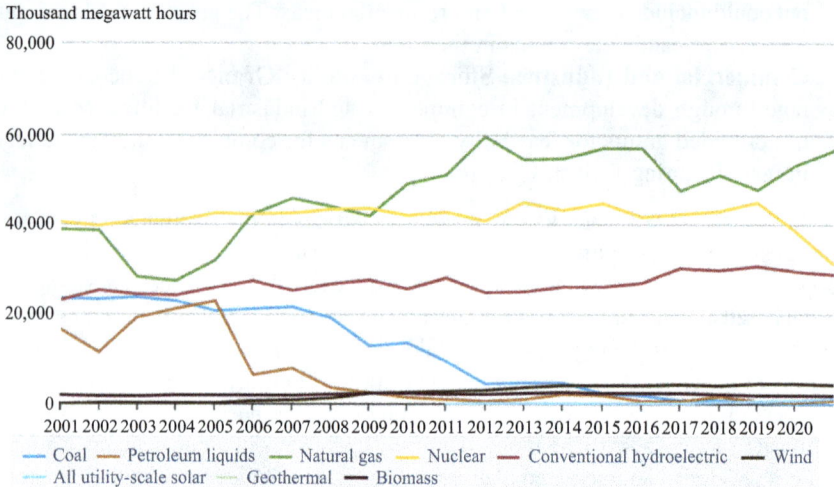

Figure 4.8   Evolution of New York electricity matrix. Source: EIA (2022).

of hydroelectric power in the United States. In addition, in 2019 the state was also the third largest producer of renewable energy in the country.

During the years 2001–2008, an increasing demand for electricity in the state it was noticed. However, in the following decade the demand for electricity decreased and according to [32] the peak demand of the state is expected to have an average annual reduction of 0.13% between the years 2018 and 2028. The state also began to see an increase in distributed renewable energy generation. This trend has increased in recent years, which contributes to the increase in the intermittence of energy resources available for system management by the NYISO operator.

In 2019, New York State passed the Climate Leadership and Community Protection Act (Climate Act), which codified some of the country's most aggressive climate and energy goals.

- 6,000 MW of solar energy in 2025.
- 70% renewable energy by 2030.
- 3,000 MW of energy storage by 2030.
- 9,000 MW of offshore wind energy by 2035.
- 100% carbon-free electricity by 2040.
- 85% reduction in GHG emissions from 1990 levels to 2050.

Energy storage will play a crucial role in meeting these ambitious state goals. Storage can help integrate clean energy into the grid, reduce costs associated with meeting peak electrical demands, and increase efficiency. In addition, energy storage can stabilize supply during peak electricity usage and help keep critical systems online during an outage. All this while projections indicate that this move will create an industry that could employ 30,000 New Yorkers by the year 2030 (NYSERDA).

With increasing investments in renewable sources and greater exposure to intermittency in electrical power generation, energy storage is critical to New York's clean energy future. Projections indicate that renewable energy sources, such as wind and solar, will supply a greater portion of the state's electricity, due to the focus presented in New York's Clean Energy Standard [30], in this way, storage will allow clean energy to be available when and where it is most needed to meet the system load [31].

New York State government launched in 2019 a plan to reach 3,000 MW (New York State Governor, 2019) in energy storage capacity by the year 2030, to accelerate the growth of the industry and decrease the costs related to technologies. A total of $400 million will be allocated to the plan for new projects that will contribute to the state's agenda of achieving the objective of producing/consuming clean electrical energy. There are different types of incentives in the state provided by the New York State Energy Research and Development Authority (NYSERDA) for storage systems.

NYSERDA is offering incentives that can help accelerate the deployment of Bulk Storage projects that provide market power, ancillary services, and/or reserve capacity.

Incentives are available at a fixed amount per usable kWh of installed storage capacity for projects above 5 MW. Projects that provide capacity services on the market, receive the stated incentive fee. Projects that only provide energy arbitrage or ancillary services receive 75% of the incentive fee. There is no maximum project size, however, the total NYSERDA incentive is no more than $25 million per project. Incentives are guaranteed by the developer at a flat rate of incentive at the time a completed form is submitted and approved by the NYSERDA.

A total of $150 million was allocated to this incentive category and it is already fully committed to new projects, totaling 639.9 MW and 2,174.1 MWh.

The NYSERDA Retail Energy Storage Incentive provides commercial customer financing for standalone systems, grid-connected energy storage systems, or in conjunction with a new or existing on-site clean generation source such as solar, fuel cells or cogeneration. In this type of incentive energy storage systems must:

• Be sized for up to 5 MW of alternating current (AC) power.
• Be new, permanent, and stationary.
• Be located in the state.
• Use commercially available thermal, chemical, or mechanical technology operated primarily for electrical load management or shifting on-site renewable generation to more beneficial time periods.
• Provide value to a customer under an investor-owned utility fee, including delivery fees or value of New York State distributed energy resources.
• Interconnect behind the customer's electrical meter or directly into the distribution system.

Incentives for all retail storage projects are provided through a network of participating contractors, approved in the retail energy storage incentive program, that contract directly with the customer. Incentive rates are available from retail incentive panels.

Consolidated Edison Inc. (Con Edson) recently signed its largest project in the state to bring a 100 MW/400 MWh lithium-ion battery system into operation. In this contract, Power Global will build the battery system and Con Edison will manage the energy bids in the New York energy market for 7 years after the start-up, which should take place in 2022 [33].

4.2.4.4   North Carolina—USA

North Carolina—USA reached in 2019 the second place in installed capacity of solar generation and approximately 6.5 GW in 2020 [34]. The state has 2 GW of hydroelectric generation capacity [35] and has very favorable conditions for off-shore renewable generation as well [36,37].

Due to the favorable characteristics of the power generation system and the growing demand of the market in North Carolina, the public commission of the state utilities started to carry out efforts to analyze storage systems to assist in the process of optimizing system resources and still as a way to reduce the need for new investments in generation, transmission and distribution. As a result, North Carolina utilities are now required to include storage systems in their investment plans [35].

The process of evaluating the need for and benefits of storage systems in North Carolina began with the 2017 state congressional order HB589, which aimed to leverage renewable energy projects in the state. Among other specific things, this order requested a technical study [38] to determine the value of storage systems to consumers.

The initial study presented in [39] showed that a storage capacity of the order of 5 GW would be economically viable for the system considering technology price projections for 2030 [40].

There are still no specific incentives for storage in the state. However, the study by [39] began to mobilize the sector in the state and utilities began to include storage in their investment plans due to the attractiveness of this type of technology for the state. As a result of this movement, in August 2020 [41], the largest lithium-ion battery system in the state was installed in the city of Asheville, with 9 MW of capacity for operation close to a Duke Energy substation. The total cost of the project was $15 million, and the company has plans to invest a further $600 million for the construction of a new 375 MW in its concession area.

4.2.5   India

As other countries, India government signed the COP-21 commitment of 40% renewable energy insertion on electricity sector by 2030. Besides the agreement, India is experiencing extreme climate events as a heat wave in 2022 summer, the third hottest in the past 122 years, consequently in this period the electricity demand reached its major demand peak. Moving for a more renewable and flexible matrix will require the intensification of storage use, as well as the diversification in used technologies.

BloombergNEF report estimates 387 W/1,143 GWh of new energy storage capacity globally. The same report predicts India as the third country in number of energy storage systems by 2040 [42]. Stationary technologies, as ion-lithium batteries, in pointed as the technology for changing the storage market, thus India needs to overcome barriers concerning to cost and lack of specific regulations.

NREL study found that BESS storage can be cost-effective in 26 of 34 South Asia regions, mostly located in India. In all the scenarios evaluated, the storage capacity grows until 2050, reaching between 180 GW and 800 GW or 10% and 25% of installed capacity. The technology most adequate for the future renewable matrix with strong presence of solar and wind generation, for providing peak capacity is 4-h battery. Despite that, the report highlight the importance of pumped hydroelectric storage for the country, because its low costs [43].

Moving from a centralized hydro and thermal electric system for a decentralized and renewable-based one requires strong regulatory framework. Central Electricity Regulatory Commission (CERC) launched in 2017 a study named "Introduction of Electricity Storage System in India" for guide the usage of storage system, and the operational and recovery aspects of storage facilities. For inclusion of different technologies, addressing the preferred location of use, the Energy Storage Sectional Committee was created in 2016, and since then is discussing and proposing standards for integrating energy storage systems on the grid [44,45].

4.3 Lessons learned and detachable initiatives

From the experiences presented, it is possible to draw some conclusions and observations about the insertion of storage systems.

With the intensification of the substitution of fossil sources for renewable sources to cope with the process of reducing greenhouse gas emissions from the 2000s onwards, electrical systems are left without dispatchable generation, making it difficult to balance load generation and the operation of electrical networks. Energy storage now has an important role in the electricity sector and the search for technologies has become a major challenge since the last decade.

With the global option for decarbonization, each country and state has created mechanisms to support the inclusion of renewable sources in the electricity generation matrix. These mechanisms appear in laws and regulations from the 2000s onwards, such as the feed-in tariff, government subsidy for financing renewable generation as tax exemption, contracts for difference to minimize price risk, etc. This same movement is not yet observed for storage technologies, but there is already discussion in several countries in this regard.

One of the regulatory doubts is regarding the classification of storage systems, which initially in several countries place them in the category of generators that have their own characteristics and regulations, often not adaptable to the storage agent.

Storage via pumped-hydro technology has existed since the beginning of the last century to monitor the load curve. This mature technology is still a viable solution when you have a favorable geography.

In specific cases like California, there is an incentive to install storage systems to improve the reliability and security of the system, that is, as a solution for the electrical system. This was more latent with what happened in the summer of 2020 when there were blackouts due to lack of power due to the occurrence of very high temperatures. Australia, for example, in the province of Queensland, held a combined energy and storage auction in 2017 which demonstrates this possibility of inserting storage.

The investment opportunity in storage mainly of the "behind the meter" type in homes through electrochemical batteries has increased mainly in countries that have "time-of-use" tariffs in low voltage. Implementation in commerce and industry is already a reality, always observing the type of business and the difference in peak and off-peak tariffs.

Lithium-ion batteries, which were initially intensely applied in the electronics industry (cell phones and other applications) and in electric vehicles, are beginning to be applied in the electrical sector, even with significant power grouping together a set of cells. It is observed that this technology is already operational leaving aside the experiences with pilot cases that still appear in other technologies.

The emergence of distributed generation with solar and wind sources has demanded distributed storage systems, which has led companies to explore these types of combined businesses. These companies have forced regulators to create regulatory mechanisms to expand the list of applications such as network services and thereby earn additional gains for investors in addition to modulating the load curve.

The transport sector has been the current target in the decarbonization process and with that has brought opportunities for investments in storage. The current cost of electric vehicles can decrease with the second use of their batteries in the electric grids or even with the use of them as an ancillary service provider for the grid (vehicle-to-grid, V2G) when integrated through a storage aggregator. With fast charging stations mainly on highways, there is a need to install batteries to minimize investments and losses in the medium voltage networks of the distributors, which represents a new type of business for these systems with the intense dissemination of electric cars.

Other locations can observe those experiences to drive their electric systems for a more sustainable, efficient, and reliable place through energy storage systems.

4.4   Issues and challenges roadmap

There are barriers to overcame in many electric systems for energy storage play the role and provide the services they are capable of. First it is urgent to create a better definition of what a storage system is. The lack of proper definition or the idea that storage is similar only to an energy generation device, limits the scope of storage. In fact, the behavior of ESS addresses energy generation and energy consumption, but it should not be mistaken with one or another. In [46], the authors advocate for a definition that clarifies energy storage as a specific step in electric system.

Other challenge, pointed by [47], mainly for stationary batteries, is regards with the storage costs per unit of energy. Despite the technology price is decreasing, there is a perception of technology high cost. It is strong mainly in underdevelopment countries, where the manufacture is not present. The diversity of types of storage systems also causes this issue, due to the market availability and technological trends that will drive the prices.

If there is no regulation about a technology use and application, each supplier is free to select between them, and in most cases the selected will be chosen by the less cost, not always for the most efficient and sustainable. The lack of regulation delays technological advances and makes difficult to control the use of safe and effective technologies, especially in small and isolated systems [48].

In electric systems, based on distributed energy resources and microgrids, the most common challenges are concentrated in technical subject: the selection of raw materials will define quality of the energy provided and life cycle of a storage appliance; the quality of the interface between renewable generation and the storage technology; and the charging discharging control system for preventing system loss and breakdown [49]. The microgrids architecture suggests the use of hybrid ESS for an optimum performance of future grids.

As important as the technical, regulatory, and economic issues, the environmental issue also figures as another challenge for storage systems. Implementation of storage technologies together with renewable sources add flexibility and safety, allied to emissions reduction to the system. Despite that, the manufacture and disposal of storage devices could no longer be environment friendly, due the use of fossil fuels, chemicals, and non-recyclable waste can be present on those processes [49].

References

[1]   J. N. Kang, Y. M. Wei, L. C. Liu, R. Han, B. Y. Yu, and J. W. Wang, "Energy systems for climate change mitigation: a systematic review," *Appl. Energy*, vol. 263, p. 114602, 2020, doi: 10.1016/j.apenergy.2020.114602.

[2]   M. Arbabzadeh, R. Sioshansi, J. X. Johnson, and G. A. Keoleian, "The role of energy storage in deep decarbonization of electricity production," *Nat. Commun.*, vol. 10, no. 1, 2019, doi: 10.1038/s41467-019-11161-5.

[3]   H. fang Lu, X. Ma, and M. da Ma, "Impacts of the COVID-19 pandemic on the energy sector," *J. Zhejiang Univ. Sci. A*, vol. 22, no. 12, pp. 941–956, 2021, doi: 10.1631/jzus.A2100205.

[4]   M. M. Mohideen, S. Ramakrishna, S. Prabu, and Y. Liu, "Advancing green energy solution with the impetus of COVID-19 pandemic," *J. Energy Chem.*, vol. 59, pp. 688–705, 2021, doi: 10.1016/j.jechem.2020.12.005.

[5]   IEA, *Russia's war on Ukraine*, 2022. https://www.iea.org/topics/russia-s-war-on-ukraine (accessed Nov. 20, 2022).

[6]   M. Nerlinger and S. Utz, "The impact of the Russia-Ukraine conflict on the green energy transition: a capital market perspective," *Swiss Financ. Inst. Res. Pap. Ser.*, 2022.

[7]   ANEEL, "Nota Técnica no 094/2020 – SRG/ANEEL – Abertura da Tomada de Subsídios para obter contribuições para as adequações regulatórias

necessárias à inserção de sistemas de armazenamento, incluindo usinas reversíveis, no Sistema Interligado Nacional," Brasília, 2020.

[8]   GTAI, *The Energy Storage Market in Germany. Germany Trade and Invest*, Berlin, 2019.

[9]   GTAI, *The Energy Storage Market in Germany, Germany Trade and Invest*, GTAI, 2019. http://www.gtai.com/.

[10]   W. E. Council, *Energy Storage Monitor: Latest Trends in Energy Storage*, 2019. https://www.worldenergy.org/assets/downloads/ESM_Final_Report_ 05-Nov-2019.pdf.

[11]   EIA, *Electric Power Monthly: Table 6.2.B. Net Summer Capacity Using Primarily Renewable Energy Sources and by State, November 2020 and 2019 (Megawatts)*," 2020. https://www.eia.gov/electricity/monthly/epm_ table_grapher.php?t=table_6_02_b.

[12]   EIA, *Battery Storage in the United States: An Update on Market Trend*, U.S. Energy Information Administration, U.S. Department of Energy, 2020. https://www.eia.gov/electricity/data.php.

[13]   W. Mackenzie, *The U.S. Energy Storage Monitor – Q4 2020 Executive Summary*, 2020. https://www.woodmac.com/reports/power-markets-us-energy-storage-monitor-q4-2020-454455/.

[14]   Yale, *In Boost for Renewables, Grid-Scale Battery Storage Is on the Rise*, 2020.

[15]   ESA, *Federal Focus: Investment Tax Credit (ITC)*, Energy Storage Association, 2021. https://energystorage.org/policies-issues/federal/itc/.

[16]   "Using the solar investment tax credit for energy storage," *Energy Sage*, 2021. https://www.energysage.com/solar/solar-energy-storage/energy-storage-tax-credits-incentives/.

[17]   E. Hossain, H. M. R. Faruque, M. S. H. Sunny, N. Mohammad, and N. Nawar, "A comprehensive review on energy storage systems: types, comparison, current scenario, applications, barriers, and potential solutions, policies, and future prospects," *Energies*, vol. 13, no. 14, p. 3651, 2020, doi: 10.3390/en13143651.

[18]   A. Castillo and D. F. Gayme, "Grid-scale energy storage applications in renewable energy integration: a survey," *Energy Convers. Manag.*, vol. 87, pp. 885–894, 2014, doi: 10.1016/j.enconman.2014.07.063.

[19]   *2020 Grid Energy Storage Technology Cost and Performance Assessment*, U.S. Department of Energy, 2020. https://www.energy.gov/energy-storage-grand-challenge/downloads/2020-grid-energy-storage-technology-cost-and-performance.

[20]   CaliforniaISO, *Root Cause Analysis: Mid-August 2020 Extreme Heat Wave*, 2021. http://www.caiso.com/Documents/Final-Root-Cause-Analysis-Mid-August-2020-Extreme-Heat-Wave.pdf.

[21]   Utility Dive, *PG&E's Total Storage Procurements Pass 1 GW with Latest Round of Projects*, 2020. https://www.utilitydive.com/news/pges-total-storage-procurements-pass-1-gw-with-latest-round-of-projects/578389/.

[22] SDG&E, *Microgrids Help Integrate Renewable Energy and Improve Community Resiliency*, 2021. https://www.sdge.com/more-information/environment/smart-grid/microgrids#:∼:text=The first of its kind,changing environmental and system conditions.

[23] University of California, "The ARPA-E project: multifunctional battery systems for electric vehicles," *Arpa Energy*, 2013. https://arpa-e.energy.gov/technologies/projects/multifunctional-battery-systems-electric-vehicles.

[24] Massachusetts Government, *Massachusetts Clean Energy an Climate Plan for 2020*, 2010.

[25] Massachusetts Department of Energy Resources, *State of Charge – Massachusetts Energy Storage Initiative Study*, 2017. https://www.mass.gov/doc/state-of-charge-report/download.

[26] Center for Sustainable Energy, *Massachusetts Offers Rebates for Electric Vehicles (MOR-EV) Program Statistics*. https://mor-ev.org/program-statistics.

[27] Massachusetts Department of Energy Resources, *ESI Goals & Storage Target*.

[28] M. P. Norton, *Eversource Battery Projects Advancing On Outer Cape, Vineyard*, 2019. https://www.wbur.org/bostonomix/2019/04/08/battery-energy-storage-eversource-cape-vineyard.

[29] I. Gheorghiu, "There once was a 48 MWh Tesla battery on Nantucket, which saved National Grid $120M in its budget," *Utility Dive*, 2019.

[30] EIA, *New York Generated the Fourth Most Electricity from Renewable Sources of Any State in 2019*, U.S. Energy Information Administration, 2019. https://www.eia.gov/todayinenergy/detail.php?id=45996.

[31] EIA, *New York State Profile and Energy Estimates*, U.S. Energy Information Administration, 2018. Administration: https://www.eia.gov/state/?sid=NY #tabs-1.

[32] NYISO, *NYISO 2018 Power Trends Report*, 2018. https://electricenergyonline.com/article/energy/category/t-d/56/699735/nyiso-2018-power-trends-report.

[33] Grenntechmedia, "Con Edison contracts its biggest battery to date in New York City," *Con Edison Contracts Its Biggest Battery to Date in New York City*, 2020. https://www.greentechmedia.com/articles/read/con-edison-contracts-new-yorks-biggest-battery-to-date.

[34] SEIA, *North Carolina State Solar Policy*, Solar Energy Industries Association, 2020. https://www.seia.org/state-solar-policy/north-carolina-solar.

[35] EIA, *North Carolina Net Electricity Generation by Source*, U.S. Energy Information Administration, 2022. https://www.eia.gov/state/?sid=NC#tabs-4.

[36] AWEA, *AWEA State Wind Energy Facts – North Carolina*, American Wind Energy Association, 2020. https://www.awea.org/Awea/media/Resources/StateFactSheets/North-Carolina.pdf.

[37] WINDExchange, *Wind Energy in North Carolina*, Wind Energy Technologies Office, 2020. https://windexchange.energy.gov/states/nc.

[38] NCLEG, *House Bill 589/SL 2017-192*. North Carolina General Assembly, 2017.

[39] NCSU, *Energy Storage Options for North Carolina*, 2018. https://energy.ncsu.edu/storage/wp-content/uploads/sites/2/2019/02/NC-Storage-Study-FINAL.pdf.

[40] Lazard, *Lazard's Levelized Cost of Storage Analysis – Version 4.0*, 2018. https://www.lazard.com/media/450774/lazards-levelized-cost-of-storage-version-40-vfinal.pdf.

[41] Duke-Energy, *North Carolina's Largest Battery System Now Operating at Duke Energy Substation*, 2020.

[42] BloombergNEF, *Global Energy Storage Market to Grow 15-Fold by 2030*, 2022. https://about.bnef.com/blog/global-energy-storage-market-to-grow-15-fold-by-2030/. Accessed Nov. 10, 2022.

[43] NREL, *Energy Storage in South Asia: Understanding the Role of Grid-Connected Energy Storage in South Asia's Power Sector Transformation*, July, 2021. www.nrel.gov/publications.

[44] A. Rose, K. Duwadi, D. Palchak, *et al.*, *India Policy and Regulatory Environment for Utility-Scale Energy Storage*, 2020.

[45] IESA, *Policy and Regulatory Framework to Accelerate Energy Storage Deployment in India*, 2022.

[46] D. Parra and R. Mauger, "A new dawn for energy storage: an interdisciplinary legal and technoeconomic analysis of the new EU legal framework," *SSRN Electron. J.*, vol. 171, p. 113262, 2022, doi: 10.2139/ssrn.4037984.

[47] ESMAP, *Deploying Storage for Power Systems in Developing Countries Policy and Regulatory Considerations*. 2020.

[48] C. S. Lai and G. Locatelli, "Are energy policies for supporting low-carbon power generation killing energy storage?" *J. Clean. Prod.*, vol. 280, p. 124626, 2021, doi: 10.1016/j.jclepro.2020.124626.

[49] M. Faisal, M. A. Hannan, P. J. Ker, A. Hussain, M. Bin Mansor, and F. Blaabjerg, "Review of energy storage system technologies in microgrid applications: issues and challenges," *IEEE Access*, vol. 6, pp. 35143–35164, 2018, doi: 10.1109/ACCESS.2018.2841407.

Chapter 5

Control and optimization of distributed energy storage systems

Filipe Perez[1] and Gilney Damm[2]

This chapter introduces control and optimization techniques for distributed energy storage systems, in the context of modern power systems. The optimization and control strategies mainly address issues of power quality and provision of ancillary services, aiming at the operation of the system with the integration of renewable sources, power converters, electric vehicles, and even microgrids. In this way, storage systems are applied for the optimal operation of the electrical network from an operation perspective, considering the restrictions of the system, such as operational and physical limits of the equipment. At the end of the chapter, the internal control loops of the converters applied to energy storage are presented.

5.1 Introduction

Energy storage system (ESS) is one of the solutions to mitigate the impact of renewables, since a number of operation modes to manage ESS are well known and have been largely applied to other applications like transportation and uninterruptible power supply (UPS) [1]. Energy storage techniques can be mechanical, electro-chemical, thermal, etc. The most popular are hydraulic in pumped storage and stored fuel for thermal power plants. ESS has been widely used to meet the energy generation with energy consumption, such as reservoirs for hydroelectric plants in Brazil, and pumped storage plants (PSP) in France, being a mature technology. Therefore, ESS has the ability to improve many aspects of power systems directly related to power quality and stability.

However, new technologies such as supercapacitors and batteries have limited power and energy capacities. In the same way, concerning efficiency, some technologies have yet to achieve a high-performance level. In particular when considering high levels of energy storage, since power losses may become considerably high [2]. All things considered, the major challenge for ESS is to find

[1]Power System Division, Lactec Institute, Brazil
[2]LISIS Laboratory, University Gustave Eiffel, France

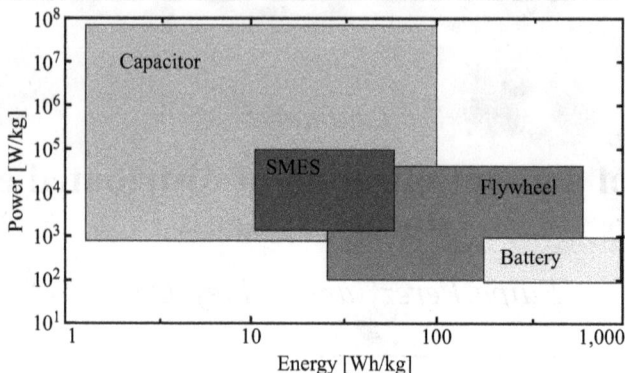

Figure 5.1   Relation between energy density and power density for different ESS

a trade-off between investment and operational costs, while fulfilling economical constraints [1].

There is a very diverse range of ESS that are distinguished by their storage capacities, technology, or the way they store energy. The different ESS technologies can be seen in Figure 5.1 adapted from [3], exposing the relation between power and energy capacity. Technologies such as batteries have been widely exploited, especially in renewables integration and distribution systems. However, the operational cost of batteries becomes harmful, since the unpredictable power generation, has a significant impact on batteries' life-cycle. On the other hand, supercapacitors are devices capable of handling large variations of power within small time intervals compared to batteries. This is due to its high power density, so it can supply much more power for a sudden demand.

At the same time, supercapacitors can inject/absorb power extremely quickly, which is a great advantage for applications with large spikes in the power range. However, applications with large amounts of stored energy turn this equipment economically less viable. For this reason, hybrid energy storage system (HESS) may be applied as an optimized solution to store energy, putting together the advantages of each technology [4]. Since each storage technology has a more suitable means of application, the combined operation of different energy storage technologies can greatly improve their application in power systems. In this chapter, it is considered the application of ESS composed of supercapacitors or batteries in the distribution context.

5.1.1   Distribution system context

ESS in the distribution system context are elements that enable strategic management of the generated energy, in order to optimize the power flow. ESS provides mitigation of the impact of renewable energy sources' intermittency and increases the reliability of the system in the event of a power supply failure. Smart grid application of ESS can be seen as a management tool on the demand side, power flow control, load–displacement, etc. The main advantage of ESS is to improve stability, power quality, and reliability of supply in power systems through different operation modes [3]. They have the ability to quickly vary power, because of the power converter characteristics without impacting the system power flow, when compared with typical synchronous generators.

To exemplify the wide application of ESS in power systems, here it is introduced the description of the main ESS application according to [2]. Its application can bring improvements in power quality, minimizing voltage and frequency variations, providing support to the network, and improving system reliability and resilience. ESS applications may be described as:

1. Ancillary services:
   - *Voltage and frequency regulation*: minimize the fluctuation impacts from generation, which can be caused by the instantaneous imbalance between demand and generation, or the intermittence of renewables. The support is made through active and reactive power injections, and brings improvements in network stability.
   - *Spinning reserve*: ability to fast respond in order to compensate a generation or transmission contingency, and maintain the system for as long as necessary.
   - *Black start*: ability to restore the network in cases of severe contingencies, as a black-out, when storage provides the power needed to restore the system.

2. Bulk storage:
   - *Arbitrage*: during low-price energy period, the storage system absorbs energy from the grid for later sale, or use during expensive times. It can also be used to store energy with renewable energy sources surplus, increasing renewables usage.
   - *Supply capacity*: characterized by annual operating hours, frequency of operation, and duration of operation for each use. For example, if the price is per hour, storage brings flexibility in supplementary hours. Supply capacity can be used for peak shaving.

3. Infrastructure service (transmission and distribution):
   - *Update deferral*: delay for improvements expenses, an example is the installation of storage systems close to the consumers relieving the transmission, and providing capacity of supply without the need to invest in transmission equipment.
   - *Line decongestion*: with the natural growth of peak demand, the line may lose the ability to deliver all the power demanded by the load, so the storage systems are allocated to supply those peaks.

4. Energy management:
   - *Power quality*: short-duration voltage variations, primary frequency regulation, power factor control, harmonic reduction, and uninterrupted service (UPS, e.g.).

- *Reliability*: greater guarantee of the energy supply, even in the face of disturbances and contingencies of the network.

5. Vehicle to grid (V2G):
   - Operating mode that benefits battery charging according to its charge level and degradation. That is, charging is carried out in a way that causes less damage to the battery, extending its useful life and maintaining the required charge level, which is essential in the context of electric vehicles.

The impact of deep-cycle in batteries is a current concern since the operation mode can drastically affect the life cycle of a battery system. Research to model the aging factor in batteries is also key because it brings many impacts to the system operation in long term. Different battery technologies have been developed to improve their operation in deep charge and discharge operation modes and to reduce the aging factors, but the main results are given by the strategies on the operation mode and optimization techniques [5].

Lead–acid batteries are the most mature of all battery technologies available, with lower-cost solutions well fit for most large applications. They are considered a safe technology, with relatively low maintenance (usually periodic water maintenance), high conversion efficiency (around 80–90%), and low self-discharge rate. However, the poor performance at low and high temperatures combined with a short lifetime due to frequent deep charge and discharge operation, and relatively low power density, handicap the use of this technology for particular applications like isolated Microgrids [6].

Lithium-ion batteries, on the other hand, have grown as prominent technology with a longer life cycle, since the deep charge operation and aging factors have minimal impact on this technology. Lithium-ion batteries also have a higher energy and power density, with higher conversion efficiency (around 95%) compared with other technologies. Other features are the lower self-discharge and maintenance, and absence of memory effect. For these reasons, lithium-ion technologies are being widely applied in low voltage distribution systems context [7]. The main concern in lithium-ion applications is the sensitivity to temperature variations, with the risk of explosion. Thus, it is necessary to use a dedicated unit for temperature control.

Supercapacitors on the other hand offer a higher surface area by using thin layers of electrolyte as a dielectric between them. The double layers capacitors used carbon electrodes with a separator between the electrolyte. The energy capacity is improved due to the large increase of surface area in the electrolyte.

The supercapacitors are still in development with few applications in power systems, since they have a high cost and limited energy density for large-scale energy applications. They are mostly used in DC power applications, for high peak power, and low energy situations, because of the high power density of this technology. So, supercapacitors have great potential, since they have a wide operation region, which can be completely discharged (unlike batteries) and operate effectively in several environments without damage (hot, cold, and moist) [1,3].

The main characteristics of supercapacitors are: high charge and discharge rates; low degradation over several hundred thousand cycles; good reversibility; less weight than the others (higher specific energy); low toxicity in the materials used for manufacturing and high cycle efficiency, approximately 95%. However, the main problem presented is the lower energy density compared with other technologies.

When combined with the use of batteries and supercapacitors, the operation field of HESS becomes even more powerful, seen as the management unit of electrical systems. Then, the hybrid storage becomes responsible for the dynamic stability (short-term supply) of the system, and also guarantees the long-term energy supply reducing damage or degradation of the storage. Batteries and supercapacitors have been one of the most popular combinations to compose hybrid energy storage system structures. In this sense, HESS can be categorised based on the number of ESS elements and converter configuration connected to the system.

5.1.2 Hybrid energy storage systems

Hybrid energy storage systems (HESS) can be configured in passive, active, or parallel and series connections. They can be categorised based on the number of ESS elements and converter configuration connected to the system. Figure 5.2 depicts the classification of battery-supercapacitor HESS topologies introduced in [8].

The passive configuration is the most simple one, where the supercapacitor and the battery are directly connected to the DC bus as depicted in Figure 5.2(a). In this case, the system works according to the devices' time constants to provide power sharing in the DC bus. The capacity of the battery and the supercapacitor cannot be fully used because the range of voltage in a DC bus is not flexible (to avoid voltage fluctuations).

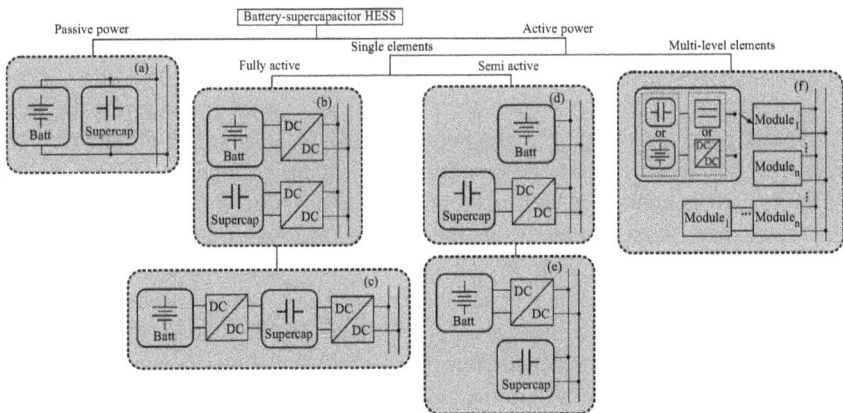

Figure 5.2 Different HESS configurations among passive, active, cascaded, and multi-level applications

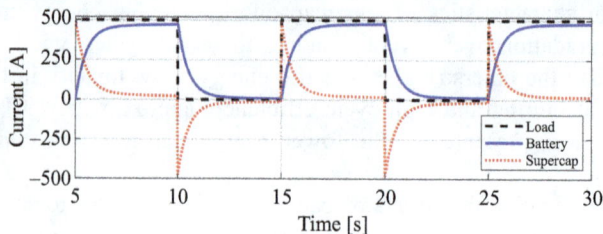

Figure 5.3   Power sharing of passive HESS where supercapacitor and battery time constant is highlighted

An example of passive HESS operation is depicted in Figure 5.3, where the power sharing between the supercapacitor and the battery under a period pulsed load is presented. In this example, the supercapacitor has a much faster response to immediately provide the total amount of the power demand. In a second step, the power of the battery slowly increases to reach the power demand while the supercapacitor reduces its contribution. Finally, in the steady-state, the battery supplies the load demand and the supercapacitor contributes only during the transients [9].

The active configuration uses a bidirectional DC/DC converter to control the power flow in the ESS, which results in a more flexible operation of the system. The active configuration of HESS can be done in parallel (Figure 5.2(b)), or cascaded (Figure 5.2(c)). In a fully active HESS application, the ESS elements are independent and can perform individual control targets allowing different control approaches at the same time, which results in much better system operation. Usually, the faster storage elements (supercapacitors) are used to act in transients and fast perturbations, while slower, higher energy density elements (batteries), are used in power flow regulation in the long term. Therefore, an active configuration has the best technical solution but can be considered an expensive and complex solution. The cascaded configuration can be used for devices with different voltage-rate operations, according to the construction of the element, but this configuration also increases the controller complexity.

Semi-active HESS topologies are explored as a good compromise between active and passive configurations (Figure 5.2(d)–(e)). In (d) configuration, the supercapacitor assures a flexible operation, improving volumetric efficiency. The supercapacitor is controlled to absorb high-frequency fluctuations in the DC bus, while the battery passively maintains the voltage variation inside the desired limits [9]. In Figure 5.2(e) configuration, the system is required to have a wider DC bus voltage range because of supercapacitor characteristics, which can be an issue for many applications. Finally, the multi-level HESS configuration in depicted in Figure 5.2(f), where the combination of passive and active elements are configured to create an unique power system management.

A common control strategy of parallel active HESS is to design low-pass filter (LPF), and moving average smoothing method (MASM), to decompose the frequency in low and high components. Thus, the supercapacitor is allocated to have fast response, dealing with high frequency perturbations, and the battery is allocated to deal with long-term supply to mitigate the low frequency impacts. The bandwidth and the cut-off frequency of the LPF are chosen to smooth the battery current, while the supercapacitor is demanded according to its capacity (or the DC/DC converter capacity), then a trade-off between supercapacitor and battery operation is necessary to assure proper operation of the HESS.

The decomposition approach is developed in [10] where the supercapacitor responds to fast power variations and the battery responds to slower variations. According to the power reserve of the proposed strategy, the voltage variation (dV/dt) is decomposed in two components: a slower one to provide battery reference and a faster component to the supercapacitor.

5.2   Optimal operation algorithm

There are many challenges that need to be faced in controlling storage systems. ESS allows the integration of energy sources in a smoother way, improving modern power system issues like bidirectional power flow problems that deteriorate the protection and control schemes of these systems. In addition, ESS helps in instability problems, related to the interaction of controls and transitions to different operation modes. There are various factors of uncertainty associated with load demand and especially with renewable energy sources that have intermittent characteristics. Thus, reliable operation of power grids involves load forecasting and weather forecasting models.

To address these challenges, the control of ESS must ensure reliable system operation. The currents and voltages must be properly controlled to reduce oscillations and frequency must be kept within the operating margins (in weak grid context[*]). The power balance must be regulated to maintain the balance between demanded load and generated power, with smooth variations for different operating modes transition and fast fault detection. In addition, economic dispatch with power sharing between generations can reduce operating costs and maintain system reliability. The optimization of operating costs can also include the provision of ancillary services together with optimal management of the power flow for the entire network. Thus, an adequate control strategy guarantees the proper operation of the electric grid.

The great challenge in the operation of ESS in distribution networks is to maintain a safe operation of the system, balancing generation and demand, where the optimal management of the system can be done through heuristic algorithms or intelligent control. Model predictive control (MPC) is also widely applied in power systems context to assure the optimal operation. MPC solves optimization problems in each sampling time, in order to determine the operational minimum (economic or technical) considering the physical limits and technical restrictions of the grid.

---

[*]Weak grids are related to systems low short-circuit ratio (SCR) values (normally $SCR < 3$), which for composed by isolated power systems, microgrids, etc.

Thus, the MPC includes a feedback feature in the optimization process to withstand uncertainties and disturbances, dealing with operational restrictions, such as the limits of storage capacity or the rate of power variation. MPC can also incorporate generation and load demand forecasts, based on the future behavior of the system. It is possible to obtain an optimal operation considering economic and technical criteria, which makes the MPC a very relevant control strategy for ESS [11].

ESS operation address different energy scenarios, where generation excess/deficit is minimized through optimization methods composed of cost functions. However, the open-loop feature of optimization systems does not allow to compensate for uncertainties and disturbances. Therefore, MPC closed-loop feature allows corrective actions using measurements to update the optimization problem, which ensures the optimal operation of the system [11].

5.2.1   Model predictive control

Considering the complex characteristics of the power system management problem, where forecast uncertainties and operational constraints appear, the MPC approach is an interesting solution. In this context, mixed integer linear programming (MILP) can be integrated into the MPC approach to solving different optimization problem. MPC strategy has gained a lot of relevance in the last years [11–15].

A formulation that integrates the MPC technique and the MILP is widely applied in distribution systems. The MPC technique will allow dealing with the power generation forecasts of renewables and the power demand of consumers, whose uncertainties, in the forecast are minimized using the concept of sliding horizon. The MILP technique will allow the solution of the linear optimization problem, in addition to continuous variables and binary variables, for example, battery charging or discharging, power flow through power converters, and connection or disconnection of the consumer [15].

The works in [16–19] are examples of successful solutions that apply MPC scheme for ESS management. These works develop MPC complemented with MILP approaches, using a mixed logical dynamical (MLD) structure to optimize the operation of ESS with AC coupling and interconnected with the utility grid. The bidirectional flow through the battery and in the utility grid is modeled, introducing binary variables, continuous auxiliary variables, and mathematical operations between vectors, performed by each bidirectional element. In these works, an economic dispatch problem is established, where the objective function is formulated to reduce the costs of energy production from distributed energy sources, and to maximize the economic benefits of energy sales to the power company.

MPC emerged not only as a control strategy but as a concept used to identify and leverage the development of a broad set of control methods, characterized by the following principles or basic elements [11,15]:

1. *Forecast model*: It represents the relationship between the output variables and the measurable or non-measurable input variables (input–output model). Its importance consists in the possibility of calculating future outputs $y(k + i|k)$ at

each instant k for a given forecast horizon (Ny) where $i \in N_1, ..., Ny$. This calculation depends on knowledge up to instant k of past values of the input and output variables; and the future control signals $u(k + i|k)$ with $i \in 1, ..., Nu$. Depending on the characteristics of the input variables, the forecast model is separated into two parts: the process model and the disturbance model.

2. *Cost function*: (or objective function) seeks to quantify the purpose/objective of the controller as a function of the values of future outputs and control signals. Generally, the cost function quantifies the error between future outputs and a given reference and the control effort incrementally $\Delta u(k + i|k)$ or directly $u(k + i|k)$. The objective function can be minimized (or maximized) considering its restrictions. However, in practical applications, when considering these restrictions, the solution of the optimization problem cannot be calculated analytically (due to very complex model), optimization techniques must be considered.

3. *Control law*: The design of the control law initially considers the calculation of the set of future control signals that optimize the objective function. Once the future control vector is calculated, the concept of *sliding horizon* is applied, where only the first element of the optimal control vector $u(k|k)$ is considered and sent to the process. The other elements are discarded; therefore, the control calculation is repeated at the following instant sampling. The robustness property of the MPC is that at the next instant of time, the output $y(k + 1)$ is already known, and new predictions are made from this updated value, obtaining a new value of the control signal $u(k + 1|k + 1)$ calculated at the instant $k + 1$, which in principle is different from the value $u(k + 1|k)$, due to this new existing information.

The MILP approach is based on the algebraic specification of a set of feasible alternatives for solving a problem, where an objective function is used as a decision criterion. The following conditions are necessary to bring a possible solution to the problem: Include integer decision variables in the modeling (often binary variables 0 or 1 are used), and continuous decision variables; express the decision criterion as a linear function of these variables; represent the set of feasible alternatives as solution variables, respecting a set of linear equations and inequalities.

Equation (5.1) defines a general formulation of the MILP approach:

$$\min \sum_{j \in J} C_j x_j + \sum_{i \in I} C_i \delta_i \tag{5.1}$$

$$s.t. \sum_{j \in J} a_{kj} x_j + \sum_{i \in I} a_{ki} \delta_i (\rho) b_k \quad \forall k \in K$$

$$x_j \geq 0 \ \forall j \in J$$

$$\delta_i \in Z_+ \forall i \in \mathbb{Z}_+$$

where C_j and C_i correspond to cost parameters of continuous x_j and binary δ_i variables, respectively; a_{kj} and a_{ki} are constraint parameters of continuous and binary variables, respectively; and b_k correspond to the requirements of the problem. J is the set of continuous variables, $\mathbb{Z}_+$ the set of positive integer variables, and K the set of constraints. The symbol ρ denotes mathematical relations of inequality and equality. A maximization model can be formulated as a minimization problem, multiplying the objective function by (-1).

In the optimization problem given in (5.1), the branch and bound method based on the divide and conquer strategy is used to deal with continuous and integer variables. The general idea is to perform successive partitioning in the space of valid solutions of the model. The problem is divided into sub-problems, considering then the solution of these sub-problems, one can reach the solution of the initial problem. The initial step of the method consists of finding the solution to a linear programming problem, which can be performed by several methods. If this solution, does not violate the constraints, the solution is optimal and the problem is solved, otherwise, a search tree is constructed [15].

Next sections introduce a number of optimization applications for ESS in distribution grid context. Different operation modes are described to improve ESS operation. The proposed strategies deal with power quality improvements and are applied to energy storage plants, where the objective function is modeled in a linear manner (MILP approach) and can be integrated into MPC solutions.

5.2.1.1   Peak shaving

The optimization for peak shaving operation seeks to reduce the peak demand of the main electrical grid, minimizing the cost of operating the distribution system. In this way, the solution to the optimization problem consists of determining the charge and discharge power of the ESS in order to reduce the peak demand with the lowest possible operating cost of the power system [13].

It is considered that the distribution system represents a cost for the system operator. Thus, energy prices have to be assumed for peak hours, intermediate hours, and off-peak hours. The peak shaving operation is composed of four objective functions that are summed to compose the complete peak shaving operation. Each part of the objective function introduces particular behavior to the optimization as detailed next.

The first the optimization function concerns the operation cost of the electrical grid according to the power level. Therefore, the more power is demanded, the higher the operating cost. This is described through a linear function:

$$\min(P_1) = \alpha_1 \sum_{i=1}^{T} (P_i^{grid} + Q_i^{grid}) \tag{5.2}$$

where α_1 is the energy price (which can be shifted according to energy price variation schedule), $P_i^{grid} \geq 0$ is the measured active power demanded from the main grid, Q_i^{grid} is the measured reactive power ($Q_i^{grid} \geq 0$ considering inductive load feature in traditional distribution systems), and T is the total time instants.

During ESS operation, the battery suffers degradation and, consequently, loss of useful life during charge and discharge cycles. Thus, there is a cost related to the use of these storage systems. In [20], the cost in $k/Wh for a lithium-ion battery is calculated. According to the authors, the cost related to battery degradation is proportional to the amount of power charged or discharged by the battery multiplied by a factor called linearized battery degradation cost coefficient (α_2) which can be obtained as:

$$\alpha_2 = \frac{C_{bat}}{2N(SOC_{max} - SOC_{min})} \tag{5.3}$$

where C_{bat} represents the price of the battery in $k/Wh, N represents the number of cycles the battery can operate during its lifetime, SOC_{max} and SOC_{min} represent the maximum and minimum limits of the battery's state of charge, respectively. The coefficient remains constant for the battery deep of discharge (DOD).

The SOC of an ESS can be calculated according to the nominal capacity of the battery (E) and the measured current in the battery terminals ($I_{bat}(t)$), which is written as follows:

$$SoC = 100\left[1 - \frac{1}{E}\int_0^t I_{bat}(t)dt\right] \tag{5.4}$$

SOC is given in percentage and the limits are between 0% and 100%.

The second objective function represents the operation cost of the ESS, it means that the price to operate the battery is proportional do the power rate from the ESS. This optimization function is calculated according to:

$$\min(P_2) \ = \ \alpha_2 \sum_{i=1}^{T}(P_i^{dc} + \ P_i^{ch} + Q_i^{dc} + \ Q_i^{ch}) \tag{5.5}$$

where α_2 is the battery degradation cost coefficient, $P_i^{dc} \geq 0$ and $Q_i^{dc} \geq 0$ are the active and reactive power discharged by the ESS, $P_i^{ch} \geq 0$ and $Q_i^{ch} \geq 0$ are the active and the reactive power absorbed by the ESS. Note that the battery charge and discharge variables have been separated to always assume non-negative values, so they assure suitable application in the optimization function.

The third optimization function is related to the desired value of SOC for the battery. It indicates the optimal operation point of the ESS. It is interesting for peak shaving operation that the ESS maintains a high amount of stored energy so that it can be used at peak hours. Thus, the third objective function is given according to the following equation:

$$\min|P_3| \ = \ \alpha_3 \sum_{i=1}^{T} |SOC^* - \ SOC_i| \tag{5.6}$$

where α_3 is the cost associated to vary the state of charge (this parameter is most used as a weight coefficient to adjust the behavior of the cost function without physical meaning), SOC^* is the desired reference for state of charge and SOC_i is measured the state of charge.

Thus, by setting a high SOC^* value, the optimization process works to charge the ESS in advance maintaining the ESS operation as close as possible to the desired reference. The charge state reference value (SOC^*) can be changed according to the proposed optimization strategy. For example, in other operation modes, it may be interesting to set the reference in $SOC^* = 50\%$, allowing maximum availability of power balance, whether for charging or discharging.

Note that an absolute value function is inserted in P_3, which results only in non-negative values for this function. This approach avoids problems with negative numbers when the optimization process is selected to minimize the function, and it is applied in the other optimization functions that may obtain negative values.

An additional cost for exceeding demand for peak shaving operation is here included, where the maximum active power value that the power system operator requires to demand from the main grid is established. So, if the power demand exceeds the given power limit (P_{lim}), the optimization function imposes an extra cost penalty (defined by substation power limits or maximum allowable demand). This maximum active power value can be determined based on the characteristic of the distribution network load curves. This is the main function to perform the peak shaving operation, so it is not possible to perform the peak shaving operation without this part. The fourth optimization function is calculated as follows:

$$\min(P_4) = \begin{cases} C_{ps}(\$), \ P_i^{grid} > P_{lim} \\ 0, \ P_i^{grid} \leq P_{lim} \end{cases} \tag{5.7}$$

with C_{ps} being a positive extra cost penalty in currency for the peak shaving operation, and P_{lim} being the maximum active power supplied by the feeder, indicating the physical limits of the substation equipment.

The optimization functions P_1, P_2, and P_3 are general in ESS application context, and can be easily implemented in different operation modes aiming at an overall optimization of the system. Therefore, the next operation modes also consider these functions as part of the complete cost function.

5.2.1.2   Renewable generation smoothing

The optimization for renewable generation smoothing operation seeks to reduce the intermittence in renewable sources (distributed generation) caused mainly by environmental factors. To reduce intermittence, the ESS is used to supply or absorb power in order to compensate for abrupt power variations at the output of the distributed generation. The idea is to produce a power reference curve, which represents the desired dispatch for the generation, considering a given profile. With this, the optimization process determines the power levels of charging and discharging of the ESS, so that the resulting generation of the generation profile is close as possible to the reference curve.

Considering a photovoltaic panel (PV), the optimization function is given by the power difference between the reference curve of solar generation and the resulting power of the solar plant and the ESS. In this way, the optimization problem seeks the ESS power values that reduce the difference between the reference curve and the solar plant generation with the lowest ESS operating cost.

The optimization function for generation smoothing can be written as follows:

$$\min|P_5| = \alpha_5 \sum_{i=1}^{T} |P_i^{ref} - (P_i^{dc} - P_i^{ch} + P_i^{pv})| \tag{5.8}$$

where α_5 is a weight coefficient, which can be sized according to the operation objective. P_i^{ref} is the reference curve at instant i, P_i^{dc} is the power discharged by the ESS, P_i^{ch} is the power absorbed by the ESS, $P_i^{pv} \geq 0$ is the measured power generated by the PV.

5.2.1.3 Voltage regulation

The voltage regulation optimization consists of using the ESS to keep the voltage at the point of common coupling (PCC) as close as possible to the rated voltage of the system, within the limits of voltage established in grid codes. To promote voltage regulation, the ESS can provide active or reactive power in order to compensate for variations in PCC voltage. Large-interconnected power systems and small grids (microgrids) may present different characteristics in voltage behavior in relation to active and reactive power. Thus, to propose a formulation for voltage regulation in the system, it is necessary to carry out a sensitivity analysis to model the relationship between active and reactive power in voltage regulation (see [21]). In these cases, a droop relation (V–Q) between voltage amplitude and reactive power may be obtained for traditional power distribution systems[†].

The objective function for voltage regulation is to minimize the difference between the nominal voltage of the system and the voltage at the PCC. This optimization function is calculated according to next equation:

$$\min |\Delta V| = \alpha_6 \sum_{i=1}^{T} |V^* - V_i^{pcc}| \tag{5.9}$$

α_6 is the weight coefficient, V^* is the rated voltage at PCC, and V_i^{pcc} is the measured voltage at the PCC.

5.2.1.4 Frequency regulation

The frequency regulation is operated in weak grids context, where the inertia is reduced and a compatible amount of power from the ESS is provided to control the frequency of the system[‡]. In this case, the ESS is used to control the frequency (obtained from phase-locked loop (PLL)) to reach the grid reference. The

---

[†]The coupling between active and reactive power can be neglected for power systems with a high X/R ratio, then the direct $V - Q$ relation is assured.

[‡]Strong grids require larger ESS to impact on frequency regulation, therefore, ESS in distribution cannot contribute, in general, to frequency control in large interconnected systems. More recently, large share of electric vehicle units operating in V2G have become able contribute to frequency regulation.

frequency control is made by active power supply in order to compensate for the frequency variations in the system. The optimization of frequency regulation is based on a higher level control approach, with time-scale from seconds to minutes. Therefore, the inertial support is not considered in the optimization algorithm, since its time-scale is slower, which can be included in the local controllers (lower level control).

So, a sensitivity analysis is necessary to model the relationship between active power and frequency, which results in the calculation of the f-P droop coefficient[§]. The objective is to minimize the difference between the nominal frequency of the system and the measured frequency of the grid. This function is calculated according to the following equation:

$$\min \ |\Delta\omega| = \alpha_7 \sum_{i=1}^{T} |\omega^* - \omega_i^{grid}| \qquad (5.10)$$

with α_7 being the weight coefficient, ω^* being the frequency reference, and ω_i^{grid} the measured frequency in the grid. The charge or discharge operation for the ESS is given in the droop relation inserted as a restriction in the model.

5.2.2 Restrictions

The restrictions delimit the physical regions to proper operate the system, providing the limits of operation of the ESS and the other equipment of the network. They establish the bounds where it is possible to operate the system optimally and safely, without damaging the electrical equipment. The basic restrictions for the operation of ESS in power systems are established as follows, considering the characteristics of each operating mode.

5.2.2.1 Active and reactive power balance

The power balance corresponds to the equilibrium between the power generated and consumed (demand) within the distribution system. Thus, the active and reactive power balance are calculated considering the power system containing the main grid, the PV system, the battery, and the load:

$$P_i^{grid} + P_i^{dc} - P_i^{ch} + P_i^{pv} - P_i^{load} = 0 \qquad (5.11)$$

$$Q_i^{grid} + Q_i^{dc} - Q_i^{ch} - Q_i^{load} = 0 \qquad (5.12)$$

where P_i^{grid} is the grid power, P_i^{dc} and P_i^{ch} are the active power provided and absorbed by the ESS, respectively. P_i^{pv} is the power generated by the PV, P_i^{load} is the power demanded by the load and losses, Q_i^{grid} is the reactive power in the main grid, P_i^{dc} and P_i^{ch} are the reactive power provided and absorbed by the ESS, respectively, and Q_i^{load} is the reactive power of the load.

[§]It is proposed the decoupling of active and reactive power, assuming high X/R ratio, then resulting in proper f-P relation.

It is extremely important to assure the active and reactive power balance in the system, to avoid power mismatch in the grid supply, so the power grid allows the balance of the whole system.

5.2.2.2   ESS charging/discharging decision and power limits

The ESS must supply active and reactive power for the system, respecting the operational limits established by the grid operators. In this way, the control algorithm must not request an active or reactive power greater than the values established for the equipment. Likewise, the control algorithm cannot request that the ESS be charged/discharged at the same time, as it would be infringing a physical limit of the equipment. These conditions can be inserted into the optimization problem as:

$$P_i^{ch} \leq \delta_i^{ch} P_{\max} \tag{5.13}$$

$$P_i^{dc} \leq \delta_i^{dc} P_{\max} \tag{5.14}$$

$$\delta_i^{ch} + \delta_i^{dc} \leq 1 \tag{5.15}$$

$$Q_i^{ch} \leq \xi_i^{ch} Q_{\max} \tag{5.16}$$

$$Q_i^{dc} \leq \xi_i^{dc} Q_{\max} \tag{5.17}$$

$$\xi_i^{ch} + \xi_i^{dc} \leq 1 \tag{5.18}$$

with $P_{\max}$ and $Q_{\max}$ being the maximum active and reactive power that can be supplied by the ESS, respectively. δ_i and ξ_i are the binary variable of decision to operate charge/discharge of active and reactive power in the ESS, respectively.

In the above restrictions, two binary variables (δ_i^{ch} and δ_i^{dc}) are used to control the charging and discharging process of the ESS, respectively. Therefore, $\delta_i^{ch} = 1$ corresponds to ESS charge and $\delta_i^{dc} = 1$ corresponds to ESS discharging. By this constraint, the two binary variables can never assume the same value of 1, that is, the ESS charge and discharge process do not occur simultaneously in the optimization problem.

The other two binary variables (ξ_i^{ch} and ξ_i^{dc}) are used to control the injection of reactive power into the grid, and they are not directly related to charge/discharge operation in the battery, i.e. reactive power does not necessarily affects the state of charge (SOC) in the the ESS. In this case, the power converter of the ESS is able to provide the required reactive power by controlling the phase angle of the current flowing into the grid. Only imaginary power is required for this operation, so no real power is expended.

5.2.2.3   State of charge limits

The state of charge (SOC) corresponds to the percentage of energy available in the ESS. The SOC must be constrained in a maximum and minimum value, in the optimization problem, in order to meet the limits of available energy in the ESS. In addition to the operational limits, the SOC varies depending on the amount of energy charged or discharged in a time interval. To cover the operational limits and the dynamics of the ESS in relation to the SOC, it is performed the discretization of the SOC equation in (5.4), to be included in the optimization:

$$SOC_i = SOC_{i-1} - \frac{P_i^{dc} - P_i^{ch}}{E} t_s \tag{5.19}$$

with SOC_i being the SOC from the storage at instant i, SOC_{i-1} is the SOC at the previous instant, and E is the total ESS capacity.

The operation limits for the SOC concerning the ESS operation are written as:

$$SOC_{\min} \leq SOC_i \leq SOC_{\max} \tag{5.20}$$

where SOC_{min} is the minimum allowed value of SOC and $SOC_{\max}$ is the maximum SOC from the ESS. This restriction allows to insert the maximum desired DOD operation for the ESS, according to the desired lifespan of the equipment. It allows safe operation of the battery.

5.2.2.4   Relation between voltage and reactive power

The relationship between the voltage at the PCC and the reactive power supplied by the ESS is inserted into the optimization problem as a constraint, which is given by V–Q droop relation:

$$V_i^{pcc} = V^* - K_v \ (Q_i^{load} - Q_i^{dc} + Q_i^{ch}) \tag{5.21}$$

where K_v is the droop coefficient, calculated according to the system sensitivity (see [13,21]).

This constraint is necessary for the optimization problem because it establishes the relationship between the variable to be controlled (V_i^{pcc}) and the control variables (Q_i^{dc} and Q_i^{ch}). The optimization problem determines the best values of Q_i^{dc} and Q_i^{ch} in order to attain voltage V_i^{pcc} in the nominal voltage of the system. In this constraint, K_R is a constant whose value depends on the reactive droop ratio of the considered distribution system.

5.2.2.5   Relation between frequency and active power

The relationship between frequency and active power in traditional power systems is given by a droop characteristic (considering $X > R$). The reference value for active power to be provided by the ESS in the grid is computed by the following f-P droop equation, inserted as a system constrain:

$$\omega_i^{grid} = \omega^* - K_f \ (P_i^{load} - P_i^{dc} + P_i^{ch}) \tag{5.22}$$

where K_f is the droop coefficient, calculated according to the system sensitivity [22].

5.2.2.6   Power converter operation limits

The ESS can either supply active or reactive power to the system. The relationship between the active, reactive, and apparent power of the ESS is given by the following equation:

$$S_i^{ess} = \sqrt{(P_i^{dc} + P_i^{ch})^2 + (Q_i^{dc} + Q_i^{ch})^2} \tag{5.23}$$

where S_i^{ess} is the apparent power of the ESS power converter. This cannot be used directly in the MILP optimization problem because it is a nonlinear equation. A linearization process is proposed in [23] to use equation in the MILP optimization problem. The equation can be linearized through a set of linear constraints to compose the behavior of the original nonlinear model in (5.23).

Therefore, the capability curve of the power converter can be represented by the following set of constraints in a linear manner.

$$\frac{-S_{max}^2 + \kappa(P_i^{dc} - P_i^{ch})}{\sqrt{S_{max}^2 - \kappa^2}} \le (Q_i^{dc} - Q_i^{ch}) \le \frac{S_{max}^2 - \kappa(P_i^{dc} - P_i^{ch})}{\sqrt{S_{max}^2 - \kappa^2}} \quad \forall \kappa \in A$$

(5.24)

where S_{max} is the maximum value of the apparent power of the ESS, κ is a discretization value inside the interval between $-S$ and S, composed by the set A. Thus, forming a linear subspace that meets the capability curve of the inverter, so that the equations generated by the constraints are linear.

The next constraint of aims to limit the phase angle of the converter connected to the ESS, based on the optimization of active power and reactive power generation. This restriction limits the phase shift angle between current and voltage between a maximum and minimum value, being an indirect restriction of the power factor. This consequently further restricts the ESS converter's operation to the following:

$$\theta_{min} \le \theta_i^{ess} \le \theta_{max}$$

(5.25)

with θ_{min} and θ_{max} being the minimum and the maximum allowed phase angle for the converter in rads, respectively, and the phase angle of the ESS being defined as:

$$\theta_i^{ess} = \tan^{-1}\left(\frac{P_i^{dc} - P_i^{ch}}{Q_i^{dc} - Q_i^{ch}}\right)$$

(5.26)

Equation (5.26) is a nonlinear restriction, which cannot be applied to the MILP approach, but this restriction can be performed indirectly by the power factor $\mathbf{FP}_i^{ess} = (P_i^{dc} - P_i^{ch})/S_i^{ess}$, limiting the operation to a defined range: $\mathbf{FP}_{min} \le \mathbf{FP}_i^{ess} \le \mathbf{FP}_{max}$.

5.3 Local controllers

ESS requires a DC/AC converter to be integrated into an AC grid bus, where the decoupling between the AC and the DC side can be advantageous. DC/AC converters uses of passive filters (L or LCL) to mitigate harmonic distortion and ripples in the AC grid caused by the switching of power converters [24,25].

Grid-connected operation mode requires a simple control strategy to inject active and reactive power into the main grid, where a phase-locked loop (PLL) structure is used to guarantee synchronism with the main network. In this case, the frequency and the voltage output are governed by the main network, represented

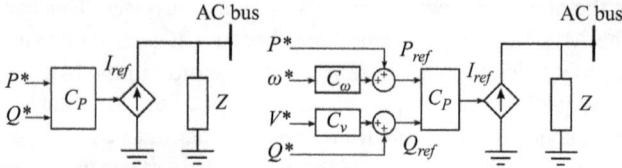

Figure 5.4 Grid-feeding (left) and grid-supporting (right) mode to connect VSC into AC grids

most of the time as an infinite bus. However, the operation of power converters in grid forming mode implies the control of frequency and voltage of the system, which is challenging for traditional control structures. The outer voltage control loop cascaded with the inner control current loop in power converters have complex tuning and suffer interference from the external parameters of the network, output converter filters, and PLL [26].

Grid-feeding converters are modeled as a current source, typically connected to the grid with high impedance, they are designed to deliver power to the grid, where active and reactive powers are controlled for this purpose. In this case, synchronization with the grid is needed and stand-alone operation is not possible. On the other hand, grid-supporting converters are modeled as controlled current sources in parallel with a shunt impedance. So, the currents and voltages of the converters are controlled to improve frequency and voltage levels of the connected grid. Extra ancillary services for grid-supporting may be implemented locally, such as inertia emulation, damping power oscillation, and unbalanced compensation [27]. Figure 5.4 introduces the model for grid-feeding and grid-supporting, where references of P^* and Q^* are given from a higher-level controller.

From the ESS point of view, it results in supplying power to the main grid following a higher control level (obtained from the optimization algorithm), ensuring power supply to the main network. The AC grid connection in simple cases is done through a three-phase AC bus connected with a VSC converter with an output L-filter. The AC grid is considered as an infinite bus, where voltage and frequency are considered stable, and the power variations are supported by the main grid. Section 6.3.2 will be presented a weak grid case[||], where frequency and voltage are taken into account for stability purposes.

5.3.1 Grid connection model

The ESS connected with the main grid is depicted in Figure 5.5. Equations (5.27) and (5.28) represent the state-space dynamics of this model:

$$\dot{I}_{ld} = -\frac{R_l}{L_l}I_{ld} + \omega_g I_{lq} + \frac{1}{2L_l}V_{dc}m_d - \frac{V_{ld}}{L_l} \tag{5.27}$$

[||]A weak grid can be composed of isolated systems with low inertia properties.

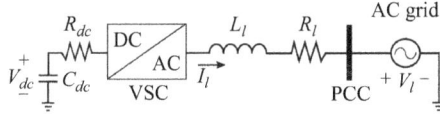

Figure 5.5 VSC connected to AC grid through L filter

$$\dot{I}_{lq} = -\frac{R_l}{L_l}I_{lq} - \omega_g I_{ld} + \frac{1}{2L_l}V_{dc}m_q - \frac{V_{lq}}{L_l} \tag{5.28}$$

where $I_{ld,q}$ and $V_{ld,q}$ are the currents and voltages on the AC grid using Park's transformation. V_{dc} is the DC bus voltage, R_l and R_{dc} are the cable losses and L_l is the line inductance. ω_g is the grid frequency and the modulation index m_d and m_q are the control inputs: they are bounded by $\sqrt{m_d^2 + m_q^2} \le 1$ to avoid over-modulation.

Here, I_{ld} and I_{lq} are the state variables, with ω_g, V_{dc}, and V_l being the system perturbations. Feedback linearization is used to meet the control target, i.e. to let the outputs reach two decoupled current references I_{ld}^* and I_{lq}^*. They are obtained by the power references P_l^* and Q_l^*. The synchronous dq reference frame is chosen such that the d-axis is fixed to the AC side of the grid, i.e., $V_{ld} = \hat{V}_l$ (voltage amplitude) and $V_{lq} = 0$. Therefore, decoupled control on P_l and Q_l can be realized considering the system synchronized [28]:

$$P_l = \frac{3}{2}V_{ld}I_{ld} \tag{5.29}$$

$$Q_l = -\frac{3}{2}V_{ld}I_{lq} \tag{5.30}$$

According to (5.29) and (5.30), it is possible to control P_l and Q_l by controlling I_{ld} and I_{lq} in the desired value. The active and reactive power references are calculated from the optimization algorithms as developed in the last section. The control objectives are explicit in (5.32) and (5.32):

$$I_{ld}^* = \frac{2}{3}\frac{P_l^*}{V_{ld}} \tag{5.31}$$

$$I_{lq}^* = -\frac{2}{3}\frac{Q_l^*}{V_{ld}} \tag{5.32}$$

5.3.1.1 Feedback linearization

According to the control objective, the control output is defined as $y = \begin{bmatrix} I_{ld} & I_{lq} \end{bmatrix}^T$, therefore the system has two outputs and two inputs. The Lie derivatives are non-singular since V_{dc} is always positive, therefore, a nonlinear feedback control input can be written as:

$$m_d = \frac{2}{V_{dc}}\left[L_l\Phi_d + R_lI_{ld} - \omega_g L_lI_{lq} + V_{ld}\right] \tag{5.33}$$

$$m_q = \frac{2}{V_{dc}} \left[L_l \Phi_q + R_l I_{lq} + \omega_g L_l I_{ld} + V_{lq} \right] \tag{5.34}$$

where $m_{d,q}$ are the control input and $\Phi_{d,q}$ are the additional inputs.

Since $I_{ld,q}^*$ is the desired trajectory for $I_{ld,q}$, it is possible to design the additional input $\Phi_{d,q}$ in a linear manner with respect to the output $I_{ld,q}$ and hence a linear stable subspace is generated:

$$\Phi_{d,q} = -K_{d,q}(I_{ld,q} - I_{ld,q}^*) - K_{d,q}^\alpha \chi_{dq} \tag{5.35}$$

$$\dot{\chi}_{d,q} = I_{ld,q} - I_{ld,q}^* \tag{5.36}$$

where $K_{d,q}$ and $K_{d,q}^\alpha$ are positive constants calculated by pole placement.

5.3.1.2 PLL synchronization

The voltages in the AC side of the VSC can be written as follows:

$$\begin{cases} V_{ld} = V_l \cos\left(\omega_g t + \theta_0 - \rho(t)\right) \\ V_{lq} = V_l \sin\left(\omega_g t + \theta_0 - \rho(t)\right) \end{cases} \tag{5.37}$$

where V_l is the peak voltage value, ω_g is the grid frequency, θ_0 is the phase angle of the fundamental component, and $\rho(t) = \omega t + \varphi$ is the angle of the synchronous system calculated by the PLL.

The adopted reference frame implies $V_{ld} = V$ and $V_{lq} = 0$ when $\rho(t) = \omega_g t + \theta_0$, which means that the converter is synchronized with the main grid. Thus, it is possible to design a controller for $\rho(t)$, such that $V_{lq} = 0$ in steady state [29,30]. The synchronous reference frame (SRF) PLL structure is shown in Figure 5.6, with $G(s)$ being a linear controller that cancels the quadrature component of the voltage ($V_{lq} = 0$) by adjusting the value of $\rho(t)$. The linear controller $G(s)$ is written as follows:

$$G(s) = k_p \frac{1 + s\tau_i}{s\tau_i} \tag{5.38}$$

where k_p is the proportional gain and τ_i the time constant.

The nonlinear dynamics of the synchronism structure is linearized around an operating point ($\bar{\rho} = \omega_n t + \theta_0$, where ω_n is the fundamental frequency), considering $\sin \mu \approx \mu$, and the grid frequency close enough to the fundamental frequency ($\omega_g \approx \omega_n$) [30]. Thus, based on the block diagram of the PLL, the tracked

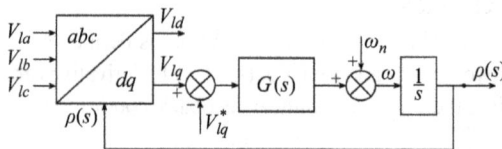

Figure 5.6   Block diagram of SRF-PLL

frequency is given as:

$$\omega_g = V_l[\omega_n t + \theta_0 - \rho(t)] \tag{5.39}$$

Therefore, the PLL is able to synchronize the VSC with the main AC grid.

5.3.2   Grid forming mode

In weak grids, voltage and frequency stability is not assured, therefore the power converter connected to this grid must be able to provide voltage and frequency support for the system. Here the VSC is interconnected with the AC grid via a LCL filter. The electrical model of VSC connected into the AC applying grid forming control is presented in Figure 5.7.

The state-pace model of the system can be written as:

$$\dot{I}_{c,d} = -\frac{R_c}{L_c}I_{c,d} + \omega_g I_{c,q} + \frac{1}{2L_c}V_{dc}m_d - \frac{V_{c,d}}{L_c} \tag{5.40}$$

$$\dot{I}_{c,q} = -\frac{R_c}{L_c}I_{c,q} - \omega_g I_{c,d} + \frac{1}{2L_c}V_{dc}m_q - \frac{V_{c,q}}{L_c} \tag{5.41}$$

$$\dot{V}_{c,d} = \frac{I_{c,d}}{C_c} - \frac{I_{l,d}}{C_c} + \omega_g V_{c,q} \tag{5.42}$$

$$\dot{V}_{c,q} = \frac{I_{c,q}}{C_c} - \frac{I_{l,q}}{C_c} - \omega_g V_{c,d} \tag{5.43}$$

$$\dot{I}_{l,d} = -\frac{R_l}{L_l}I_{l,d} + \omega_g I_{l,q} + \frac{V_{c,d}}{L_l} - \frac{V_{l,d}}{L_l} \tag{5.44}$$

$$\dot{I}_{l,q} = -\frac{R_l}{L_l}I_{l,q} - \omega_g I_{l,d} + \frac{V_{c,q}}{L_l} - \frac{V_{l,q}}{L_l} \tag{5.45}$$

with $V_{c,dq}$ being the voltage on the filter capacitor C_c and $I_{c,dq}$ the current on the filter inductor L_c. $I_{l,dq}$ are the line currents in the AC grid and V_l is the voltage on the main AC bus.

5.3.2.1   Voltage and current control loop

The voltage control loop is designed via cascaded proportional–integral (PI) controllers with an external voltage control loop and an inner current control loop. The time-scale separation artificially makes the inner currents' dynamics much faster than the voltages' dynamics to avoid control interaction and assure stability. The voltage control scheme is depicted in Figure 5.8.

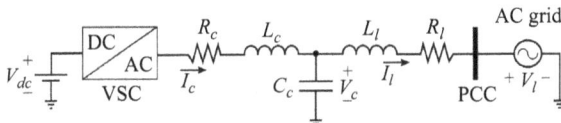

Figure 5.7   VSC connected to the AC grid through LCL filter

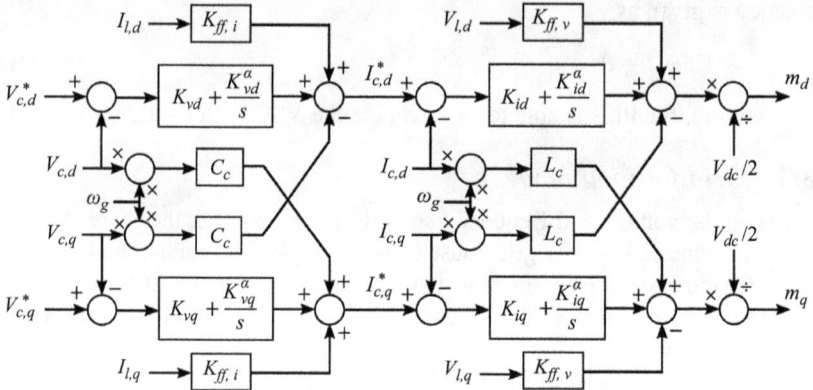

Figure 5.8   Voltage control with inner current control loop

The first step is to design a current control loop, in which the currents $I_{c,d}$ and $I_{c,q}$ track their respective references $I_{c,d}^*$ and $I_{c,q}^*$. Applying a PI controller with feed-forward and decoupling terms ($\omega_g L_c I_{c,q}$) to currents $I_{c,d}$ and $I_{c,q}$, the control inputs are calculated as:

$$m_d = \frac{2}{V_{dc}}[\Phi_{id} - \omega_g L_c I_{c,q} + K_{ff,v} V_{c,d}] \tag{5.46}$$

$$m_q = \frac{2}{V_{dc}}[\Phi_{iq} + \omega_g L_c I_{c,d} + K_{ff,v} V_{c,q}] \tag{5.47}$$

where $K_{ff,v}$ are the voltage feed-forward gains, and the PI controllers are written as follows:

$$\Phi_{id} = -K_{id}(I_{c,d} - I_{c,d}^*) - K_{id}^{\chi}\chi_{id}$$
$$\dot{\chi}_{id} = I_{c,d} - I_{c,d}^*$$
$$\Phi_{iq} = -K_{iq}(I_{c,q} - I_{c,q}^*) - K_{iq}^{\chi}\chi_{iq}$$
$$\dot{\chi}_{iq} = I_{c,q} - I_{c,q}^*$$

χ_{dq} are auxiliary dynamics representing integral terms to ensure zero error in steady state. The proportional gains are given by K_{id} and K_{iq}, with K_{id}^{χ} and K_{iq}^{χ} being the integral gains.

The dq currents' references $I_{c,d}^*$ and $I_{c,q}^*$ will be used as virtual control inputs, such that the voltages $V_{c,dq}$ reach their references $V_{c,dq}^*$. This procedure is completed by inserting the PI controllers with the feed-forward and decoupling terms ($\omega_g C_c V_{c,q}$) into the voltage control loop, leading to:

$$I_{c,d}^* = \Phi_{vd} - \omega_g C_c V_{c,q} + K_{ff,i} I_{l,d} \tag{5.48}$$

$$I_{c,q}^* = \Phi_{vq} + \omega_g C_c V_{c,d} + K_{ff,i} I_{l,q} \tag{5.49}$$

with $K_{ff,i}$ being the current feed-forward gain and the PI controller is expressed as:

$$\Phi_{vd} = -K_{vd}(V_{c,d} - V_{c,d}^*) - K_{vd}^\chi \chi_{vd}$$

$$\dot{\chi}_{vd} = V_{c,d} - V_{c,d}^*$$

$$\Phi_{vq} = -K_{vq}(V_{c,q} - V_{c,q}^*) - K_{vq}^\chi \chi_{vq}$$

$$\dot{\chi}_{vq} = V_{c,q} - V_{c,q}^*$$

K_{vd}, K_{vq}, K_{vd}^χ and K_{vq}^χ are positive gains computed by pole placement, such that the explicit time-scale separation is ensured. $\chi_{v,dq}$ are the integral terms to eliminate steady-state errors.

Here the voltage references $V_{c,d}^*$ and $V_{c,q}^*$ are provided to the controller by the secondary control level, since the system is modeled as a voltage source, allowing grid-forming operation mode, to control the frequency and voltage amplitude. The ESS can provide primary frequency and voltage support with power-sharing properties.

5.3.2.2   Frequency and inertial support

The stability management in power electronics dominated systems has become a relevant issue in the last years, with the integration of renewable energy and the growth of modern electronic loads. In this context, the inertia of power systems is systematically reducing, affecting the stability margins and causing blackouts. Weak grids, such as microgrids in remote areas and the stand-alone operation of these systems, also have as main feature low inertia, based on power converters generators [31,32].

The use of the virtual inertia concept was raised as a solution to provide inertial support and frequency regulation, mitigating the inertial issues of weak grids [26]. Virtual inertia consists in emulating the inertial behavior of synchronous machines, where the virtual synchronous machine (VSM) case introduces the swing equation of a traditional synchronous generator given by:

$$\dot{\omega}_{vsm} = \frac{1}{2H}[P^* - P - D_p(\omega_{vsm} - \omega^*)], \tag{5.50}$$

with H being the virtual inertia coefficient and D_p being the damping factor, ω_g the measured grid frequency, P^* is the active power reference and P is the AC grid measured power. The power angle of the VSM is computed by frequency integration $\delta = \omega_o \int \omega_{vsm} dt$, where the active power injection by the ESS is done [33].

The inertia coefficient H is computed to reduce power oscillations and improve transient stability, it involves a second-order system modeled to result in the desired transient behavior of frequency [34]. The transfer function is expressed as:

$$\frac{\delta(s)}{P^*(s)} = \frac{\omega_n}{2Hs^2 + D_p s + \omega_n P_{max}} \tag{5.51}$$

P_{max} is the maximum available power in the converter, computed as $P_{\text{max}} = E.V_c/X_{eq}$, with E being the output voltage on the VSC and X_{eq} being the system equivalent reactance.

The active power reference P^* given by the second level controller (optimization algorithm) from the previous section is inserted in the swing equation to compose the virtual inertia scheme. Therefore, the inner loop current reference is calculated as:

$$I_{ld}^* = \frac{2}{3} \frac{V_{ld}P^* - V_{lq}Q^*}{V_{ld}^2 + V_{lq}^2} \tag{5.52}$$

considering $V_{lq} = 0$ when the VSC is connected to the main grid.

5.4 Conclusions

Current energy transition is bringing heavy strain to the power grid, by the intermittence inherent to renewable energy sources and electric vehicles. The fact that most new power generation is interfaced to the grid through power converters, creates even more serious issues because of the reduction of the system's inertia. To cope with such new problems, storage appears as one of the main solutions the system has in the short to medium future. This chapter addresses the optimal operation of such energy storage systems (ESS) in the context of distribution networks. First, it discussed the interest of hybrid energy storage systems (HESS), mixing different types of storage (batteries and supercapacitors) or different technologies of batteries (lead-acid and lithium-ion). MPC-based control algorithms are then introduced, where an objective function is formulated to guarantee the optimization of battery operation. Different modes of ESS operations are proposed, based on improvements in power quality and provision of ancillary services. It considers battery operation costs, network operation costs at critical moments, SOC control, renewable injection profile control, voltage and frequency regulation, and even peak demand reduction. In this way, the optimization algorithm is considered a high-level control structure, which provides the active and reactive power references to be dispatched by the ESS. Then, the power references are transformed into voltage and current references for the internal control loops of the converters. It is important to remark that ESS includes private-owned systems, household or condominium-wide batteries, bidirectional charging electric vehicles, and also DSO's systems as large ESS. There are several ways to optimize ESS operation, even though the main physical issues and mechanisms are those discussed in this chapter. Future mechanisms will be developed, either in the optimization of the usage, as well as in the control algorithms to be used.

References

[1]　Khan N, Dilshad S, Khalid R, *et al.* Review of energy storage and transportation of energy. *Energy Storage*. 2019;1:e49.

[2] Akhil AA, Huff G, Currier AB, *et al. DOE/EPRI 2013 Electricity Storage Handbook in Collaboration with NRECA.* Sandia National Laboratories, Albuquerque, NM; 2013.

[3] Ribeiro PF, Johnson BK, Crow ML, *et al.* Energy storage systems for advanced power applications. *Proceedings of the IEEE.* 2001;89(12):1744–1756.

[4] Beardsall JC, Gould CA, and Al-Tai M. Energy storage systems: a review of the technology and its application in power systems. In: *2015 50th International Universities Power Engineering Conference (UPEC)*, 2015. p. 1–6.

[5] Omar N, Monem MA, Firouz Y, *et al.* Lithium iron phosphate based battery – assessment of the aging parameters and development of cycle life model. *Applied Energy.* 2014;113:1575–1585.

[6] Divya and Østergaard J. Battery energy storage technology for power systems – an overview. *Electric Power Systems Research.* 2009;79(4):511–520.

[7] Kousksou T, Bruel P, Jamil A, *et al.* Energy storage: applications and challenges. *Solar Energy Materials and Solar Cells.* 2014;120:59–80.

[8] Jing W, Lai CH, Wong WS, *et al.* A comprehensive study of battery-supercapacitor hybrid energy storage system for standalone PV power system in rural electrification. *Applied Energy.* 2018;224:340–356.

[9] Jing W, Lai CH, Wong WS, *et al.* Dynamic power allocation of battery-supercapacitor hybrid energy storage for standalone PV microgrid applications. *Sustainable Energy Technologies and Assessments.* 2017;22:55–64.

[10] Siad SB, Malkawi A, Damm G, *et al.* Nonlinear control of a DC MicroGrid for the integration of distributed generation based on different time scales. *International Journal of Electrical Power & Energy Systems.* 2019;111:93–100.

[11] Bordons C, Garcia-Torres F, and Ridao MA. *Model Predictive Control of Microgrids.* Springer, New York, NY; 2020.

[12] Fuchs L, Tortelli OL, Perez F, *et al.* Operação Ótima de uma Microgrid Conectada à Rede Elétrica utilizando Controle Preditivo e Modelo de Previsão baseado em Redes Neurais para Peak Shaving. In *2021 14th IEEE International Conference on Industry Applications (INDUSCON).* IEEE, 2021. p. 439–446.

[13] Fuchs L. Metodologia para a otimização de microrrede conectada ao sistema elétrico de distribuição utilizando a abordagem de controle preditivo baseado em modelo. *Federal University of Paraná.* 2022. p. 1.

[14] Perez F, López-Salamanca HL, Medeiros Ld, *et al.* Optimal operation of an urban microgrid using model predictive control considering power quality improvements. *Brazilian Archives of Biology and Technology.* 2021;64.

[15] Salamanca HLL. Controle e otimização de microrredes em baixa tensão no contexto brasileiro. *Universidade Tecnológica Federal do Paraná.* 2018. p. 1.

[16] Aguilera-Gonzalez A, Vechiu I, Rodriguez RHL, *et al.* MPC energy management system for a grid-connected renewable energy/battery hybrid power

plant. In *2018 7th International Conference on Renewable Energy Research and Applications (ICRERA)*. IEEE; 2018.p. 738–743.

[17]   Carli R, Cavone G, Pippia T, *et al.* A robust MPC energy scheduling strategy for multi-carrier microgrids. In *2020 IEEE 16th International Conference on Automation Science and Engineering (CASE)*. IEEE; 2020. p. 152–158.

[18]   Venkat AN, Hiskens IA, Rawlings JB, *et al.* Distributed MPC strategies with application to power system automatic generation control. *IEEE Transactions on Control Systems Technology.* 2008;16(6):1192–1206.

[19]   Antonopoulos S, Visser K, Kalikatzarakis M, *et al.* MPC framework for the energy management of hybrid ships with an energy storage system. *Journal of Marine Science and Engineering.* 2021;9(9):993.

[20]   Shi Y, Xu B, Wang D, *et al.* Using battery storage for peak shaving and frequency regulation: joint optimization for superlinear gains. *IEEE Transactions on Power Systems.* 2017;33(3):2882–2894.

[21]   Zad BB, Hasanvand H, Lobry J, *et al.* Optimal reactive power control of DGs for voltage regulation of MV distribution systems using sensitivity analysis method and PSO algorithm. *International Journal of Electrical Power & Energy Systems.* 2015;68:52–60.

[22]   Patino J, Valencia F, Espinosa J. Sensitivity analysis for frequency regulation in a two-area power system. *International Journal of Renewable Energy Research.* 2017;7(2):700–706.

[23]   Moreno R, Moreira R, Strbac G. A MILP model for optimising multi-service portfolios of distributed energy storage. *Applied Energy.* 2015;137:554–566.

[24]   Engler A and Soultanis N. Droop control in LV-grids. In *2005 International Conference on Future Power Systems*. IEEE, New York, NY, 2005. 6pp.

[25]   Sao CK and Lehn PW. Control and power management of converter fed microgrids. *IEEE Transactions on Power Systems.* 2008;23(3):1088–1098.

[26]   Zhang H, Xiang W, Lin W, *et al.* Grid forming converters in renewable energy sources dominated power grid: control strategy, stability, application, and challenges. *Journal of Modern Power Systems and Clean Energy.* 2021;9(6):1239–1256.

[27]   Rocabert J, Luna A, Blaabjerg F, *et al.* Control of power converters in AC microgrids. *IEEE Transactions on Power Electronics.* 2012;27(11):4734–4749.

[28]   Chen Y, Damm G, Benchaib A, *et al.* Feedback linearization for the DC voltage control of a VSC-HVDC terminal. In *European Control Conference (ECC)*, 2014. p. 1999–2004.

[29]   Golestan S, Monfared M, Freijedo FD, *et al.* Performance improvement of a prefiltered synchronous-reference-frame PLL by using a PID-type loop filter. *IEEE Transactions on Industrial Electronics.* 2013;61(7):3469–3479.

[30]   Pereira HA, Cupertino AF, da SG Ribeiro C, *et al.* Influence of PLL in wind parks harmonic emissions. In *2013 IEEE PES Conference on Innovative Smart Grid Technologies (ISGT Latin America)*. IEEE, New York, NY, 2013. p. 1–8.

[31] ENTSO-E. *High Penetration of Power Electronic Interfaced Power Sources, Guidance Document for National Implementation for Network Codes on Grid Connection*, 2017.

[32] Perez F. Control of ac/dc Microgrids with Renewables in the Context of Smart Grids: Including Ancillary Services and Electric Mobility. *Université Paris-Saclay.* 2020. p. 1.

[33] Perez F, Damm G, Lamnabhi-Lagarrigue F, *et al.* Adaptive variable synthetic inertia from a virtual synchronous machine providing ancillary services for an AC MICROGRID. *IFAC-PapersOnLine.* 2020;53(2):12968–12973.

[34] Xiao J, Jia Y, Jia B, *et al.* An inertial droop control based on comparisons between virtual synchronous generator and droop control in inverter-based distributed generators. *Energy Reports.* 2020;6:104–112.

Chapter 6

Considerations for EV usage for distributed energy storage employment & charging infrastructure

Manav Giri[1] and Sarah Rönnberg[1]

The proliferation of electric vehicles (EVs) goes hand in hand with the proliferation of the charging infrastructure. The charging infrastructure forms an essential backbone for the proliferation of EVs. The charging infrastructure is usually classified based on the type of charging connector, the type of output (AC or DC), or the associated power levels. This chapter extends the discussion on the charging infrastructure to the electrical installation requirements derived from the functions incorporated in the EV supply equipment (EVSE) and presents an overview of the electrical installation requirements for safe charging infrastructure. EVs represent a nonlinear load on the grid. As with all new technologies, there are challenges and opportunities. The challenges associated with the increase in the adoption of EVs and their impact on the grid are briefly discussed. EVs represent a unique load on the grid as it provides flexibility in terms of demand management and the possibility of energy storage. The primary function of an EV is transportation and hence the inclusion of grid support functions in an EV will require both regulatory support by updating the building and electrical installation codes as well as support at the policy level for incentivizing the different stakeholders. As outcomes, readers should be able to get a brief overview of the installation requirements, the challenges posed by the proliferation of EVs on the grid, and the unique opportunities and possibilities that the embrace of such a new technology brings.

6.1 History of EVs

EVs have been around for over a hundred years and back in the 1800s, they were powered by lead acid batteries. The hazards of carrying the liquid electrolyte, bulky lead acid batteries, underpowered motors, and lack of charging infrastructure were among the reasons which saw the demise of these experiments [1]. The advent of lithium-ion batteries and the tightening of pollution norms saw many a company

[1]Luleå University of Technology, Sweden

Euro emission norms for Diesel Engines

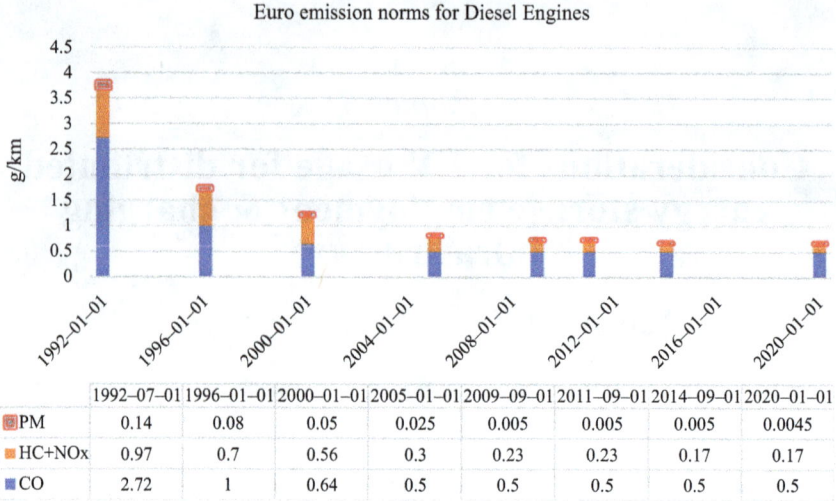

	1992–07–01	1996–01–01	2000–01–01	2005–01–01	2009–09–01	2011–09–01	2014–09–01	2020–01–01
PM	0.14	0.08	0.05	0.025	0.005	0.005	0.005	0.0045
HC+NOx	0.97	0.7	0.56	0.3	0.23	0.23	0.17	0.17
CO	2.72	1	0.64	0.5	0.5	0.5	0.5	0.5

Figure 6.1   Progression of diesel engine emission norms over the years. Data from [2].

like Tesla, Mitsubishi, Nissan, etc. re-initiate the development of EVs in the early 2000s [1].

The tightening of the emission norms as shown in Figure 6.1 has forced the automotive industry to invest in cleaner fuels and the development of better engines [3]. These have eventually led to an increase in development costs [4] and made EVs increasingly more appealing to the original equipment manufacturers [1]. The year 2010 saw the launch of the Nissan Leaf, which was the first commercially successful EV that was developed for the mass market by a major manufacturer [5].

6.2   Types of EVs

The road to the adoption of a fully EV has been a gradual one. A survey of customer perception towards EVs brought forth the following issues [6]:

- Range anxiety
- Availability of charging facility
- Repair and assistance during breakdown especially in remote areas
- Long charging times
- High cost of battery

Besides the automotive original equipment manufacturers (OEMs) not wanting to let go of the skills developed in the development of the internal combustion engine, the above issues also prompted them to take a gradual approach toward the introduction of a fully EV. Over the years, OEMs have tried different solutions to address the above issues. This gradual approach first saw the introduction of mild

Table 6.1   Classification of EVs

Vehicle attributes	Mild hybrid	Full hybrid	Plug-in hybrid	Battery EV
Drive	IC engine + Starter/ generator	IC engine + Motor	IC engine + Motor	Electric motor
Fuel	Hydrocarbon	Hydrocarbon	Hydrocarbon + Battery	Battery
Typical battery voltage	12 V, 24 V and 42 V	~42 V	~300 V	300–800 V
Typical battery capacity	< 1 kWh	~1.5 kWh	~15 kWh	>30 kWh
Electrical assistance	10%	20%	50%	100%

and full hybrid vehicles. Mild hybrids are vehicles with an integrated starter generator powering up a small battery, which provides additional assistance to the primary internal combustion powertrain as detailed in Table 6.1. The aim is to improve the efficiency of the internal combustion (IC) engine by using the additional power from the batteries as a boost [1].

Next in line has been the fully hybrid vehicles, which carry both an IC engine and an electric motor. The size and weight of the batteries in such vehicles are higher than those in the mild hybrids. This helps in allowing the IC engine to work in the region of peak efficiency and thus helps in improving the efficiency of the vehicle. The plug-in hybrid EV (PHEV) has a larger battery pack that can be charged from the power grid. In plug-in hybrids, the traction effort is shared by both the electric drive and the combustion engine. A PHEV's battery pack is smaller than an all-EV for the same vehicle weight as it still accommodates a combustion engine and a hybrid drivetrain. Fully EVs are EVs with the highest battery capacity and the entire traction effort being supported by the on-board battery. Another technology that is still at a nascent stage is a fuel cell-based EV which converts chemical energy from a fuel like hydrogen into electrical energy [1].

6.3   EV charging infrastructure

The EV charging infrastructure is the crucial backbone for the dissemination and promotion of electromobility. The most common topology of the power converters used in charging an EV is shown in Figure 6.2 [7]. The topology consists of a three phase/single phase active power factor correction (PFC) converter followed by a DC–DC converter for shaping the battery voltage in accordance with the current requested through the battery management system. Depending upon the mode of charging selected, the interface with the grid and the interface with the EV are selected and identified in the relevant IEC standards. Accordingly, the topology shown in Figure 6.2 may be on-board or off-board. The relevant IEC standards and

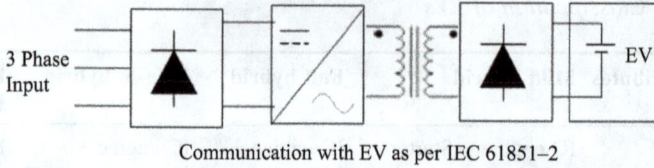

Communication with EV as per IEC 61851–2

Figure 6.2   General power architecture of an EV charger

Table 6.2   EV charging modes

Mode 1	Standard power lead plugged into normal outlet On-board charger in vehicle converts the AC input to DC output and controls the battery charging This mode of charging is used only for two and three wheelers	
Mode 2	In-line EVS control box (blue) is part of lead and provides leakage detection, over current protection, earth connection detection Lead can be plugged in normal outlet (15 A) On-board charger in vehicle converts the AC input to DC output and control the battery charging	
Mode 3	Dedicated wall box with in-built controller for monitoring EV charging with leakage current protection, over current protec- tion, earth connection detection Possible to control rate of charging through upstream On-board charger in vehicle converts the AC input to DC output and controls their battery charging	
Mode 4	Off-board AC to DC conversion Responds to requested current by the EV by modulating output voltage Ultra-fast DC charger up to 360 kW is being installed Communication protocols and connecters are defined in IEC 61851-24	

the connection interfaces will be discussed in Section 6.3. The EVSE thus chosen works as a mode to connect/disconnect the EV to the electric grid. The different modes of EV charging have been described in the IEC 61851 series of standards and are briefly described in Table 6.2.

The primary functions of the EVSE are as follows:

(i)   To provide a reliable and robust connection to the grid for the EV.
(ii) To establish an equipotential ground plane to facilitate energy transfer from the grid to EV.
(iii) To ensure that the EV requested current does not exceed the current safely supplied by the underlying grid architecture and thus protect the installation.

Over a period, some additional functions have been added to the EVSE to further increase the robustness of the installation and improve the safety of the customers using the EVSE. These functions are:

(A)   Provide protection against leakage currents both AC and DC to protect the customer from electric shock.
(B)   Monitor the input voltage within the ± 10% threshold of nominal voltage, i.e., from 207 V to 250 V for a 50 Hz system to ensure that the EV is protected from grid faults like a broken PEN conductor.
(C)   Provide flexibility for demand management by altering the rated current that the EVSE can provide.

In this chapter, the focus will be on the infrastructure requirements and not on the features provided inside the EVSE and thus the focus is on functions A, B, and C. Each of the above listed functions can be further decomposed into further subfunctions and we will list these as we go further in this discussion.

6.3.1   Provide reliable and robust connection to the grid for the EV

The provision for reliable and robust connection is accomplished through the connection interface and the underlying grid infrastructure supporting the charger.

6.3.1.1   Connection interface

Traditionally, the charging infrastructure has been classified based on the connector used for connecting the EV to the electric grid. Many publications identify the connectors and the geographical areas where these connectors are prevalent. For example, the IEC 62196 type I connector for single-phase charging is prevalent in geographies like the USA, Japan, and South Korea. Japan is the home of the CHAdeMO protocol for DC fast charging, while the USA and South Korea are increasingly adopting the combo connector based on IEC 62196 type 1 also referred to as the IEC 62196-3 EE connector (CCS1). Europe and China adopted the IEC 62196 type 2 connector for AC charging. For DC charging Europe has adopted the combo connector based on IEC 62196 type 2 also referred to as IEC 62196-3 type FF connector (CCS2) whereas China has GB/T 20234.2 connector for DC charging. India has been a recent entrant in adopting electromobility and provides a unique opportunity where the market is expected to be dominated by two and three-wheeled vehicles and commercial vehicles unlike other markets like Europe, America, Japan, and China which are expected to be dominated by passenger cars. India has adopted the IEC 62196 type 2 connector for three phase AC charging while for low voltage DC charging (< 100 V DC) it has decided to use the Bharat DC EV charging standard which uses the GB/T 20234.2 connector. The efficacy of this new charging standard for high current and high voltage applications, i.e., for commercial vehicles and passenger cars, though may seem limited but may turn out to be useful for two and three-wheeled automobiles that are expected to dominate this market as illustrated in Figure 6.3.

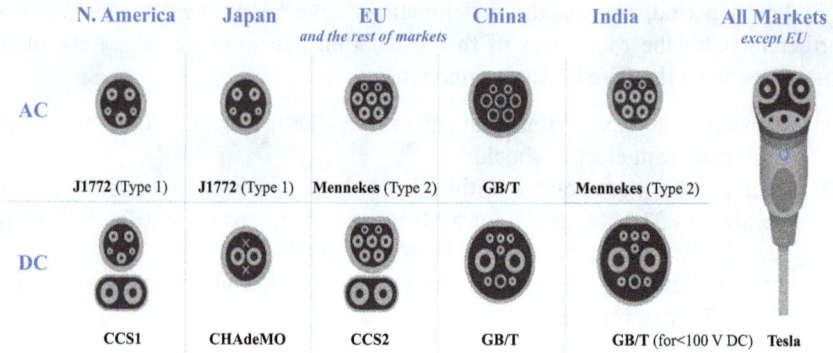

Figure 6.3   Connectors employed (EVSE interface) in the charging infrastructure

The common features associated with the above set of connectors are the following:

(i)   Providing a robust interface between vehicle and charger for at least 10,000 connect/disconnect cycles.
(ii) Ensure that the conductive earth pin is the first pin getting connected while connecting and the last pin during disconnection.
(iii) Monitor the connection between the EV and the EVSE through the proximity pin.
(iv) Provide communication between the EV and the EVSE.

6.3.1.2   The underlying grid infrastructure supporting the charger

The backbone of any successful charging infrastructure is the governing installation requirements and recommendations set by the local governing bodies responsible for the electrical safety of the installations. In Table 6.3, the IEC standards applicable for a standard EVSE mode 3 and mode 4 installation are listed, which is derived from a survey of different EVSE manufacturers and their installation documents available online. These standards have been notified by different countries in different regulations like the NEC in the USA, BS7671 in the UK, etc., and the list of such documents is given in Table 6.4. The installation guidelines and the technical standards/regulations specified in Tables 6.3 and 6.4 provide the requirements essential for building a secure charging infrastructure. The focus of these guidelines is on the following two points:

(i)   Protection from electrical shock.
(ii) Providing a secure and robust electrical infrastructure.

Protection from electrical shock:
The recommendations for protection from electrical shock focus on the following parameters:

Table 6.3 Standards governing the EV charging infrastructure requirements

#	Description	International standards	
		Name	Title
1	Cable sizing	IEC 60227-1	Polyvinyl chloride insulated cables of rated voltages up to and including 450/750 V – Part 1: General requirements
2		IEC 60287-1-1	Electric cables – calculation of the current rating – Part 1-1: Current rating equations (100% load factor) and calculation of losses – General
3		IEC 60287-3-1	Electric cables – calculation of the current rating – Part 3-1: Sections on operating conditions – Reference operating conditions and selection of cable type
4		IEC 60364-5-52	Part 5-52: Selection and Erection of Electrical Equipment – Wiring Systems (IEC 60364-5-52:2001, IDT)
5		IEC 60228: 2004-11	Conductors of insulated cables
6	Power quality	IEC 60038	IEC standard voltages
7		IEC 61000-2-2	Electromagnetic compatibility (EMC) – Environment-Compatibility levels for low-frequency conducted disturbances and signaling in public low-voltage power supply systems
8		IEC 60364-4-44	Low-voltage electrical installations – Part 4-44: Protection for safety – Protection against voltage disturbances and electromagnetic disturbances
9	MCB (overcurrent protection)	IEC 60364-4-43	Low-voltage electrical installations – Part 4-43: Protection for safety – Protection against overcurrent
10		IEC 60898-1	Electrical accessories – Circuit-breakers for overcurrent protection for household and similar installations – Part 1: Circuit-breakers for a. c. operation
11		IEC 60947-1	Low-voltage switchgear and control gear – Part 1: General rules
12		IEC 60947-4-1	Low-voltage switchgear and control gear – Part 4-1: Contactors and motor-starters – Electromechanical contactors and motor-starters
13		IEC 60947-5-1	Low-voltage switchgear and control gear – Part 5-1: Control circuit devices and switching elements – Electromechanical control circuit devices
14		IEC 61439-2	Low-voltage switchgear and control gear assemblies – Part 2: Power switchgear and control gear assemblies

(Continues)

Table 6.3   Standards governing the EV charging infrastructure requirements
(continued)

#	Description	International standards	
		Name	**Title**
15	RCD (leakage protection)	IEC/TR 60755	General requirements for residual current operated protective devices
16		IEC 61008-1	Residual current operated circuit-breakers without integral overcurrent protection for household and similar uses (RCCBs) – Part 1: General rules
17		IEC 61009-1	Residual current operated circuit-breakers with integral overcurrent protection for household and similar uses (RCBOs) – Part 1: General rules
18		IEC 61543	Residual current-operated protective devices (RCDs) for household and similar use – Electromagnetic compatibility
19		IEC 62955	Residual direct current detecting device (RDC-DD) to be used for mode 3 charging of EVs
20		IEC 60364-5-53	Electrical installations of buildings – Part 5-53: Selection and erection of electrical equipment – Isolation, switching, and control
21	Safety and protection of electric shocks	IEC 61140	Protection against electric shock – Common aspects for installation and equipment
22		IEC 60529	Degrees of protection provided by enclosures (IP Code)
23		IEC 62262	Degree of protection provided by enclosures for electrical equipment against external mechanical impacts (IK code)
24		IEC 60364-4-41	Part 4-41: Protection for safety – Protection against electric shock
25	Electromagnetic compatibility	IEC 61000-6-1	Electromagnetic compatibility (EMC) – Part 6-1: Generic standards – Immunity for residential, commercial and light-industrial environments
26		IEC 61000-6-3	Electromagnetic compatibility (EMC) – Part 6-3: Generic standards – Emission standard for residential, commercial and light-industrial environments
27	Surge protection device (optional)	IEC 61643-11	Low-voltage surge protective devices – Part 11: Surge protective devices connected to low-voltage power systems – Requirements and test methods
28		IEC 61643-12	Low-voltage surge protective devices – Part 12: Surge protective devices connected to low-voltage power distribution systems – Selection and application principles
29		IEC 62305-1	Protection against lightning – Part 1: General principles

(Continues)

Table 6.3   Standards governing the EV charging infrastructure requirements (continued)

#	Description	International standards	
		Name	Title
30		IEC 62305-2	Protection against lightning – Part 2: Risk management
31		IEC 62305-3	Protection against lightning – Part 3: Physical damage to structures and life hazard
32		IEC 62305-4	Protection against lightning – Part 4: Electrical and electronic systems within structures
33		IEC TR 60664-2-2	Insulation coordination for equipment within low-voltage systems – Part 2-2: Interface considerations – Application guide
34	Earthing	IEC 60364-5-54	Low-voltage electrical installations – Part 5-54: Selection and erection of electrical equipment – Earthing arrangements and protective conductors
35	Plugs and sockets	IEC 60884-1	Plugs and socket-outlets for household and similar purposes – Part 1: General requirements
36		IEC 60906-1	IEC system of plugs and socket-outlets for household and similar purposes – Part 1: Plugs and socket-outlets 16 A 250 V a.c.
37		IEC 62196-1	Plugs, socket-outlets, vehicle connectors, and vehicle inlets – Conductive charging of electric vehicles – Part 1: General requirements
38		IEC 62196-2	Plugs, socket-outlets, vehicle connectors, and vehicle inlets – Conductive charging of electric vehicles – Part 2: Dimensional compatibility and interchangeability requirements for a.c. pin and contact-tube accessories
39		IEC 60309-1	Plugs, socket-outlets, and couplers for industrial purposes. Part 1: General requirements
40		IEC 60309-2	Plugs, socket-outlets, and couplers for industrial purposes. Part 2: Dimensional interchangeability requirements for pin and contact – tube accessories.
41	EV supply equipment	HD 60364-7-722	Low voltage electrical installations: Part 7-722: Requirements for special installations or locations – Supply of EV
42		IEC 61851-1	EV conductive charging system – Part 1: General requirements
43		IEC 61851-21	EV conductive charging system – Part 22: AC EV charging station
44		IEC 61851-22	EV conductive charging system – Part 22: AC EV charging station
45	Installation requirements and verification	IEC 60364-1	Low-voltage electrical installations – Part 1: Fundamental principles, assessment of general characteristics, definitions

(Continues)

Table 6.3   Standards governing the EV charging infrastructure requirements (continued)

# Description	International standards	
	Name	**Title**
46	IEC 60364-5-51	Electrical installations of buildings – Part 5-51: Selection and erection of electrical equipment – Common rules
47	IEC 60364-6	Low-voltage electrical installations – Part 6: Verification
48	IEC 60950-1	Information technology equipment – Safety – Part 1: General requirements
49	IEC 61386-1	Conduit systems for cable management – Part 1: General requirements

Table 6.4   Regulations quoting the implementation of the governing standards for EV charging

	Sweden	Netherland	France	USA	UK	Israel
Regulations for Electrical Installations	ELSÄK-FS 2022:1	NEN1010	NFC 15-100	NEC	BS 7671	SI 1419
	SVENSK STANDARD SS 436 40 00 + R1	NEN 3140	NF C 32-321			

Providing a secure Earthing connection for the EV

The earthing systems used in the distribution network are described in Table 6.5. The earthing scheme employed provides a path for the conduction of the leakage current from the device and away from the user. Based on the potential hazards associated with the failure of the earth connection, the different earthing schemes have been rated in Table 6.5. It is to be noted that the potential risks of disconnection of the earthing conductor are the highest in the TN-C scheme of earthing and it is this scheme that has been expressly prohibited for its use in EVSE installations, as the failure of the PEN conductor can lead to potentially dangerous voltages to appear between the installation and the user. Additionally, the over-current protection devices and RCDs employed for the protection of the users against the leakage currents are ineffective in the event of an open PEN conductor fault.

The most common earthing scheme employed in Europe is the TN-C or the TN-C-S. For such schemes, the standard BS 7671 recommends the installation of a localized earth electrode and additional measures to ensure that the touch potential does not exceed the threshold of 70 V RMS (Figure 6.4). The steps are enumerated in Table 6.6.

Choice of a residual current detection device

In the ideal case of a properly working converter, the issue of DC injection from an EV does not arise while the EV is working as a load on the grid. However, in the

Table 6.5 Earthing schemes employed

Earthing scheme	Description	Potential Hazard of Earth conductor failure	Countries
TT	The neutral point of the transformer is connected to the ground. Provision of localized earthing through an earthing pit in the vicinity of the installation	Less	India, Japan, Italy, UK, Belgium, Spain
IT	The neutral point of the transformer and the protective earth is isolated	Less	Norway
TN-S	The neutral point of the transformer is connected to the ground and a separate conductor is taken from the transformer to the installation	Less	
TN-C	The neutral point of the transformer is connected to the ground and a PEN conductor is routed to the installation	Highest	UK, Sweden
TN-C-S	Combination of TN-C and TN-S with a connecting link in the distribution box	High	Sweden, Norway, USA

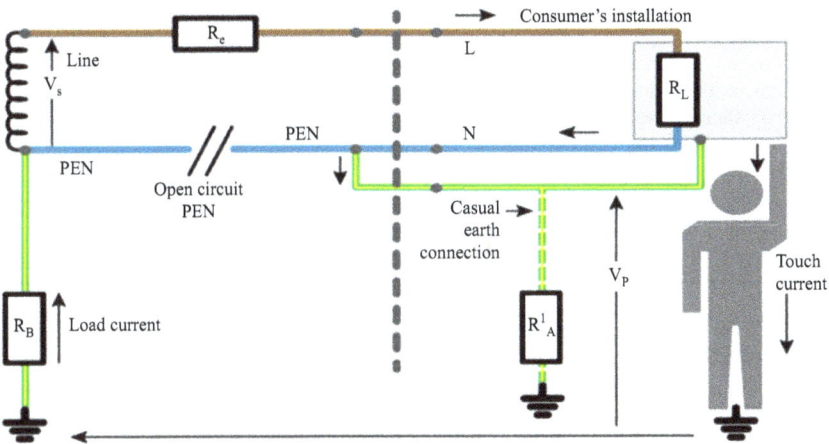

Figure 6.4 Conducted leakage current paths in the event of an open PEN conductor fault

event of a malfunction in the converters or especially in the event that the EV is used as a source of power, i.e., in the event of vehicle to grid (V2G) or vehicle to home (V2H) applications), DC injection into the grid is a distinct possibility. DC injection may cause saturation of transformer cores, overheating of grid-connected equipment, and acceleration of cable corrosion. Therefore, it is important to ensure that its value stays

Table 6.6   Recommended steps for earth conductor sizing

Step 1	Compute neutral current for a three phase installation after neglecting triple harmonics	$I_m = \sqrt{I_{L1}^2 + I_{L2}^2 + I_{L3}^2 + I_{L1}I_{L2} + I_{L2}I_{L3} + I_{L1}I_{L3}}$
Step 2	Compute the local installation Earth electrode resistance	Refer Annexure D of IEC 60364-5-54 for computing the electrode resistance in accordance with shape and type of electrode
Step 3	Compute the resistance of the cable used for connecting the protective earth to the earth electrode	The cable cross-section for protective earth shall be such that the cable cross section S shell not be less than the value derived from the formula (referred from IEC 60364-5-54). Wherever this computation is not used refer Table 54.2 from IEC 60364-5-54: $$S = \frac{\sqrt{I^2 t}}{k}$$ S is cross-section area mm^2; I is the r.m.s. value expressed in amperes of protective fault current, for a fault of negligible impedance, which can flow through the protective (see IEC 60909-0); t is the operating time in second of the protective device for automatic disconnection; k is the factor dependent on the material of the protective conductor, the insulation and other part the initial and final temperatures (for calculation of k, sec Annex A)
Step 4	Compute the total resistance of the total resistance of the earth electrode and protective conductor (R_{Aev})	
Step 5a	For single-phase operation, ensure the value of (R_{Aev}) is less than the value generated by the following computation	$R_{Aev} = \frac{70 U_0 C_{max}}{I_{inst} U_0 C_{max} - 70}$ ($C_{max} = 1.1$ for most installations) I_{inst} is the device current for single-phase installations
Step 5b	For three-phase operation, ensure the value of (R_{Aev}) is less than the value generated by the following computation	$R_{Aev} = \frac{70 U_0 C_{max}}{I_m U_0 C_{max} - 70(I_{L1} + I_{L2} + I_{L3})}$ ($C_{max} = 1.1$ for most installations)

within the allowable limit stated in the relevant clauses from the IEEE 1547, IEC 61727, and DIN VDE 0126 standards. Although the power converter is responsible for limiting the injected DC current on the AC lines, additional verification of DC injection by the EVSE is a cost-effective additional safety measure, allowing different EVs to be checked by one EVSE. The human body also needs to be protected from pulsating or pure DC components and the selected RCD needs to detect those.

The recommendations for the introduction of the protective scheme in an EVSE will be discussed in subsequent sections; in this section, we will only focus on the electrical infrastructure side recommendations as suggested by regulations like BS 7671. The recommendations from BS 7671 regarding the choice of RCDs are as follows:

(A)   The RCD shall disconnect all the live conductors in the installation.
(B)   The detection threshold shall be 30 mA.

(C) The RCD employed shall be a type "B" in the event the EVSE does not have any residual current protection built in.

(D) In the event, the EVSE has a built-in RCD protection of the type RDC-DD with a threshold of 6mA as per IEC 62955, then a type A or type F RCD shall be used in conjunction. The threshold of 6 mA is chosen so as to not interfere with the operation of type A or type F RCD with a detection threshold of 30 mA.

Additional precautions that are recommended in the installation of these RCDs are as follows:

(i) A type AC RCD should not be fitted upstream of a Type A, F, or B RCD as the load characteristics might impair the operation of a type AC RCD.

(ii) A type A or F RCD should not be connected upstream of a type B RCD as the load characteristics might impair the operation of the type A or type F RCD.

(iii) A type A RCD should not be connected upstream of a type A or type F RCD for similar reasons as cited above.

Protection from accidental discharge from energy stored in passive elements like capacitors
Capacitors are used in power electronics and other circuits for blocking the DC component, for providing a low impedance path to noise as in filters, or as a local reservoir of energy to stabilize the voltage and thus decouple the system from instantaneous variations in DC voltage. Thus, in such instances, an accidental touch may lead to a potential discharge through the human body even when the power electronic device is de-energized. The protection devices installed in such instances may provide no protection to the human in contact with the stored energy element. The UL2202 has additional recommendations to protect the human body from coming in contact with such stored energy elements. These recommendations are:

(i) Rules to limit the accessibility of such stored energy elements through appropriate mechanical design.

(ii) Specification of maximum permissible voltage available on the terminals of the capacitor with values measured after 5 sec from the time when the capacitor has become accessible. These limits can be referred to in Section 6 of UL 2202.

Providing a secure and robust electrical infrastructure
The recommendations for providing a secure and robust electrical infrastructure focus on the following aspects:

(a) Providing an overload protection device
(b) Cable cross-section identification
(c) Physical infrastructure-related requirements
(d) In this chapter, we will keep the focus limited to the electrical requirements and not on the physical infrastructure requirements, which specify parameters like the height of the installation and other physical attributes like parking space.

The specification of the overload protection device is specified in accordance with clause 443.1 of IEC 60364-4-43, wherein

$$I_B \leq I_n \leq I_Z \tag{6.1}$$

$$I_2 \leq 1,45 \times I_Z \tag{6.2}$$

where I_B is the design current for that circuit; I_Z is the continuous current-carrying capacity of the cable; I_n is the rated current of the protective device; I_2 is the current ensuring effective operation in the conventional time of the protective device.

To ensure that the requirements (6.1) and (6.2) above are satisfied, the following procedure may be followed:

(i)   Ascertain IB from the peak power to be discharged from the EVSE (e.g., 11 kW). Therefore, for a three-phase installation with a nominal phase to phase voltage of 400 V, the value of IB = 16 A.
(ii) The rating of the overcurrent device is chosen such that

$$In = k * IB$$

where k can be a value ranging from 1.1 to 1.6 depending upon the type of overcurrent protection device used.
(iii) Next, follow the steps as per IEC 60354-5-52 refer and clause 543 of IEC 60364-5-54 to ascertain the suitable cable cross-section.

6.4   Power electronics in EV

Power electronics in an EV can be used for both traction applications and battery charging applications. In this chapter, we will limit the discussion to the application for battery charging and V2G applications as this is the only mode wherein the EV is connected to the electricity grid. Several topologies have been proposed in the literature for these applications. However, the application of a power converter topology for a particular application depends upon factors such as:

(a)   Efficiency of power conversion
(b)   Available space constraints
(c)   Scalability of design
(d)   Associated bill of material cost

The parameters (a) and (b) listed above can be quantified together in the power-density of the applicable converter. An improvement in power-density entails a reduction in power losses such as conduction and switching losses in the converter and this reduction in turn helps the designer to package the components of a power converter module within a smaller footprint. Other factors that influence the packaging of the components are the applicable creepage and clearance restrictions. The creepage and clearance restrictions on a printed circuit board on which the power electronics components are assembled is a function of the over-voltage category which is defined by the point of connection of the device, the pollution degree which is defined by the operating environment of the device and the material group which is defined by the comparative tracking index (CTI). Based on the above parameters, standards for insulation coordination like UL840, EN60601, UL840, IEC60664, IPC2221, and EN60950-1 identify the creepage and clearance distances to be maintained. The scalability of the design is a function of the possibility of re-use of the design for different applications. It is typically observed that only the control and feedback circuits have the possibility of re-use (shown in orange) whereas the power circuits which include the snubbers and other reactive elements have to be re-designed in accordance with the switching and load requirements (shown in green in Figure 6.5).

Depending upon whether the converter is used for on-board or off-board applications in an EV, different topologies can be chosen for the AC–DC converter. Typically for on-board applications, the bridgeless and semi-bridgeless topology or the conventional 6-pulse converter can be used. However, the 12-pulse rectifier and matrix converter topologies are typically suited for off-board applications. The Vienna rectifier could be used in both on-board and off-board applications, but acting as a unidirectional active AC/DC converter and lack of regeneration may limit its application for off-board systems only. Generally, a PWM control scheme with constant switching frequency is employed in these converters to reduce the complexity of the input filter.

DC–DC converter topologies can be classified as non-isolated and resonant converter or phase shift topologies. The non-isolated topologies are the typical buck, boost, and buck–boost topologies used in low voltage-low power applications like two and three wheeled vehicles. Typically, EV charging of passenger cars and commercial vehicles requires operation over wide input and output voltages, and hence the resonant or phase shift converter topologies are utilized. The resonant converter topology can be further classified into different types in accordance with

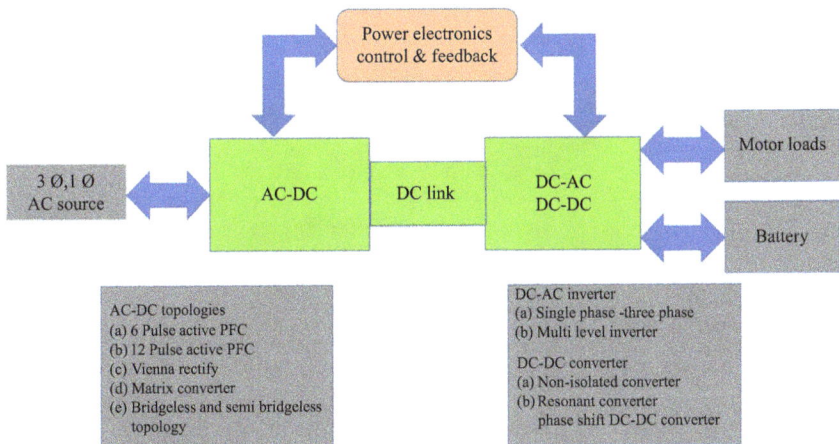

Figure 6.5 Block diagram of power electronic converters in EVs

the composition of the resonant tank (L, L–C, L–L–C, L–C–C, etc.) utilized in the construction of the DC–DC converter. The desired resonant tank topology is chosen to have a high permissible voltage gain variation and narrow frequency variation. Additional requirements to achieve zero voltage switching (ZVS) which include inductive input impedance, sufficient energy stored in the resonant tank, and sufficient dead time. The use of interleaved phase shifted DC–DC converter topology brings benefits like low output voltage and current ripple but may suffer from hard switching at low load conditions and thus in an increased harmonics/supraharmonics emission.

6.5   Impact of EVs on power quality

As discussed in the previous section, EVs represent a power electronic load on the grid. A review of published literature on the stability of power electronic converters on the electric grid suggests that power electronic loads with regulated output voltage feature negative incremental input impedance [8–10]. This is a cause of instability in the power system. Various instances of sub-synchronous resonance and resonances at higher frequencies associated with dynamic interactions of power converters and passive elements in the power system have been reported [11–13]. However, not all of these have been associated with EVs. A review of the literature associated with the proliferation of EVs suggests the following power quality events as described in Figure 6.6 [14,15]. Events in grey are those that can be attributed to the large penetration of EVs in the grid, while those in orange can be observed even at lower levels of penetration in the grid. The discussion at hand will be limited to such events that are observable even at lower penetration levels only.

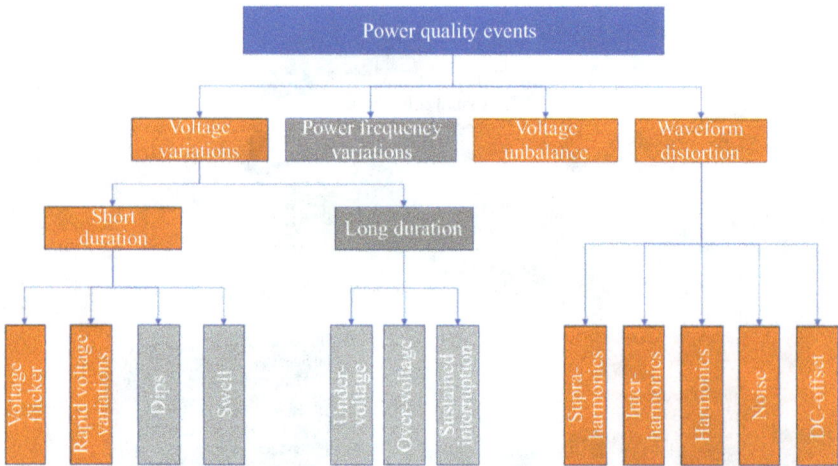

Figure 6.6   Power quality events associated with EV proliferation in the power system

In this section, the collective term emission will be used to describe voltage and current distortion in the form of harmonics, interharmonics, and supraharmonics.

6.5.1 Harmonics

The increase in power electronic loads in the power system helps improve the efficiency of operation, however, introduces new issues in the power system. A classic diode bridge rectifier with a DC link capacitor leads to a discontinuous input current which is present only when the output voltage of the rectifier exceeds the capacitor voltage. The non-sinusoidal and discontinuous nature of current leads to a high value of total harmonic distortion (THD) and a poor power factor. This led to the inclusion of active power factor correction topologies in power electronics devices like the boost converter at the output of the diode bridge rectifier [15]. However, the load current is sourced through the inductor and the semiconductor switch and thus such a topology is only suited for low power applications like LED lamps. High power loads like EVs require an active front end power factor correction module. The increase in penetration of such non-linear devices and the corresponding waveform distortion led to the introduction of the concept of distortion power factor [16]. The true power factor is the product of the distortion power factor and the displacement power factor [16]. The input current of a 6-pulse or 6P active rectifier will consist of $NP\pm1$ harmonic components. For $N = 1$, the input current is expected to be rich in the 5th and the 7th harmonic [17]. The EV is expected to be compliant with the IEC 61000-3-2 and IEC 61000-3-12 (standards applicable to electrical loads on the grid specifying the harmonic content of the load), and the magnitude of the injected harmonics is expected to be within the specified limits in the standard. Harmonics in the power supply are introduced through the interaction of the harmonic content of the current consumed by non-linear loads with the source and line impedance of the system. However, the harmonic current of the load is known to be a function of the background disturbances of the voltage waveform [17].

Typically, the power electronic loads have been modeled as a current source with an admittance in parallel [18]. The current source acts as the source of primary emissions from the device and the admittance is responsible for the secondary emissions. The input admittance of a power electronic device comprises of the input admittance of the filter and the converter. The admittance of the converter is decided by the control system. As described in [8], each of the converter control loops adds a parallel admittance to the converter and thus the overall converter input admittance is a function of the operating point of the converter. For EVs, especially in mode 4 charging, this operating point keeps changing. For example, if we take the CHAdeMO protocol as an example, the charger receives a new value of desired charging current from the vehicle every 100 ms [19]. To meet this varying charging current request, the operating point of the charger and thus the input impedance of the charger keeps changing. The operating point is in turn a function of the estimate of the battery open circuit voltage, as illustrated in Figure 6.7. The estimation of the battery open circuit voltage is based on parameters like the state of charge, the rate of charging, ambient temperature, battery temperature, etc. [20].

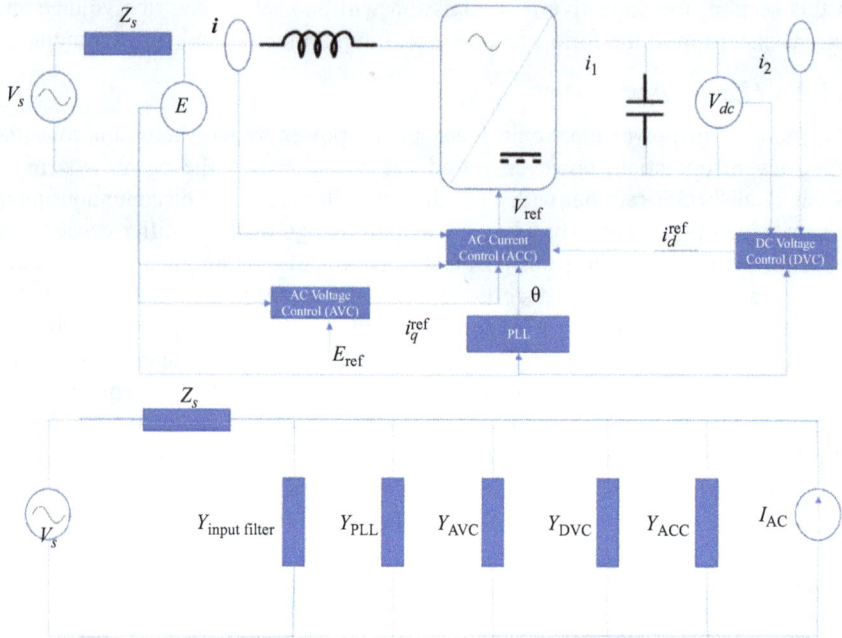

Figure 6.7   *Block diagram of front-end converter and equivalent model used in power system studies*

The input impedance and the emission from the charger are expected to vary for different vehicles, different battery sizes, and chemistries as the operating point are expected to be different. Figure 6.8 shows the variation of harmonic current emission from an EV charging at a fast charger for the 5th and the 7th harmonic. The emission from the EV can be seen to change depending upon the operating stage of the charging and to a lesser degree within each operating stage. The amplitude of the harmonic current emissions is a function of the grid impedance. A strong grid is defined as one with quasi-zero impedance while a weak grid can be emulated using reference impedances identified in IEC/TR 60725. From measurements conducted in a lab environment for the same EV, in the same charging mode, connected to a strong and then subsequently to a weak grid, it is seen that the recorded harmonic current amplitude changes. For a strong grid, the amplitude of the dominating harmonics originating from the EV is higher compared to when the EV is connected to a weak grid. The results are shown in Figure 6.9.

6.5.2   Supraharmonics

Supraharmonics are frequency components in the range from 2 kHz to 150 kHz. EVs, like most modern power electronic-based devices, are known sources of supraharmonics as residues from the active switching occurring in this frequency range are injected into the grid [21]. It was discussed in Section 6.4 that the active

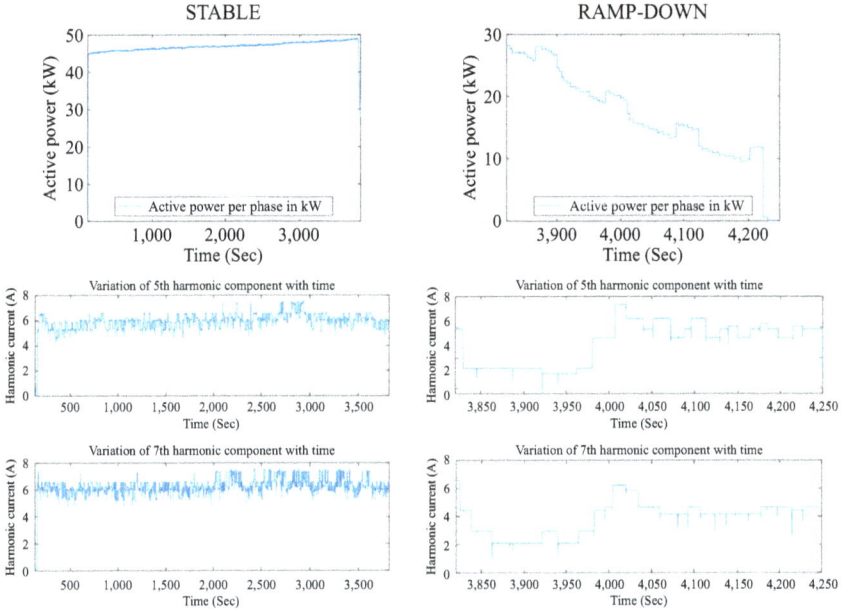

Figure 6.8   Variation in 5th and 7th harmonics components of current observed in a DC charging session

front end PFC is expected to have a constant switching frequency to reduce the complexity of the input filter; however, a resonant DC–DC converter can have a varying switching frequency to meet the operational output voltage and current requirements. An interleaved phase shifted DC–DC converter is expected to have a constant switching frequency but a varying pulse width. It is these switching frequency components and their harmonic multiples that are expected to be visible in the spectrum of the recorded emissions from the charger. These switching artefacts couple with the common mode capacitances and are conducted to the protective earth conductor [22]. The protective earth could be shared by other loads on the same grid and thus the protective earth can work as a bridge for the spread of such high frequency components [23]. Such propagation of harmonics is, however, contingent upon the type of earthing employed. Such propagation of supraharmonic frequencies has been reported in the United States and not in Europe where a resonant earthing is employed on MV installations. The supraharmonic current emissions also have a dependence on the grid impedance with higher values reported for stronger grids as shown in Figure 6.9.

6.5.3   Noise

The IEC 61851-24 standard introduced power line carrier (PLC) as a means of secure communication between the EV and the EV charger for DC fast charging. The communication protocol has been described in detail in the ISO 15118 series of

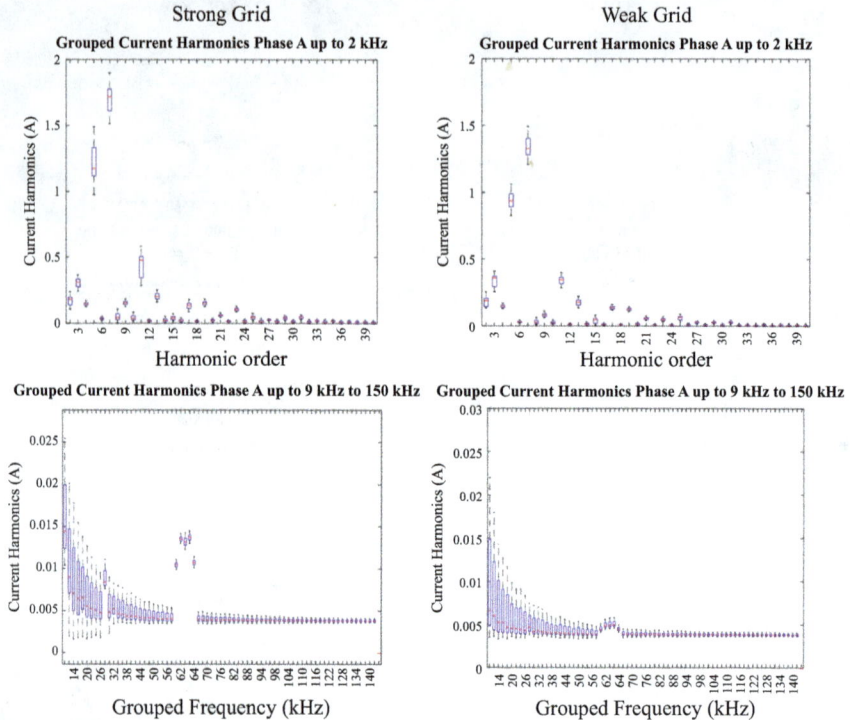

Figure 6.9   Variation in harmonic and supraharmonic components of current observed in an AC charging session for the same vehicle charged on a strong and a weak grid

standards and is implemented through the HomePlug Green Phy Standard. The frequency range of the communication is between 2 and 30 MHz. The PLC communication takes place on the control pilot line and the protective earth in the connection interface between the EV and the DC fast charger. The protective earth is also connected to other loads and thus the protective earth can act as a bridge for the spread of these signals.

6.5.4   Interharmonics

Interharmonics are non-integral multiples of the fundamental frequency. Studies have shown double conversion systems (systems with DC link) to be sources of such frequencies [24]. The emission of interharmonic frequencies can be estimated from variations in the active power of the load [25] that there exists a linear relationship between the magnitude of the emitted sideband and the change in the active power of the load. Besides as discussed above, the switching frequency components, another phenomenon that has been discussed in [26] is the deviation of the switching frequency from its integral value due to component nonlinearities and the grid frequency interacting with each other and causing a constructive or destructive interference, with the net result being a beat frequency that will not be an integral multiple of the grid frequency.

6.5.5   Unbalance

Unbalance is an issue commonly associated with single-phase charging of EVs. Most users prefer charging their EVs at home. Countries like South Korea, the United States, Japan, and India have predominantly single-phase charging installations. Mode 2 and Mode 3 chargers are usually installed in residential installations and single-phase installations may lead to greater instances of unbalance on the grid. The impact of unbalance on active front end converters is characterized by increased stress on switching elements [27], an increase in voltage ripple causing increased stress on passive elements [27], and an increase in zero sequence harmonics and causing an increase in the neutral to earth (N-PE) potential. Excessive neutral-earth voltage may influence communication lines in network equipment and cause the increase of error rate of data transmission but also has the potential to damage some of the network equipment. Some network equipment like servers have a built in neutral-earth voltage detection circuit. Once neutral-earth voltage exceeds a certain specified value, the equipment cannot start up [28]. Such features have also been integrated in EVSE's and may prevent the EVSE from charging the EV.

6.5.6   DC offset

The presence of a DC voltage or current in the grid is called a DC offset. DC offset is a phenomenon commonly associated with inverter modules such as those observed in photovoltaic systems [29–31]. The DC current is not a fault current but is caused due to asymmetry between the positive and the negative half wave of the current fed into the grid. The source of such asymmetry is attributed to minor asymmetry in pulse width modulation signals caused by truncation errors in the calculation of switching instances, non-linearity in switching devices, and small errors and offset drift in the voltage and current measurement sensors used to provide a feedback signal for the control system [29–31]. Inverter modules introduce a small DC voltage which is injected into the grid and the low network impedance of an AC network can result in a large DC current injection. The injected DC current from an inverter module is regulated through standards like IEEE 1547.1, IEC 61727, and DIN VDE 0126. The injected DC currents can cause saturation of the distribution transformers which causes waveform distortion, excessive losses, overheating, and hence an overall diminished life expectancy [32]. Increasingly EVs are looked upon as sources of energy storage. With concepts like V2G and V2H gaining traction [33–35], such issues that have been commonly seen in PV systems could also get attributed to EVs.

6.5.7   Rapid voltage variations and voltage flicker

EV charging can create fast voltage fluctuations whose magnitude depends on the transient changes in current and impedance of the grid [36]. The variation in current is associated most commonly with DC fast charging of EVs. During the DC fast charging, there are two stages associated generally with rapid variations in the current sourced from the grid. The first stage is associated with the initial ramp up in requested power from the EV at the initial stages of charging and the second is associated with the ramp down of requested power commonly associated with the end of EV charging or also as identified in [37] as the constant voltage (CV mode) or trickle charge modes of EV charging. This fast voltage fluctuation could create light flicker (repetitive change in light intensity) that the human eye can perceive [36].

6.6   EVs as a flexible load

The global EV outlook data from 2015 to 2020 suggests a marked increase in the adoption of EVs (both plug-in hybrids and battery EVs). The share of new EV registrations in Europe has increased from less than 2% in 2015 to about 11% in 2020 [38]. With the increase in numbers, EVs charging without any kind of load management functionality could lead to grid constraints and increased transmission and distribution costs that prompt the construction of more peak load delivery plants, unplanned grid upgrades, and other costly solutions to meet the peak load demand. The EVs (or more correctly their batteries) could essentially be used in two ways, as a solution for loading issues and as a solution for voltage issues [39]. When it comes to voltage issues, the charging of the EV could potentially be controlled in such a way that it can help mitigate both overvoltage and under-voltage (in case of V2G) as well as voltage unbalance. For loading issues, there are different ways that the EV could provide flexibility:

- Curtail the charging at times of high demand. This is commonly referred to as peak shaving.
- Shift the charging to times with less demand. This is commonly referred to as load shifting.
- Support the system by feeding in power when needed. This requires an EV with V2G capabilities.

Additionally, the report by the Idaho National Laboratory (INL) in 2012 on V2G power flow regulations and building codes review suggests the following applications for the vehicle to grid application [40]:

- Electric energy time shift
- Electric supply capacity
- Load following
- Area regulation
- Voltage support
- Time-of-use energy cost management
- Demand charge management
- Renewables energy time shift
- Renewables capacity firming
- Wind generation grid integration.

A detailed discussion on each of the listed applications is beyond the scope of this chapter but some simplified examples can be used to evaluate the potential of EVs as a solution during times of high demand.

Imagine that a large production unit of e.g., 1.2 GW gets disconnected leading to an instant deficit of available power. To replace this would require 400,000 EVs that feed in 3 kW each. This corresponds to about 8% of Sweden's total fleet of cars that could potentially replace this production unit for about 10 h.

Sweden is a net exporter of energy but there are instances when the country as a whole need to import electricity, usually during cold days when the demand for heating is high. Such an instance occurred for example on January 6, 2021, when Sweden had to import 0.51 GW during an hour. If about 5% of all cars in Sweden reduced their charging at this time with 2 kW, this would have kept the Swedish system in balance and no import would have been necessary.

From a customer perspective, the benefits of demand management can be seen in lower costs of energy for charging the EV and the monetary benefits of using EVs as frequency containment reserves for grid strengthening. The other applications may require longer durations of operation of the EVs in a grid support role. This may not be amenable to most customers as this would translate into vehicles being immobile for a longer duration of time. Another reason for the lack of enthusiasm can be associated with the aging of batteries. Lithium-ion batteries use the process of intercalation, wherein the Li^+ ion is transferred from the cathode to the anode during charging and vice versa during discharging. Over a period, some of the Li+ ions do not get absorbed into the anode and accumulate on the surface of the electrode. These accumulated Li^+ ions have the effect of inhibiting the absorption of new Li^+ ions and thus lead to an aging effect on the batteries.

For the demand management, time of use tariffs for residential home charging has been demonstrated to be an effective tool for managing peak loads [41]. The regulatory and technological requirements facilitating demand side management are also in place [41]. However, from an energy storage perspective, the INL report identifies challenges on the regulatory front in building and electrical installation codes that need to be addressed to facilitate the uptake of V2G/V2H. From a web survey of available technology demonstrators of V2G/V2H functionality, two possible configurations have been reported [42]:

- Configuration A: Vehicle with bi-directional converter.
- Configuration B: External off-board converter with EV as a battery bank.

Configuration A will require the introduction of additional functions related to synchronizing with the grid, anti-islanding protection, and limits against DC current injection in the V2G mode. The V2H mode will require special attention to be paid towards the safety of the installation and earthing, as has been reflected upon in Section 6.3.1.2.

Configuration B: The architecture amenable to most OEMs is likely to be that of an external inverter with the EV being the DC source as the additional functions listed in configuration A will have to be incorporated in the external converter. Such a system has been developed and deployed by Nissan and demonstrated in the European commission's joint research center report [42,43].

6.7   Conclusion

In this chapter, an overview of the grid infrastructure requirements necessary for the installation of EVSE's has been presented. This is followed by a discussion on the power electronic converters associated with EV charging and the possible impact of EVs on the grid has also been briefly discussed. The proliferation of EVs presents a unique opportunity where-in the lessons learnt from the implementation of photovoltaic systems can be implemented. However, the primary function of an EV is not grid support but transportation. And hence automotive OEM's, users and utility companies will need to be incentivized to embrace the unique opportunities of utilizing EV's for grid support functions.

References

[1]   A. Ajanovic, 'The future of electric vehicles: prospects and impediments', in *Wiley Interdisciplinary Reviews: Energy and Environment*, vol. 4, no. 6, pp. 521–536, 2015 doi: 10.1002/wene.160.

[2]   IARC Working Group on the Evaluation of Carcinogenic Risks to Humans. Diesel and Gasoline Engine Exhausts and Some Nitroarenes. Lyon (FR): International Agency for Research on Cancer; 2014. (IARC Monographs on the Evaluation of Carcinogenic Risks to Humans, No. 105.) ANNEX: EMISSION STANDARDS FOR LIGHT- AND HEAVY-DUTY VEHICLES. https://www.ncbi.nlm.nih.gov/books/NBK294251/.

[3]   N. Hooftman, M. Messagie, J. van Mierlo, and T. Coosemans, 'A review of the European passenger car regulations – real driving emissions vs local air quality', *Renewable and Sustainable Energy Reviews*, vol. 86, pp. 1–21, 2018. doi: 10.1016/j.rser.2018.01.012.

[4]   W. Knecht, 'Diesel engine development in view of reduced emission standards', *Energy*, vol. 33, no. 2, pp. 264–271, 2008. doi: 10.1016/j.energy. 2007.10.003.

[5]   F. Carranza, O. Paturet, and S. Salera, 'Norway, the most successful market for electric vehicles', in *2013 World Electric Vehicle Symposium and Exhibition, EVS 2014*, October 2014. doi: 10.1109/EVS.2013.6915005.

[6]   M. O. Tupe, 'Consumer perception of electric vehicles in India', *European Journal of Molecular & Clinical Medicine*, vol. 7, pp. 4861–4869, 2020.

[7]   D. Aggeler, F. Canales, H. Zelaya-De La Parra, A. Coccia, N. Butcher, and O. Apeldoorn, 'Ultra-fast DC-charge infrastructures for EV-mobility and future smart grids', in *2010 IEEE PES Innovative Smart Grid Technologies Conference Europe (ISGT Europe)*, October 2010, pp. 1–8. doi: 10.1109/ISGTEUROPE.2010.5638899.

[8]   L. Harnefors, M. Bongiorno, and S. Lundberg, 'Input-admittance calculation and shaping for controlled voltage-source converters', *IEEE Transactions on Industrial Electronics*, vol. 54, no. 6, pp. 3323–3334, 2007. doi: 10.1109/TIE.2007.904022.

[9]   C. Dang, X. Tong, W. Song, and J. Huang, 'Stability analysis of high power factor Vienna rectifier based on reduced order model in d-q domain', *Journal of Modern Power Systems and Clean Energy*, vol. 7, no. 1, pp. 200–210, 2019. doi: 10.1007/s40565-018-0463-8.

[10]   K. Pietiläinen, L. Harnefors, A. Petersson, and H. P. Nee, 'DC-link stabilization and voltage sag ride-through of inverter drives', *IEEE Transactions on Industrial Electronics*, vol. 53, no. 4, pp. 1261–1268, 2006. doi: 10.1109/TIE.2006.878308.

[11]   E. Mollerstedt and B. Bernhardsson, 'Out of control because of harmonics: an analysis of the harmonic response of an inverter locomotive', *IEEE Transactions on Control Systems Technology*, vol. 20, no. 4, pp. 70–81, 2000. doi: 10.1109/37.856180.

[12]   A. Emadi, 'Modeling of power electronic loads in AC distribution systems using the generalized state-space averaging method', *IEEE Transactions on Industrial Electronics*, vol. 51, no. 5, pp. 992–1000, 2004. doi: 10.1109/TIE.2004.834950.

[13]   L. Harnefors, 'Analysis of subsynchronous torsional interaction with power electronic converters', *IEEE Transactions on Power Systems*, vol. 22, no. 1, pp. 305–313, 2007. doi: 10.1109/TPWRS.2006.889038.

[14]   A. Ahmadi, A. Tavakoli, P. Jamborsalamati, *et al.*, 'Power quality improvement in smart grids using electric vehicles: a review', *IET Electrical Systems in Transportation*, vol. 9, no. 2, pp. 53–64, 2019. doi: 10.1049/iet-est.2018.5023.

[15]   E. F. El-Saadany and M. M. A. Salama, 'Effect of interactions between voltage and current harmonics on the net harmonic current produced by single phase non-linear loads', *Electric Power Systems Research*, vol. 40, pp. 155–160, 1997.

[16]   E. B. Makram and S. Varadan, 'Analysis of reactive power and power factor correction in the presence of harmonics and distortion', *Electric Power Systems Research*, vol. 26, pp. 211–218, 1993.

[17]   J. Arrillaga, D. A. Bradley, and P. S. Bodger, *Power System Harmonics*, Wiley, New York, NY, 1985.

[18]   M. Bollen, J. Meyer and H. Amaris, *et al.*, 'Future work on harmonics – some expert opinions Part II – supraharmonics, standards and measurements', in *Proceedings of International Conference on Harmonics and Quality of Power, ICHQP*, 2014, pp. 909–913. doi: 10.1109/ICHQP.2014.6842871.

[19]   S. Svensk Elstandard, 'Standarder underlättar utvecklingen och höjer elsäkerheten', 2014. www.elstandard.se.

[20]   J. Gomez, R. Nelson, E. E. Kalu, M. H. Weatherspoon, and J. P. Zheng, 'Equivalent circuit model parameters of a high-power Li-ion battery: thermal and state of charge effects', *Journal of Power Sources*, vol. 196, no. 10, pp. 4826–4831, 2011. doi: 10.1016/j.jpowsour.2010.12.107.

[21]   S. K. Rönnberg, M. Bollen, H. Amaris, et al., 'On waveform distortion in the frequency range of 2 kHz–150 kHz—review and research challenges',

Electric Power Systems Research, vol. 150, pp. 1–10, 2017. doi: 10.1016/j.epsr.2017.04.032.

[22]   J. Sutaria, S. Rönnberg, and Á. Espín-Delgado, 'Analysis of supraharmonics in a three-phase frame', *Electric Power Systems Research*, vol. 203, 109213, 2022. doi: 10.1016/j.epsr.2021.107668.

[23]   G. Singh, T. Cooke, J. Johns, L. Vega, A. Valdez, and G. Bull, 'Telephone interference from solar PV switching', in *Proceedings of International Conference on Harmonics and Quality of Power, ICHQP*, 2022, May 2022. doi: 10.1109/ICHQP53011.2022.9808805.

[24]   V. Ravindran, T. Busatto, S. K. Ronnberg, J. Meyer, and M. H. J. Bollen, 'Time-varying interharmonics in different types of grid-tied PV inverter systems', *IEEE Transactions on Power Delivery*, vol. 35, no. 2, pp. 483–496, 2020. doi: 10.1109/TPWRD.2019.2906995.

[25]   J. Böhler, F. Krismer, and J. W. Kolar, 'Analysis of low frequency grid current harmonics caused by load power pulsation in a 3-phase PFC rectifier; analysis of low frequency grid current harmonics caused by load power pulsation in a 3-phase PFC rectifier', in *Conference: 2019 2nd International Conference on Smart Grid and Renewable Energy (SGRE)*, 2019.

[26]   S. Sakar, S. K. Ronnberg, and M. Bollen, 'Interharmonic emission in AC–DC converters exposed to nonsynchronized high-frequency voltage above 2 kHz', *IEEE Transactions on Power Electronics*, vol. 36, no. 7, pp. 7705–7715, 2021. doi: 10.1109/TPEL.2020.3047862.

[27]   M. Makoschitz, M Hartmann, and H. Ertl, 'Effects of unbalanced mains voltage conditions on three-phase hybrid rectifiers employing third harmonic injection', *2015 International Symposium on Smart Electric Distribution Systems and Technologies (EDST)*, Vienna, Austria, pp. 417–424, 2015.

[28]   L. Huijuan and T. Yanfeng, 'Brief analysis on reasons and solutions of excessive neutral-earth voltage in machine room', Weather Bureau, Beihai, 2013.

[29]   A. Omar, M. Fouad, A. El-Rfaey, and Y. Gaber, 'DC offset compensation technique for grid connected inverters', in *2018 9th International Renewable Energy Congress, IREC 2018*, May 2018, pp. 1–7. doi: 10.1109/IREC.2018.8362473.

[30]   M. Kim, S. K. Sul, and J. Lee, 'Compensation of current measurement error for current-controlled PMSM drives', *IEEE Trans Ind Appl*, vol. 50, no. 5, pp. 3365–3373, 2014. doi: 10.1109/TIA.2014.2301873.

[31]   T. Ahfock and L. Alan Bowtell, 'DC offset elimination in a single-phase grid-connected photovoltaic system', in *PV Invertor Research*, 2006. https://www.researchgate.net/publication/228984688

[32]   A. O. Elghareeb, A. M. Elrefaey, M. F. Moussa, and Y. G. Dessouky, 'Review of DC offset compensation techniques for grid connected inverters', *International Journal of Power Electronics and Drive Systems (IJPEDS)*, vol. 9, no. 2, p. 478, 2018. doi: 10.11591/ijpeds.v9.i2.pp478-494.

[33]   W. Kempton and J. Tomić, 'Vehicle-to-grid power implementation: from stabilizing the grid to supporting large-scale renewable energy', *Journal of*

Power Sources, vol. 144, no. 1, pp. 280–294, 2005. doi: 10.1016/j. jpowsour.2004.12.022.

[34] C. D. White and K. M. Zhang, 'Using vehicle-to-grid technology for frequency regulation and peak-load reduction', *Journal of Power Sources*, vol. 196, no. 8, pp. 3972–3980, 2011. doi: 10.1016/j.jpowsour.2010.11.010.

[35] H. Turton and F. Moura, 'Vehicle-to-grid systems for sustainable development: an integrated energy analysis', *Technological Forecasting and Social Change*, vol. 75, no. 8, pp. 1091–1108, 2008. doi: 10.1016/j.techfore.2007.11.013.

[36] S. Shimi, S. Letha, and M. Bollen, 'Impact of electric vehicle charging on the power grid' Technical Report, 2021.

[37] S. Habib, M. M. Khan, F. Abbas, L. Sang, M. U. Shahid, and H. Tang, 'A comprehensive study of implemented international standards, technical challenges, impacts and prospects for electric vehicles', *IEEE Access*, vol. 6, pp. 13866–13890, 2018. doi: 10.1109/ACCESS.2018.2812303.

[38] International Energy Agency, *Global EV Outlook 2022 Securing Supplies for an Electric Future*, 2022. www.iea.org/t&c/.

[39] K. Knezovic. *Active Integration of Electric Vehicles in the Distribution Network – Theory, Modelling and Practice*. Technical University of Denmark, Department of Electrical Engineering, 2017.

[40] A. Briones, J. Francfort, P. Heitmann, M. Schey, S. Schey, and J. Smart, *Vehicle-to-Grid (V2G) Power Flow Regulations and Building Codes Review by the AVTA*, 2012. http://www.inl.gov.

[41] E. Dudek, 'The flexibility of domestic electric vehicle charging: the electric nation project', *IEEE Power and Energy Magazine*, vol. 19, no. 4, pp. 16–27, 2021. doi: 10.1109/MPE.2021.3072714.

[42] V. Barranco and P. Covrig, 'Vehicle-to-grid and/or vehicle-to-home round-trip efficiency – a practical case study', *JRC Technical Report*, 2021. doi: 10.2760/997207.

[43] C. B. Robledo, V. Oldenbroek, F. Abbruzzese, and A. J. M. van Wijk, 'Integrating a hydrogen fuel cell electric vehicle with vehicle-to-grid technology, photovoltaic power and a residential building', *Applied Energy*, vol. 215, pp. 615–629, 2018. doi: 10.1016/j.apenergy.2018.02.038.

Chapter 7

Standards and grid codes for distributed energy storage employment

Danny Pham[1], Maira R. Monteiro[1] and Yuri R. Rodrigues[1]

Recent changes in the generation resource mix due to the significant introduction of variable renewable resources are leading to increasing interest and rapid development of energy storage systems (ESS), including multiple technologies, sizes, and applications. Aware of this challenging perspective, international organizations, regulatory agencies, systems operators, and utilities worldwide have been continuously working toward developing standards and grid codes for the establishment of good practices for the multiple aspects associated with ESS application, including deployment, operation, maintenance, design, safety, cybersecurity, among others.

In this sense, this chapter seeks to provide an overall perspective of the main efforts toward the establishment of standards and grid codes for distributed ESS employment. As outcomes, readers should be able to identify and put into context the key standards, grid codes, and opportunities for distributed ESS (DESS) employment, as well as the expected challenges based on the prior lessons learned.

7.1 General considerations: development process for standards and grid codes

ESS are critical components for modern distribution systems conceptualizations such as automated distributed networks (ADN), microgrids, and smart grids. A non-exhaustive list of ESS application for power systems operation and management support includes ancillary service, renewable resources integration, energy management, as well as short- and long-term congestion and stability (Figure 7.1) [1].

In this perspective, the establishment of standards defining clear requirements that ensure the interoperability of ESS in these diverse environments play a key role in the effective utilization of these resources [1]. This perspective is especially critical given the different types of distributed energy resources (DER) available in modern power systems and their distinct abilities to respond to power systems management requests, which in turn can demand and allocate additional

[1]Seattle Pacific University, USA

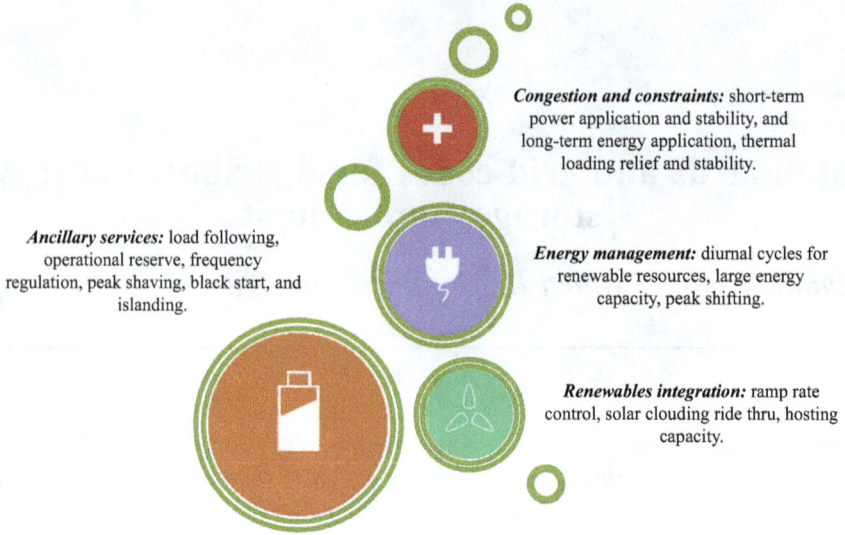

Figure 7.1   ESS application for power systems operation and management support

responsibilities for ESSs. Therefore, well-developed codes and standards with clear pathways to the deployment of DESSs are necessary. This process can be led by different organizations such as private companies, non-profits, government agencies, etc. (Figure 7.2) [2].

Standards development organizations (SDOs) exist in the United States as private sector entities with the mission of developing and publishing standards and model codes to address specific issues, technologies, and design/construction solutions. The developed documents from SDOs are focused on specific issues, and impact other stakeholders who want to deploy the technology. The provisions published by the SDO are not developed by the SDO, but instead are developed by all interested and affected parties under a process administered by the SDO. Any revisions of standards from SDO will have to be developed and supported with safety-related research and other documentation. The revised documents then will be fostered through the process by the revision proponent, which is the ESS industry. For new standards, it is advantageous for the ESS industry to develop a pre-standard, protocol, or guideline that can address any immediate need for safety-related criteria and also serve as the first draft of a standard [3].

Regulatory bodies, e.g., federal, state, or local government Public Utility Commission, Indian Tribes, etc., can also develop their requirements instead of adopting the standards and codes from SDOs. Ideally, established requirements from SDOs meet the needs of all adopters from technical, administrative, and time standpoints. Thus, enabling regulatory bodies to adopt SDOs requirements in full or selecting what they feel as appropriate for their needs [3].

The general workflow process related to standards development is depicted in Figure 7.3. Following, key characteristics related to distributed energy storage (DES)

Pathways for Adoption of Codes and
Standards

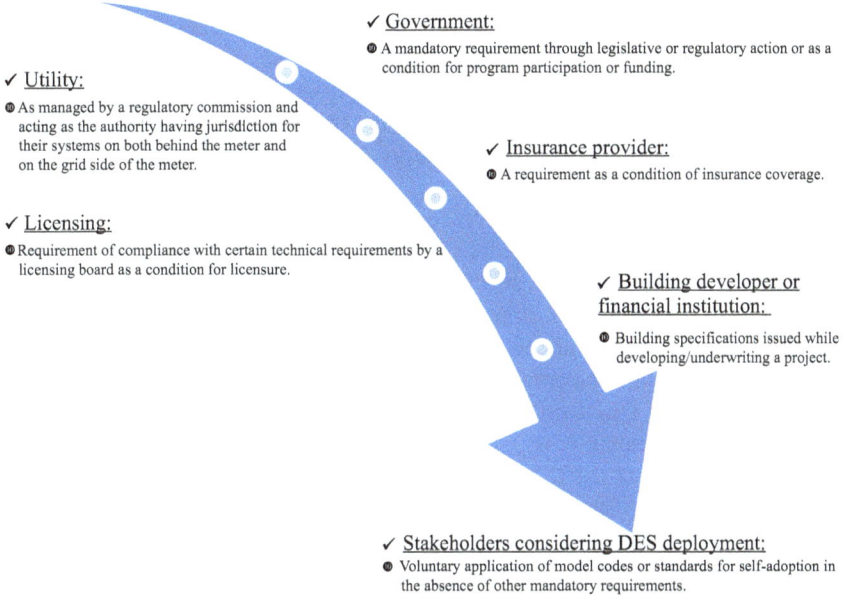

✓ Government:
  ◉ A mandatory requirement through legislative or regulatory action or as a
    condition for program participation or funding.

✓ Utility:
  ◉ As managed by a regulatory commission and
    acting as the authority having jurisdiction for
    their systems on both behind the meter and
    on the grid side of the meter.

✓ Insurance provider:
  ◉ A requirement as a condition of insurance coverage.

✓ Licensing:
  ◉ Requirement of compliance with certain technical requirements by a
    licensing board as a condition for licensure.

✓ Building developer or
  financial institution:
  ◉ Building specifications issued while
    developing/underwriting a project.

✓ Stakeholders considering DES deployment:
  ◉ Voluntary application of model codes or standards for self-adoption in
    the absence of other mandatory requirements.

Figure 7.2   Pathway for standard and grid codes development and adoption

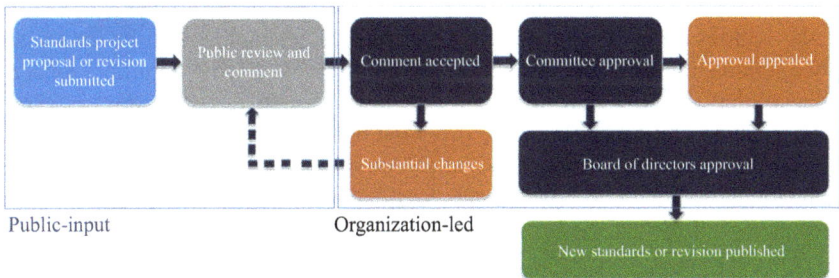

Standards project proposal or revision submitted	Public review and comment	Comment accepted	Committee approval	Approval appealed
		Substantial changes	Board of directors approval	
Public-input		Organization-led		
			New standards or revision published	

Figure 7.3   General standards development process [3]

systems application are identified and contextualized, including must-have capabilities to ensure the adequate interconnection of these units with the power grid [4].

Application: Determining the desired role of DES systems is one of the considerations as regulators may wish to see DESs systems primarily serve a customer's own energy demand or for individual customer backup purposes during grid outages. Other roles could be the provision of grid services to the power system or preserving those services for the future without expensive retrofits. Ranking these intended roles in priority order can help guide a range of decisions related to compensation mechanism design, metering and technical configuration requirements, and technical interconnection processes and requirements.

Market: Customizing rules and requirements based on the characteristics of the DES system is a key strategy for promoting fairness, which ensures appropriate regulation for various segments of the market. As mentioned above about the commercial market of DES system's deployment, the market context strongly influences compensation mechanisms design and its timing. For instance, higher penetrations of photovoltaics may motivate the utilization of more granular, cost-effective rates, and alternatives to net energy metering (NEM) to better align customer and grid system needs.

Jurisdictions: Jurisdictions with financially insolvent utilities or significant cross-subsidy are more concerned with utility revenue sufficiency and avoiding cost-shifting. To align the interests of DES customers with a broader power system, tariff design is a crucial tool. A paired behind-the-meter storage system, time-of-use (TOU) energy rates, and coincident demand-based charges can help customers to have better control of electricity consumption from the grid. They also help ease the management of DERs on the distribution system, and lead to a reduction in power system operational costs, serving as a grid-friendly incentive for customers to install DES systems.

Regulators: Regulators can help balance common utility concerns with consumer interests, ensuring the implementation cost of solutions to fit with the scale of the issue being created. Regulators also can enable innovation of business model for DES systems by defining how the participation of aggregators are allowed to operate. Therefore, standardizing interconnection application forms, processes, and requirements may offer a variety of benefits and reduce barriers to market entry for developers. Standardization can also reduce the number of incomplete applications for the utilities without redundancy of streamlines interconnection processes and utility administration. These considerations must be balanced with utility concerns to ensure that the provision of certain services from DES systems does not compromise the safe and reliable operation of the power system.

Here it is important to note that additional research, analysis, and documentation are important to standards development and ESS deployment. Without proper documentation, it is difficult to secure approval to progress into the public review stage and comment on the proposed criteria for the standards. The needs for the basis and documentation for the criteria will guide the transparency and deliverability of the standards development literature in SDO consideration. Furthermore, if criteria of those standards seem to be controversial or supported with marginal documentation, they are likely to be withdrawn or go through significant revision. As well, it is expected that manufacturers and accredited third parties conduct testing, analysis, and documentation of their systems, yielding robust, defensible, and uniform products that can be compliant with grid code and standards [3].

7.2   Standards: developed and under development

In the previous section, we reviewed several paramount factors relating to DES system deployment, ability to manage power grid capabilities, industrial standards, and interconnections to DERs. We also discussed potential pathways to adopt standards and grid codes, as well as several tactic considerations for DES system standards. In this section, we will expand on this perspective and discuss key standards in terms of conventional standards developed and under development for DES systems.

Electric utility systems by their nature are complex, and the addition of DER systems to the power grid introduces another layer of complexity. Standards are essential to the safe interconnection of DER systems to the larger grid, through establishing technical design rules. DER standards also enable the deployment of modular hardware and software components enabling greater competition and for cost-effective equipment to be deployed in an interoperable fashion between different vendors [5].

Therefore, we must agree and carefully follow such conventions to consolidate and foster equipment interoperability, preventing disruptions and catastrophic events from happening within equipment interface operations. In this perspective, the following standards present key considerations on the main areas for DES systems applications, including interconnection, operation, safety, and interoperability.

The IEEE 1547-2018 is the industry standard for interconnection and operation aspects of the DERs to the grid [2]. Safety aspects are governed by the Standard for Energy Storage Systems and Equipment (UL 9450). The UL 1741 is a certification standard for equipment testing, covering inverters and other interconnection system equipment for both off-grid and grid-connected systems. Both IEEE 1547-2018 and UL 1741 have been adopted as binding interconnection and equipment requirements by most of the states in USA. Additionally, the Standard for Interconnection and Interoperability of Inverter-Based Resources Interconnecting with Associated Transmission Power Systems (IEEE P2800) provides guidance and key recommendations for safe integration of inverter-based technologies to the bulk power system. In the following section, a timeline presenting the development of key DER-related standard series by the IEEE is presented.

7.2.1 DER standard series development timeline

The IEEE Standard 1547 (2003) was the first in the standard series to be developed concerning distributed generators and ESSs interconnection. This standard was not initially focused on types of DER technologies, but on technical specifications and testing for the interconnection. The standard requirements included the performance, operation, testing, safety, and maintenance of the interconnection, e.g., aspects such as responses to abnormal conditions, power quality, islanding, and test specifications for design, production installation evaluation, commissioning, and periodic tests. The 1547 standard considered that the DERs were a 60 Hz source [6].

IEEE 1547 Std-2003 established criteria and requirements for interconnection, yet it was not a design or application guideline. IEEE 1547 Std-2003 did not address DER self-protection, planning, design, operation, or maintaining the customer/local facilities and the utility grid. Additionally, it did not provide the procedure for the interconnect tests. Due to these deficits, multiple versions of the 1547 standard have been published until now to review and fit the standard case-by-case.

IEEE 1547 Std-2005 specified the type, production, and commissioning tests for the interconnection function and equipment of DERs. In 2008, the update provided background on requirements, tips, techniques, and rules of thumb with technical descriptions, schematics, application guidance, and interconnection examples. Meanwhile, the 2007 version addresses guidelines for monitoring, information exchange, and control for DER interconnections.

In 2011, the IEEE Std 1547.4 provided good approaches for the design, operation, and integration of microgrids, addressing the capability to separate from and reconnect to the grid while providing power to grid customers. Later, 1547.6 updated recommended practices addressing spot and grid distribution secondary networks as well as an overview of considerations and potential solutions for interconnecting DER with network distribution systems. Two years later, the 1547.7 added criteria, scope, and extent for engineering studies of the DER impact on the distribution grid.

In the next year, IEEE Std P1547.8 addressed advanced controls and communications for inverters and practices for multiple inverters and microgrids. It provided innovative information for DER behavior, interactions with grid equipment, and interconnection system response in abnormal conditions. The practices identified in P1547.8 led to the development of advanced hardware and software, resulting in higher penetration levels of DER.

At the same time, the IEEE Std 1547a (2014) Amendment 1 to 1547 allowed DER to support grid voltage regulation and provide voltage and frequency ride-through. This amendment was applicable to static power inverters and converters, induction machines, and synchronous machines [6].

Remarkably, IEEE Standard 2030.2 tackled both distribution and transmission levels of discrete and hybrid ESSs. It is the updated version of standard 2030, which provided guidance in understanding the technical characteristics of ESSs and their implementation with electric power infrastructure. IEEE Std 2030.2 also provided some examples for ESSs applications, including frequency regulation, voltage/Var support, distributed energy services, and customer-located, ESSs for microgrids, renewable energy integration combining power smoothing and peak shifting, and multiple services combined to support the grid.

In November 2013 and September 2014, Federal Energy Regulatory Commission (FERC) revised the small generator interconnection standards for DERs up to 20 MW. The conclusion stated that the reforms adopted in the final rule will reduce the time and costs required to process small generator requests for interconnection customers, maintain reliability, increase energy supply, and remove barriers to the development of new DERs (FERC 2013) [4]. Key changes in FERC standards include:

- Provision of a pre-application report for the interconnecting customer
- Revision of the 2-MW threshold for qualification in the fast-track process
- Revision of customer options for supplemental review
- Inclusion of energy storage devices as part of DERs interconnecting to the grid
- Revision of the SGIP Facilities Study Agreement

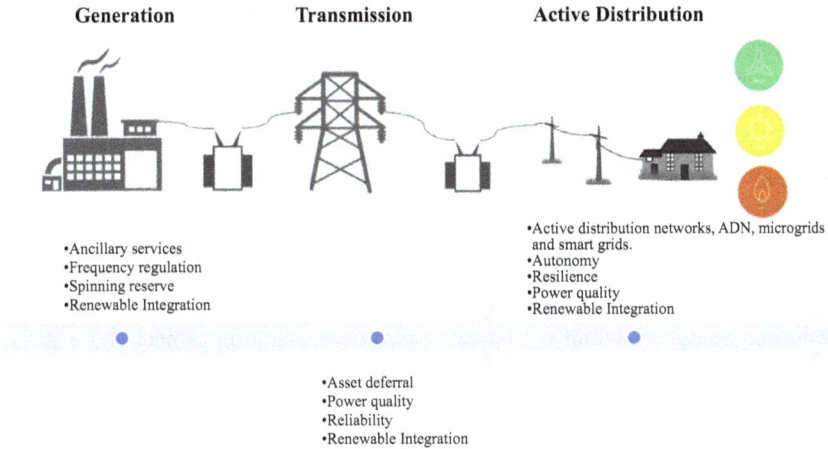

Generation   **Transmission**   **Active Distribution**

•Ancillary services
•Frequency regulation
•Spinning reserve
•Renewable Integration

•Active distribution networks, ADN, microgrids
 and smart grids.
•Autonomy
•Resilience
•Power quality
•Renewable Integration

•Asset deferral
•Power quality
•Reliability
•Renewable Integration

Figure 7.4   Examples of DES systems integration with the power grid (IEEE 2030.2) [6]

On the basis of ESS safety standards, standards typically apply to the smallest parts of the system-like wires, relays, switches – to address their design, construction and safety features. Standards for components of the ESS – cells/modules, inverters, management systems, etc. – for the resultant assembly of those components ensure that the components safely serve the intended purpose. In looking at the integration of the ESS, necessary criteria of standards will also be provided to ensure the safety of the ESS in relation to its environment, such as fire department access, fire alarms, and suppression, clearances to combustibles, ventilation of spaces, etc. [3]. Examples of DES systems integration with the power grid are depicted in Figure 7.4.

7.2.2   Aspects under development

Although published standards for DES systems present sophisticated considerations on several key aspects, there are still aspects that need further development and/or clarifications. For example, even though IEEE 1547 defines DER as a small-scale electric generator located in and connected to the local electric power system, it does not specify a distinction between energy storage devices and generators within the DER portfolio. In the standard P1547.4, there is no standardization for functioning during islanding and no voltage support specification. There is also no ramp rate specification that would enable hybrid generation-storage to mitigate intermittency of renewables. The trip point specifications do not enable renewables or storage to avoid tripling under moderate grid transients, and there are inconsistencies between the anti-islanding requirements of IEEE 1547 and the ride-through requirements defined by FERC's Large Generator Interconnection Procedure (LGIP) [1]. Despite many increases in storage adoption, safety codes, and standards for storage are still under development, and questions have been raised about safety risks. According to [7], multiple efforts are underway to ensure that safety codes and standards address ESSs; however, these types of standards tend to delay the development of storage technologies.

Optimization benefits for energy deployment require innovative and futuristic policies. With more energy storage deployments and the experience of the states, foundational policy actions and solutions have had more clearly defined paths to address barriers, and some are still under development or ready for further policy innovation. A few utilities and commissions are updating their resource planning processes to accurately model advanced storage [8] such as: In 2015, Missouri's utility submitted an integrated resource plan (IRP) that includes the required discussion of storage potential with a smart grid demonstration project and a storage component; Oregon's commission and utility addressed storage in its 2016 IRP to meet the utility's requirement to procure 5 MWh of energy storage under Oregon's storage procurement mandate; utility Hawaiian Electric Company updated its Power Supply Improvement Plan with recent estimates of energy storage cost and performance data and accounting for ancillary services benefits. Many regulatory methods tackling storage difficulties have already existed and have been proven effective; though, some of them are still under development.

The duty cycles and other parameters are derived from digital modeling, limited operational experience with energy storage projects, or technical experts' predictions. However, the ESS vendors do not offer a standard product for standard performances and test protocols, which matches these parameters. Hence, the industry has offered existing products that closely match the customer's stated needs. Although it has worked reasonably until now, vendors and users still need the codes and standards for performance requirements and test procedures to stimulate the widespread use of energy storage in the grid. Formulating a standard is a lengthy process, and it requires consensus from a broad base of stakeholders, but the DOE ESS program is facilitating the development of protocols to precede the formulation of the needed standards. This effort for standard development allows storage system vendors, utilities, and other storage users to evaluate the performance of storage technologies on a uniform basis with transparency. Furthermore, these protocols will also differentiate technologies and products for specific services [9].

As referred to innovative technologies, standards and codes are also pressed to set orders for many new emerging storage technologies. They are either mature and fully deployed or still in the research and development (R&D) phase. Some storage technology examples in [9] and their statuses are Underground Pumped Hydro (new concepts are under development), Advanced Flywheels (standards and codes for deployment are under development), Advanced Lead-Acid Battery (modules are still under tests with some early trials), and Advanced Li-ion (under laboratory).

7.3   Grid codes for distributed energy storage systems

Well-established grid codes help improve the grid overall reliability and resilience, as well as create a united defense against cybersecurity threats [10]. There are five categories of codes that should be noted, including the definition of documents, grid codes/interconnection requirements (our central idea in this chapter), certification/compliance/acceptance standards, performance tests, and internal manufacturer/vendor testing. These codes impact the core requirements for grid performance, such as voltage and frequency control, protection to ensure safety, preventing cascading events and damage to equipment, system stability, and continuity of service [11]. In this perspective, general reviews of the technical requirements dedicated to DER and ESS control in cooperation with the power system are available in [12–13], following the main requirements for these units' applications are discussed.

7.3.1   Grid codes for DERs

The general standard performance required for DERs operation under normal and fault conditions are depicted by the IEEE 157 standard [13–14].

- **IEEE 1547:** Standard for Interconnection and Interoperability of Distributed Energy Resources with Associated Electric Power Systems Interfaces defining functional requirements for equipment and resource types.
- **Scope:** Establishes criteria and requirements for the interconnection of DERs with electric power systems (EPS) and associated interfaces.
- **Purpose:** Provides a uniform standard for interconnection and interoperability of DERs to EPS, to provide relevant requirements to DER–EPS connection performance, operation, testing, safety, maintenance, and security considerations.

Overall, these requirements seek to support the bulk-power system (BPS) operation and its reliability. Here one should note that, despite the minimum impact of DERs on BPS at low penetration levels, issues regarding transmission line loading, grid voltage, and system frequency can still occur as the penetration level increases during both normal and disturbed operations. Therefore, constant revisions and lessons learned on the following key characteristics have been performed.

- Active power control
- Reactive power supply and voltage support
- Power quality[*]
- Frequency and voltage levels[†]
- Short-circuit current contribution[‡]
- Fault-ride-through capability[§]
- Ride-through capability[‖]
- Protection concepts[#]

---

[*]Certain limits on harmonics, voltage change, and so on.
[†]Usually associated with requirements for reactive power regulation, as well as frequency control issues.
[‡]Which exhibits itself in the requirements for protection devices and settings.
[§]As in response to over-frequency conditions.
[‖]Response to grid faults.
[#]Unintentional islanding protections.

These changes available in the IEEE 1547-2018 standard seek to establish settings and requirements that protect DERs while enabling support capabilities to be harnessed by BPS's utility and operators. Selected key aspects are discussed below.

7.3.1.1   Interconnection application

- In the FERC technical screen, the interconnection application must be on the utility's distribution system. This screen is used to direct projects connecting networked transmission systems to the study process because the remaining screens were not designed to evaluate transmission system impacts.
- Penetration screen requires supplemental review when the circuit contains a sum of DER nameplate capacity equal to or greater than 15% of the peak demand on that feeder or line section.
- IEEE 1547-2018 as well as IEEE 1547.6 offer guidance for DER systems to interconnect onto a secondary spot network distribution system or an area network distribution system. However, it may be inappropriate to receive fast-track interconnection approval, so this screen requires either supplementary review or a detailed impact study of the proposed DER.

7.3.1.2   Hosting capacity

- On the distribution circuit, the small generating facility shall not contribute more than 10% to the distribution circuit's maximum fault current due to the concern over protection of utility equipment and the coordination of the protection system.
- By evaluating the total fault current, the utility sources and the DERs must not exceed 87.5% of the short-circuit equipment rating on the circuit. This is because many modern inverters have fault durations that are shorter than the durations of synchronous machines, causing them to drop offline more quickly than traditional synchronous machines.

7.3.1.3   Infrastructure

- To pass screening criteria, primary distribution line types must match their types of interconnections. For instance, three-phase, three-wire must be connected in three-phase or single-phase, phase-to-phase, and three-phase, four-wire type must be connected in effectively grounded three-phase or single-phase, line-neutral.
- To prevent causing overloads or voltage imbalances on the secondary conductors or the transformer windings, or at the customer/DER system point of common coupling (PCC), the aggregate generation capacity on the shared secondary shall not exceed 20 kW if small generating facility is interconnected on single-phase shared secondary.
- Inverters that produce power at 120 volts should not be connected to 240-volt electric service (usually on a center tap neutral) if the connection create an imbalance between the two sides of the 240-volt service being more than 20% of the service transformer rating.

7.3.1.4 Ride-through capacity

- Bulk-level grid codes (approved by NERC Reliability Standard PRC-024-1), bulk generation and transmission elements are compulsory to have a degree of disturbance tolerance. The elements must not be tripped at all times unless to clear a fault to prevent equipment damage or preserve system stability. Disturbance tolerance is a required element to prevent cascading outages caused by voltage or frequency excursions.
- For bulk power systems connected to generators, IEEE Std 1547 requires DERs to disconnect within a brief period of time when voltage or frequency falls outside a certain range. For instance, a voltage drops to 0.5 p.u. or lower is required to trip/disconnect within 10 cycles.
- Figures 7.5 and 7.6 present mandatory tripping requirements, as well as voltage and frequency ride-through requirements.

7.3.1.5 Safety, power quality, and protection

- IEEE Std 1547 is not a safety standard and was written to minimize DERs risk to workers. It includes performance requirements that can be coordinated utility protection equipment.
- Anti-islanding protection and response to short-circuit and open-phase conditions are part of the core requirements.
- Requirements for output power quality are quantifiable and include limits on: harmonics causing distortion; voltage fluctuations, including contributions to overvoltage at the point of common coupling; new synchronization tolerances.

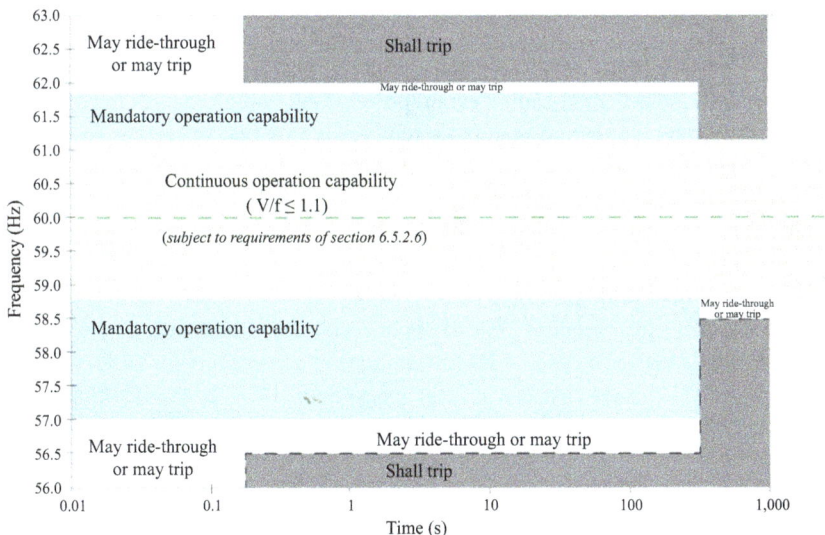

Figure 7.5 Frequency ride-through requirements for all DERs (IEEE 1547)

Illustrative comparison of voltage ride through capabilities – Categories I, II, III

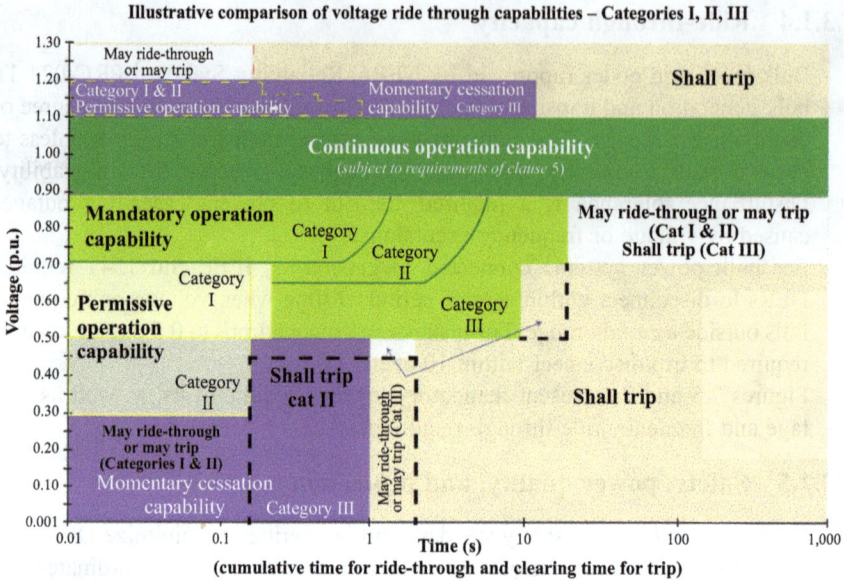

Figure 7.6   Voltage ride-through requirements for all DERs (IEEE 1547)

7.3.1.6   Operation support

- Rule 21 in California allows non-exporting standalone storage to be reviewed through the expedited fast-track process. A developer can also specify the storage charging behavior of an interconnecting application with three modes: no grid charging, peak shaving, and unrestricted charging.
- In Colorado PUC 2016, Xcel Colorado provides Net-metering compensation for a DER storage system only when the storage is charged by the NEM generator alone; thus, the generator cannot export power if the storage charges from the grid.
- Nevada's interconnection standards require one of two operating restrictions on storage systems paired with a NEM generator. One operating restriction is that the storage system cannot export power. The alternative restriction is that the storage system can only be charged by the NEM generator.

7.3.1.7   Frequency regulation

- IEEE Standard 1547 requires DER to disconnect within 160 ms when the frequency is above 60.5 Hz or below 59.8 Hz (upper range of adjustability for DER >30 kW).
- When a major event happens in high penetration DER cases, IEEE Std 1547 requires DERs to wait 5 minutes to restart automatically – a requirement for system restoration.

7.3.1.8   Stability

- If a substation or utility area has posted transient stability limitations, detailed impact studies are required, including transmission-level transient and

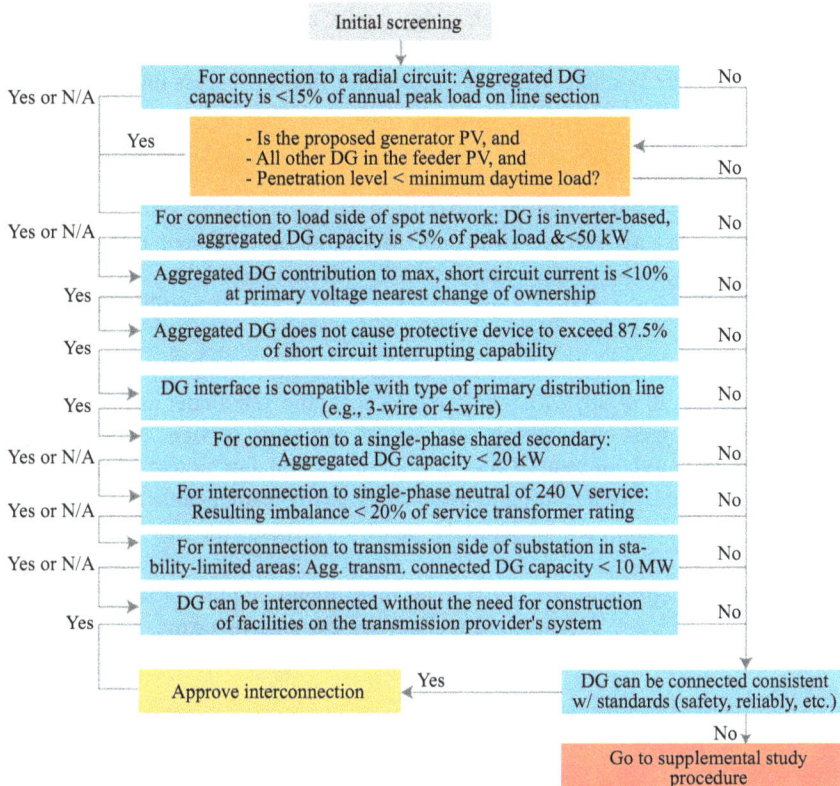

Figure 7.7   FERC SGIP technical screens summary, including the additional considerations for PV (yellow box) from Codding 2012

dynamic studies. These screening standards are designed to sustain the need for efficiency and technical accuracy for all DERs and distributed generation. Even though the interconnection screening procedures have been developed with high credentials in order to serve the industry, they also need to adapt innovative rectifications to remain with evolving standards, technology, and practical experience. Figure 7.7 presents the summary and key notation for FERC screening standards.

7.4   Challenges for interconnection to urban grids (DESS)

The DOE/EPRI Electricity Storage Handbook indicates that the biggest challenges hindering the adoption of energy storage technology are cost, the ability to deploy ESS, and lack of standards. Standards and codes have a direct impact on the cost of an ESS and its installation, and administrative burdens and time to approval issues affect the ability to deploy the technology and cost. It is difficult to uniformly document safety criteria and determine what can be approved in a uniform and timely manner. For example, the lack of specifics limits progress until appropriate standards and codes are available; for the outdated standards and codes can be conservatively applied to the technology, affecting the cost of the installation, or limiting its application [3]. In this sense, DOE recognizes four key challenges to the widespread deployment of ESS [15]:

Performance and safety: Grid operators must be confident that ESSs will perform as intended within the larger network. Advanced modeling and simulation tools can facilitate acceptance – particularly if they are compatible with utility software.

Cost-competitive systems: Actual energy storage technology (e.g., the battery) contributes 30%–40% to total system cost; the remainder is attributed to auxiliary technologies, engineering, integration, and other services.

Regulatory environment: ESSs provide divergent functions to their owners and the grid at large, often leading to uncertainty as to the applicable regulations for a given project. Regulatory uncertainty poses an investment risk and dissuades adoption.

Industry acceptance: Energy storage investments require broad cooperation among electric utilities, facility and technology owners, investors, project developers, and insurers. Each stakeholder offers a distinct perspective with distinct concerns.

Grid operators and utilities have limited experience with storage and face technical challenges integrating storage into existing systems [7]. Grid operators may not have experience considering planning for the integration and operation of storage, and they typically use models based on traditional resources with better-understood capabilities to help decide about investments. Moreover, storage can be more challenging to integrate than other resources because of changes in the system function from charging to discharging to generating electricity [7]. Some other challenges of storage deployment are related to uncertainty about the performance of storage technologies over time and their operating conditions. They are expected to last for a decade or more, but some like battery storage do not have a predictable degrading duration. Because standards and codes for safety in storage deployment tend to lag the development of storage technologies, energy storage owners may face challenges with applying existing codes and standards to storage deployment and safety verification. In the grid, the fire hazard is a remarkable challenge to storage system installers. Local entities like fire departments may not allow the deployment of storage on certain sites, or the owners may not have a complete understanding of the appropriate fire protection measures [7].

In addition, it is also important to point out that changes in reliability requirements can lead to modifications in previously established standards and influence DES regulations in the short and medium term. Table 7.1 highlights identified standards within NERC with potential for change due to DES systems.

A general description presenting the complexity of DES systems integration is illustrated in Figure 7.8.

Table 7.1   Standards with potential for change due to resilience requirements

Resilience requirement	Standards
Workforce	PER-006-1 – Specific Training for Personnel
	EOP-006-3 – System Restoration Coordination
Physical security	CIP-006-6 – Physical Security of BES Cyber Systems
	CIP-014-2 – Physical Security
Natural events	EOP-011-2 – Emergency Preparedness
	IRO-010-4 – Reliability Coordinator Data Specification and Collection
	TOP-003-5 – Operational Reliability Data
	TPL-007 – Transmission System Planned Performance for Geomagnetic Disturbance Events
Critical infrastructure	PRC-024-3 – Frequency and Voltage Protection Settings for
Resource mix	Generating Resources
Planning	TPL-001-4 – Transmission System Planning Performance
Resource adequacy	Requirements
	EOP-004-3 – Event Reporting
	EOP-005-2 – System Restoration from Blackstart Resources
	EOP-006-2 – System Restoration Coordination
	EOP-011-1 – Emergency Operations
	CIP-013-1 – Cyber Security – Supply Chain Risk Management
Supply chain	CIP-013-1 – Cyber Security – Supply Chain Risk Management
	CIP-014-2 – Physical Security

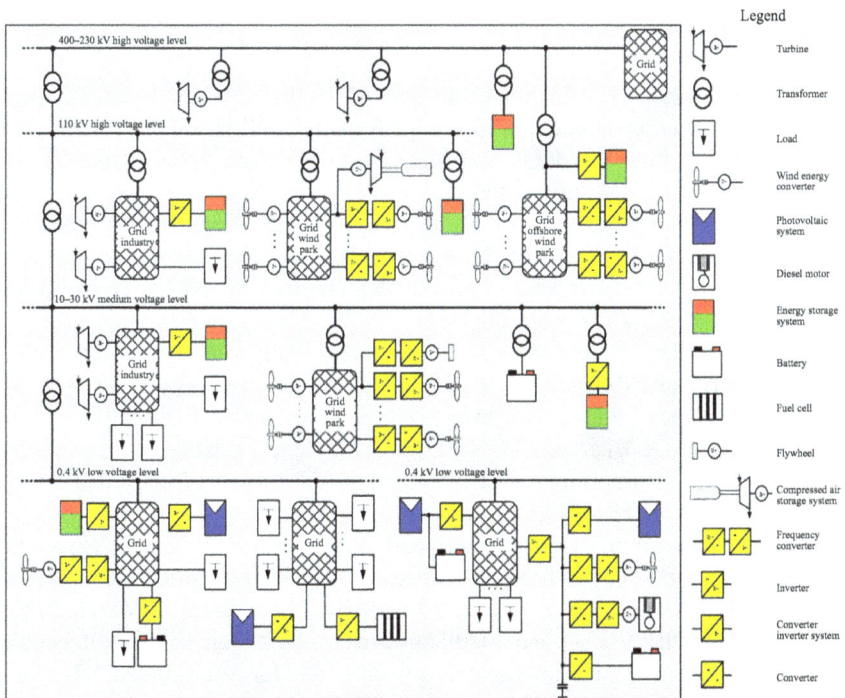

Figure 7.8   Integrating different DES systems to the electric grid [16]

7.5 Conclusion

To integrate ESSs into conventional electric grids, specially designed topologies and control are required. Thereof, costly design and debugging time of each component and control of the system are added during the deployment. However, the present and future modern power systems require extra flexibility and the importance of the DES integration has increased more than ever. Still, storage devices, standardized architectures, and techniques for distributed intelligence and smart power systems as well as planning tools and models to support the integration of ESSs are still lagging [16]. The current situation of electrical ESSs is characterized by:

- Disagreement on the role and design of ESSs
- Common use of storage only by large pump hydro or small batteries
- New technologies still under demonstration
- No widely recognized planning tools/models to aid understanding of storage devices
- System integration including power electronics must be improved

In this sense, even though meaningful steps toward standardization and integration of DES have been taken, there still exist significant issues for energy storage managers to overcome to harness the full potential of DES systems.

References

[1]   F. Cleveland, M. McGranaghan, and Al Hefner, "What: Energy Storage Interconnection Guidelines (6.2.3)," Report to NIST on the Smart Grid Interoperability Standards Roadmap, National Institute of Standards and Technology. p. 24, July, 2009.

[2]   C. Gokhale-Welch and S. Stout, "Key Considerations for Adaptation of Technical Codes and Standards for Battery Energy Storage Systems in Thailand," *Advanced Energy Partnership for Asia*, no. NREL Transforming Energy, p. 23, 2021.

[3]   D. Conover, "Overview of Development and Deployment of Codes, Standards and Regulations Affecting Energy Storage System Safety in the United States," Pacific Northwest National Laboratory, p. 53, 2014.

[4]   M. Ingram, A. Bhat, and D. Narang, "A Guide to Updating Interconnection Rules and Incorporating IEEE Standard 1547," National Renewable Energy Laboratory, p. 59, 2021.

[5]   I. S. Association, "IEEE SA Beyond Standards," IEEE SA, 9 November 2021. https://beyondstandards.ieee.org/ieee-standards-for-the-evolving-distributed-energy-resources-der-ecosystem/ [Accessed 24 March 2022].

[6]   T. Basso, "IEEE 1547 and 2030 Standards for Distributed Energy Resources Interconnection and Interoperability with the Electricity Grid," National Renewable Energy Laboratory, p. 22, 2014.

[7]   U. S. G. A. Office, "Energy Storage: Information on Challenges to Deployment for Electricity Grid Operations and Efforts to Address Them," United States Government Accountability Office, 2018.

[8]   S. Stanfield, J. "Seph" Petta, and S. B. Auck, "Charging Ahead: An Energy Storage Guide for State Policymakers," Interstate Renewable Energy Council, p. 59, 2017.

[9]   A. A. Akhil, G. Huff, A. B. Currier, *et al.*, *DOE/EPRI Electricity Storage Handbook in Collaboration with NREC*, *Sandia National Laboratories*, New Mexico, 2015.

[10] D. Narang, "NREL Transforming Energy," NREL Transforming Energy. www.nrel.gov/grid/standards-codes.html [Accessed 26 March 2022].

[11] J. Johnson, "Photovoltaic and Distributed Systems Integration," *Changing Grid Codes Around the World*, Sandia National Laboratories, p. 31, 2015.

[12] T. Sikorski, M. Jasiński, E. Ropuszyńska-Surma, *et al.*, "Determinants of Energy Cooperatives' Development in Rural Areas Evidence from Poland," Energies, vol. 14, no. 2, p. 319, 2021.

[13] M. Ingram, "Grid Code Essentials and Streamlining Process for Interconnections," NREL Transforming ENERGY, p. 30, 2020.

[14] K. Horowitz, Z. Peterson, M. Coddington, *et al.*, "An Overview of Distributed Energy Resource (DER) Interconnection: Current Practices and Emerging Solutions," National Renewable Energy Laboratory, Golden, CO, 2019.

[15] O. O. T. Transitions, "Spotlight: Solving Challenges in Energy Storage," U. S. Department of Energy, p. 51, 2019.

[16] A. Mohd, E. Ortjohann, A. Schmelter, N. Hamsic, and D. Morton, "Challenges in Integrating Distributed Energy Storage Systems into Future Smart Grid," IEEE International Symposium on Industrial Electronics, June 2008.

Chapter 8

Monitoring distributed energy storage for power quality analysis

Carlos Augusto Duque[1], Leandro Rodrigues Manso Silva[1] and Paulo Fernando Ribeiro[2]

Energy storage systems (ESSs) have been gaining significant importance with the insertion of renewable energy sources in the electrical systems. Monitoring these systems is of paramount importance for their control and protection, and for understanding their behavior when interacting with other system components. This interaction may degrade the power quality (PQ), so monitoring PQ in both DC and AC sides is necessary. In this way, this chapter will show the definition of some PQ parameters in the DC side and the concept of novelty detection for waveform recording, that can be used in both DC and AC signals. Some new aspects of PQ in AC systems will be presented, such as the increase of supraharmonic distortion and concerning with the time varying harmonic phasor. The description of a monitoring system based on the substation edge device (SED) will be discussed as the way to unify the monitoring of both DC and AC side. Some results of a SED implemented in an real transmission systems will be presented.

8.1 Introduction

Energy storage systems (ESSs) are becoming increasingly important in power system networks with the increasing penetration of renewable generation sources, especially photovoltaic and wind generation. Storage technologies play a fundamental role in maintaining the reliability and quality of energy due to the characteristics of intermittence typical of these renewable sources. However, as it is a new technology, the interactions between the various components are not yet known, which can cause unstable electrical resonances that can disconnect several branches of the system, or electrical resonances at the limit of instability that can deteriorate the power quality (PQ) of the system.

[1]Federal University of Juiz de Fora, Brazil
[2]Federal University of Itajuba, Brazil

In this sense, monitoring systems for electrical and non-electrical quantities are essential to ensure the correct functioning of the networks. These quantities must be monitored at different sampling rates, depending on the level of application required. Some of the quantities in DC/AC systems are necessary at the control level of inverters and converters and, therefore, require high sampling rates. Other quantities are necessary at the supervision level, such as the generated power and stored energy in batteries may be acquired and transmitted at rates in the range of seconds, minutes, and hours.

This chapter is focused on monitoring systems for the analysis of electrical PQ, waveform recorders and harmonic phasor estimation (HPE). Initially, an overview of PQ in DC systems is presented. PQ for DC systems is an emerging area, there is still no clear definition of the phenomena that should be monitored and their respective quality parameters, but some current works have already presented some preliminary results. This gap between the evolution of DC networks and standardization reinforces the need for waveform acquisition systems based on the novelty concept. Next, this chapter goes on to consider some important aspects for monitoring PQ on the AC side. Clearly, the problem of voltage fluctuation and supraharmonics come to the fore with the increasing penetration of generation based on intermittent sources with power electronics inverters and grid-connected electric vehicles. Another aspect that attracts attention is the concept of dynamic harmonic phasor, necessary for the characterization of the harmonic contribution and estimation of the transfer impedance. Finally, the chapter ends with the presentation of a monitoring system using the fully digitized substation philosophy. The concept of substation edge device (SED) is presented as a possibility of integration between AC and DC monitoring systems.

8.2 PQ on DC systems

Traditionally ESS and solar panel arrays were directly connected to AC networks through inverters. While concerns about PQ in AC systems were and continue to be extremely relevant, the same was not true for DC systems, since the PV modules or storage batteries presented practically constant impedance to the inverters, causing no serious problems on the DC side, concerning to PQ [1].

However, in more modern configurations, several DC blocks, with their respective DC/DC converters, are connected to a single power inverter, as illustrated in Figure 8.1. Historically, as mentioned before, inverters were designed for DC sources with constant impedance, as in the case of solar panels or storage batteries. The new connections with several DC/DC converters have generated interactions and resonances not observed in the past, causing instabilities or oscillations at the instability limit. While instabilities generate increasing oscillations until they exceed the limits of hardware operation, forcing the shutdown of entire generation blocks, sustained resonances are generally not detected by the protection and can lead to reduced efficiency, aging of components and loss of PQ.

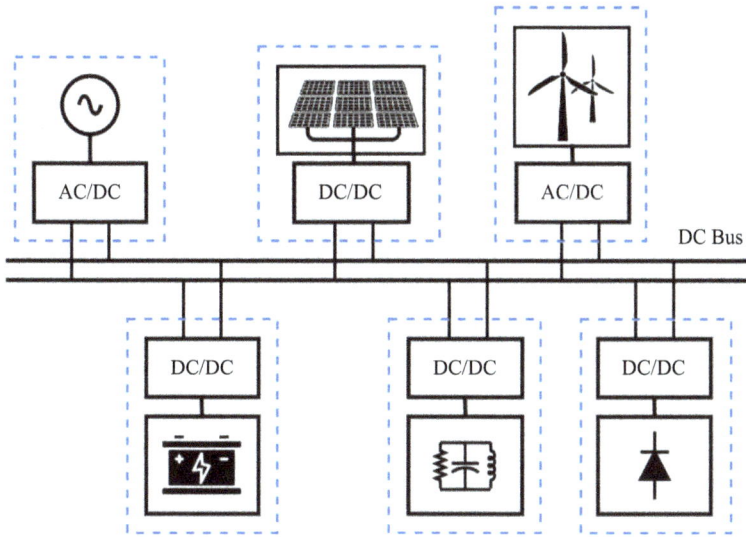

Figure 8.1   Typical topology of DC microgrid

The resonances that appear due to interactions are usually of high frequency, therefore, high sampling rates must be used and big-data can be generated if a continuous monitoring is taking place. To get around these problems, the definition of PQ parameters for DC system is a way out to avoid the generation of big data. Parameters can be calculated locally, even if the processing requires a high sampling rate, thus reducing the need for large data storage or the transfer of large amounts of information. The challenge, then, lies in defining the PQ parameters for DC systems that can be implemented on low-cost hardware. Clearly, monitoring systems that allow interoperability and has open source is one of the aspects that should be looked for, since the number of nodes to be monitored in the distribution system tends to grow exponentially over time.

8.3   DC PQ parameters

Clearly, the definitions of PQ applied to AC systems must be adapted to DC systems. However, the definition given by IEC 61000 is quite generic and can be applied to both DC and AC systems: "Power quality encompasses the characteristics of the electricity at a given point on an electrical system, evaluated against a set of reference technical parameters" [2]. The great challenge lies in defining the set of parameters that characterize the PQ of DC systems, since some problems of PQ in DC are still poorly known and studied. But overall, this quality can be associated with voltage stability [3]. Thus, many of the DC PQ indicators reflect the stability of the DC bus. Below is a brief summary based on [4,5].

8.3.1   DC voltage fluctuation

The voltage fluctuation can be defined by (8.1):

$$\delta = \frac{U_{\max} - U_{\min}}{U_N} \times 100\% \tag{8.1}$$

where $U_{\max}$ and $U_{\min}$ are the maximum and minimum voltage during the measured interval, and U_N is the rated voltage.

The voltage fluctuation on DC systems is mainly caused by the power fluctuation on the grid, even in the generation as in the load side. For example, in PV generation, the power is dependent of the sun shining. The common control strategy for PV model will not be able to keep the voltage constant and a certain amount of fluctuation on DC voltage is expected. The limit for δ is suggested in [4] as $\pm 10\%$.

8.3.2   DC voltage ripple

Ripple is a repetitive phenomenon superimposed to the DC steady value. However, the oscillation is not necessarily related with fundamental frequency, so the ripple concept must be reformulated to a more general definition taking into accounting for non-harmonically related, possibly non-stationary, components. For example, in MIL-STD-704F [6], ripple is defined as "the maximum absolute difference between an instantaneous value and the steady value V_{dc}": this is an instantaneous definition, with $V_{\max}$, $V_{\min}$, and V_{dc} taken over a predetermined time interval, according to (8.2):

$$R_{p,1} = \max\left\{\frac{V_{\max} - V_{dc}}{V_{dc}}, \frac{V_{dc} - V_{\min}}{V_{dc}}\right\} \tag{8.2}$$

In [7], the ripple is defined using only the max and min value of the voltage,

$$R_{p,3} = \frac{V_{\max} - V_{\min}}{V_{\max} + V_{\min}} \tag{8.3}$$

In [4], the total ripple distortion (TRD) is defined as

$$TRD_U = \frac{U_H}{U_d} \tag{8.4}$$

where U_H is the square root of the sum of ripples and U_d is the root mean square value of DC bus voltage. This definition is closely related to the classical definition of total harmonic distortion; however, no fundamental frequency is assumed, so the term U_H corresponds to the sum of the energy of all components, except the DC component. If no fundamental frequency is assumed in this definition, then it is not clear how to compute U_H. Thus, to use this PQ index, more information must be added to its definition, which requires further investigation.

8.3.3   Voltage deviation

The voltage deviation of DC system is quite similar to the definition used for AC grid. Equation (8.5) gives the voltage deviation θ_u:

$$\theta_u = \frac{U_{re} - U_N}{U_N} \times 100\% \tag{8.5}$$

In this equation U_{re} is the measured voltage and U_N is the rated voltage. Figure 8.2 illustrates the voltage deviation in DC systems. At time instant t_1, there is a power imbalance caused by a load switching. If the system is not able to support the new power, for example, if in the islanding operation, there is no sufficient energy storage module to feed the new load, then a voltage deviation will appear at the DC bus.

8.3.4   Voltage sag and interruption

Voltage sags and interruptions in DC systems have the same meaning as in AC systems. The sag intensity may be defined as

$$A = \frac{U_d - U_d'}{U_N} \times 100\% \tag{8.6}$$

where U_d is the measured voltage at DC bus before sag, U_d' is the measured voltage after sag, and U_N is the rated voltage. The time duration of the sag is the time the voltage U_d' holds below 10% U_d. Figure 8.3 shows the typical voltage sag in DC systems.

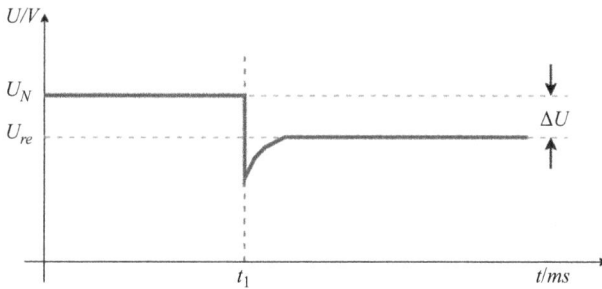

Figure 8.2   Voltage deviation caused by switching loads. Adapted from [4].

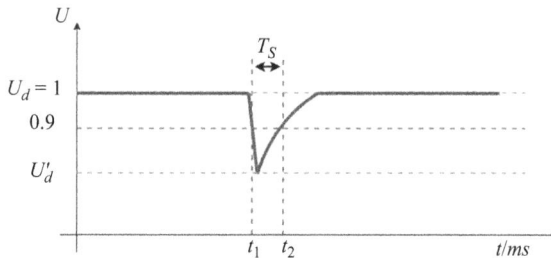

Figure 8.3   Voltage sag. Adapted from [4].

Voltage sag in the DC system can be caused by short circuit to the bus ground, switching of the high power load and sudden changes in the power of the DGs. Of course, if the ESS has an efficient control, sag, and interruption can be avoided. However, if the control system is deficient or the electrical charge stored in the batteries is not sufficient to supply the other momentary variations, sag and interruption will occur.

8.3.5   Oscillatory load transient

Oscillatory transient may happen in DC system as a consequence of load switching and power variation. Frequency bandwidth higher than 1 kHz and duration of a few milliseconds are common in DC systems. This oscillation may cause insulation breakdown or system resonances.

8.4   Novelty detection for waveform capture

The new microgrid topologies with their various DC/DC converters connecting different DC sources, battery banks and loads, in addition to power inverters connecting DC/AC systems, tend to generate new situations not yet fully studied and understood. In the meantime, intelligent monitoring of waveform from various points in the network is the safe option to ensure a reliable transition. Within the concept of intelligent monitoring, the methodology based on the detection of novelties in the monitored signals stands out. This methodology was proposed in the context of electrical power systems for AC signals, however, as it does not focus on a specific type of disturbance, but on any type of anomaly present in the signal, it can be directly applied to DC signals.

The novelty detection methodology consists of partitioning a signal into frames and comparing each frame with a reference frame according to some metric. When the result of the comparison is above a pre-established threshold, that frame is indicated as a novelty in the signal. Figure 8.4 shows an example of a signal where the novelty frames are highlighted in gray.

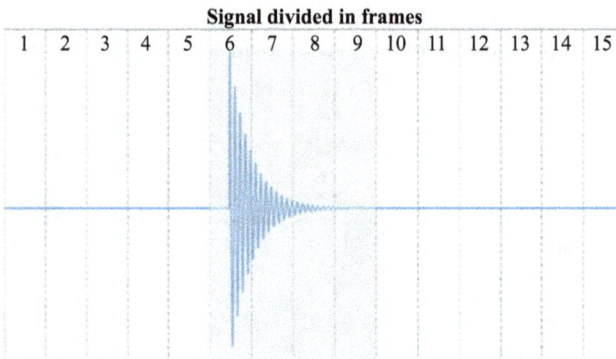

Figure 8.4   Example of the application of novelty detector in an oscillatory transient signal

Novelty detection works as follows: it is necessary to define a reference frame. At the beginning, frame 1 is the reference and each frame will be compared to it. Thus, frames 2, 3, 4 and 5 are nothing new; when comparing frame 6 to frame 1, a novelty is detected and frame 6 becomes the new reference; when frame 7 is compared to frame 6, a novelty is also detected and the same happens in the comparisons of frame 8 with 7 and of frame 9 with 8; in this way, frame 9 becomes the reference and the following frames will be compared to it and nothing new will be detected in frames 10, 11, 12, 13, 14, and 15.

To detect novelties in AC signals, several comparison metrics have already been used, both in the time domain and in the frequency domain and are found in the literature. For example, Difference of Frame Energies [8], Ruzicka Dissimilarity and Dynamic Similarity Metrics: Dynamic Time Warping (DTW), Edit Distance on Real signal (EDR), and Time Warp Edit Distance (TWED).

8.4.1   DTW

The signals can be modeled as one-dimensional vectors **x** and **y**:

$$\mathbf{x} = \begin{bmatrix} x_1 & x_2 & \cdots & x_m & \cdots & x_M \end{bmatrix}$$
$$\mathbf{y} = \begin{bmatrix} y_1 & y_2 & \cdots & y_n & \cdots & y_N \end{bmatrix} \tag{8.7}$$

where M and N are the dimensions of x and y, respectively. Note that M and N may be different.

Unlike classical similarity distance metrics (such as Euclidean distance), which analyzes the similarity between two signal vectors through "point-to-point" correspondence, the DTW similarity metric does it through "one point to many points" or "many points to one point" correspondence, as shown in Figure 8.5, in such a way that the total distance, given by the value of DTW, can be minimized between the two signals. If the signals are identical, the distance is null [9,10].

Since the DTW works with a multi-point distance calculation, the distance between each sample of x and all samples of y must be calculated. This can be done using several metrics, such as Euclidean Distance, absolute distance, or squared distance, among others. In this chapter, the Euclidean distance will be used, as shown in (8.8).

$$d_{m,n}(\mathbf{x},\mathbf{y}) = \sqrt{(x_m - y_n)^2} \tag{8.8}$$

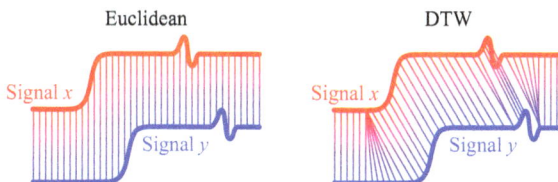

Figure 8.5   Euclidean distance and DTW

The DTW distance calculation is based on a cost matrix $\mathbf{D}$ with M rows and N columns. Each element of this matrix $D_{m,n}$ can be calculated using the following equation:

$$D_{m,n} = d_{m,n}(\mathbf{x},\mathbf{y}) + \min \begin{cases} D_{m-1,n}, & \text{or} \\ D_{m,n-1}, & \text{or} \\ D_{m-1,n-1} \end{cases} \tag{8.9}$$

This means that, starting from the upper left first element $D_{1,1}$, the matrix $\mathbf{D}$ is constructed by adding to the distance $d_{m,n}(\mathbf{x}, \mathbf{y})$ the smallest adjacent value in the upper ($\uparrow$), left ($\leftarrow$) or upper left ($\nwarrow$) directions of the respective element. After constructing matrix $\mathbf{D}$ as described above, the desired DTW distance corresponds to the last element at the bottom right of matrix $\mathbf{D}$, that is $D_{M,N}$.

8.4.2   EDR signal

The EDR metric, in addition to having a multipoint correspondence, aims to convert a vector to another vector through three operations: addition, substitution, and removal of points from the vector. In relation to the DTW metric, the EDR metric is more robust to noise [11].

The EDR metric is an extension of the Levensthein [12] distance metric. The EDR (8.10) algorithm follows to determine the degree of similarity between the vectors:

$$D_{m,n} = \min \begin{cases} D_{m-1,n} + 1, & \text{ou} \\ D_{m,n-1} + 1, & \text{ou} \\ D_{m-1,n-1} + \begin{cases} 0 \Leftarrow d_{m,n}(x, y) \le \varepsilon \\ 1 \Leftarrow d_{m,n}(x, y) > \varepsilon \end{cases} \end{cases} \tag{8.10}$$

where ε is the tolerance, also known as the elasticity parameter, and $d_{m,n}(x, y)$ is the Euclidean distance (8.8).

The processing of the EDR algorithm works as follows: the vectors (7) will be responsible for building the matrix of order $M \times N$. Each sample of mth row and nth column will be mathematically represented as $D_{m,n}$, which will be the smallest value found from the previous adjacent samples, that is, sample previous row $D_{m-1,n} + 1$, previous column sample $D_{m,n-1} + 1$, or previous diagonal sample, this one will depend on the elasticity parameter ε, that is, $D_{m-1,n-1} + 0$, or $D_{m-1,n-1} + 1$. The EDR similarity metric value will be $D_{M,N}$. For more details on the processing of the EDR algorithm, see [11].

8.4.3   TWED

The TWED metric, developed by Marteau [13], is a combination of the DTW and EDR metrics. In addition to the advantages of DTW and EDR, TWED is also capable of handling time series of different sample rates, including the reduced sample time series [14]. The TWED algorithm is described by:

$$D_{m,n} = \min \begin{cases} D_{m-1,n} + \Gamma_x, & \text{ou} \\ D_{m,n-1} + \Gamma_y, & \text{ou} \\ D_{m-1,n-1} + \Gamma_{xy} \end{cases} \tag{8.11}$$

where,

$$
\begin{aligned}
\Gamma x &= D_{m-1,n} + d_{m,m-1}(x,x) + v + \lambda \\
\Gamma y &= D_{m,n-1} + d_{n,n-1}(y,y) + v + \lambda \\
\Gamma x, y &= D_{m-1,n-1} + d_{m,n}(x,y) + d_{m-1,n-1}(x,y) + 2v|m - n|
\end{aligned} \tag{8.12}
$$

where v and λ are elasticity and penalty parameters, respectively, and, $d_{m,n}(x,y)$, $d_{m-1,n-1}(x,y)$, $d_{m,m-1}(x,x)$ and $d_{n,n-1}(y,y)$, are the cost functions.

The processing of the TWED algorithm is practically similar to the EDR algorithm, what changes are the calculations in relation to the nth row with the mth column, present in the TWED algorithm.

8.4.4 Ruzicka similarity

A similarity metric S determines how similar two vectors x and y are to each other [15]. Generally, the similarity is unity when the two vectors are identical and null when they are completely different. In this work, we used the Ruzicka similarity metric [16], expressed by (8.13).

$$
S(x_n, y_n) = \frac{\sum_{n=1}^{N_s} \min\{x_n, y_n\}}{\sum_{n=1}^{N_s} \max\{x_n, y_n\}} \tag{8.13}
$$

where the index n represents the nth samples of the vectors $\mathbf{x}$ and $\mathbf{y}$ and N_s is the total number of samples contained in each cycle of the analyzed signal. To calculate the detection threshold, the dissimilarity metric D is used, complementing the similarity metric. For the Ruzicka metric, the expression of dissimilarity is given by (8.14):

$$
D(x_n, y_n) = 1 - S(x_n, y_n) \tag{8.14}
$$

For the detection threshold calculation, a sensitivity constant γ and a size for the reference window l_r are defined by the user, which can contain one or more reference cycles for the comparison. For the ith cycle, the threshold is defined by:

$$
\eta_i = 1 - \gamma \times \text{median}(D_{ref}), \tag{8.15}
$$

where D_{ref} is a vector that contains the dissimilarity values of the cycles $i - 1_r$ to the ith cycle. A detection is defined at the moment when the similarity value becomes less than the detection threshold:

$$
S_i < \eta_i \tag{8.16}
$$

8.4.5 The difference of frame energies

The *DFE* is calculated as the absolute value of the subtraction of the current frame energy from the energy of the reference frame, as shown in (8.17). If *DFE* is higher than a given threshold, a novelty frame is detected, and this frame is labeled as the new reference frame:

$$DFE = |EF_c - EF_r| \tag{8.17}$$

The energy of each frame is calculated according to the following equation:

$$EF = \sum_{n=0}^{M-1} |x[n]|^2 \tag{8.18}$$

where $x[n]$ is the nth sample of the frame, and M is the number of samples per frame.

8.4.6 Metrics comparison

In order of comparing the performance of the metrics described above in the novelty detection, a set containing various signals with different disturbances were generated. Although the disturbances generated are typical AC PQ disturbances, the objective is only to compare the performance of the metrics, remembering that the novelty approach can be used in DC systems. The results are shown in Figure 8.6 for a scenario with signal-to-noise ratio (SNR) = 60 db and in Figure 8.7 for a scenario with SNR=30 db. The figures show the receiver operating characteristic (ROC) curves. The ROC curve compares the probability of detection

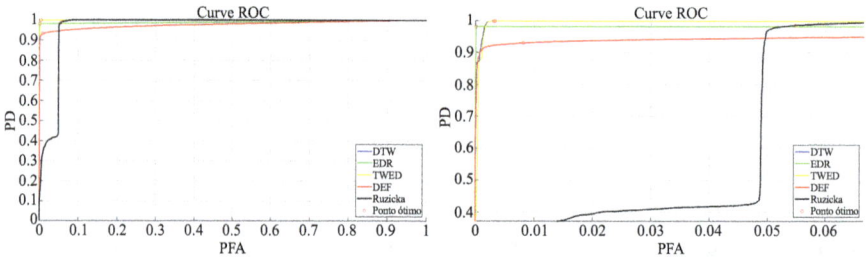

Figure 8.6   *ROC curves for SNR = 60 dB scenario*

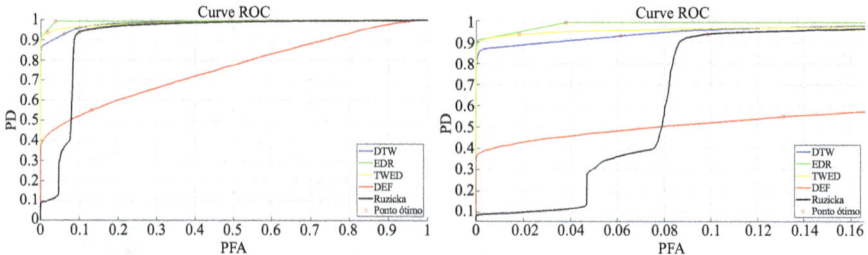

Figure 8.7   *ROC curves for SNR = 30 dB scenario*

(PD) with probability of false alarm (PFA). The closer the corner is of (0,1), the better is the detector. Note the plot on right side is just the zoom of the left side plot.

It is possible to note that for the signals with SNR = 60 db DTW and TWED presented the best performances (P_D approximately one and P_{FA} approximately zero), unlike Ruzicka which did not present a satisfactory performance (low P_D and high P_{FA}). For the scenario of SNR = 30 dB, EDR and TWED presented the best performances (P_D approximately one and P_{FA} approximately zero), in contrast to the other techniques adopted, mainly DEF and Ruzicka that did not present good performances (low P_D and high P_{FA}).

8.5 PQ in the AC grid side

PQ in the AC network is already a consolidated field in electric power systems, with standardization for the main disturbances that affect the quality of the product and services related to the supply of electric energy. Nonetheless, with the constant modernization of electrical networks, especially with the increase of the penetration of power generation from renewable sources and the inclusion of ESSs, there is an increasing concern about the emergence of new disturbances or even potentiating some existing disturbances. This shows that the concern with PQ must be constant and follow the evolution of power grid. However, the philosophical aspects of PQ monitoring remain the same and can be summarized according to the following objectives presented in JWG C4.112 [17,18]:

(i)   Compliance verification when a set of PQ parameters must be compared with limits given by standards, rules, or regulatory specifications. The lack of compliance may result in penalties or incentives associated with PQ compliance and improvements.

(ii) Performance analysis/benchmarking of a site, feeder, substation, etc., used mainly for internal purposes (planning, asset management, among others).

(iii) Site characterization that is used for characterizing the PQ at a specific site in a detailed way.

(iv) Troubleshooting because poor PQ may lead to malfunction of equipment connected to the systems. For customers, PQ disturbances can be expensive if they lead to interruption of production.

(v)   Advanced applications and studies covering more specific measurements and analyses. These studies are becoming more necessary in the utility side as the consequence of implementing smart grid concepts. This kind of monitoring is characterized by higher time resolution of data acquired and consequent demand for more complex processing and faster communication channel for data transferring.

(vi) Active PQ management including all applications where a system operation and control is related to PQ parameters.

Taking into account the increased penetration of renewable generation, ESSs, and electric vehicle, two points deserve attention from the agents involved in grid operation and PQ maintenance:

(i)   the increase of harmonic and interharmonic distortions in the system;
(ii)   the increase of supraharmonic components.

8.5.1   *The increase of harmonic and interharmonic distortions in the system*

The increase in harmonic and interharmonic distortion has been reported in the literature due to the inclusion of renewable sources and new electronic loads. In this context, the location of harmonic sources and the attribution of responsibility is a relevant research topic [19–22]. For a fair attribution of responsibility for harmonic distortion, it is necessary to estimate the harmonic phasor (HP) dynamically, as well as the harmonic transfer impedance from the harmonic sources to the point where the harmonic contribution is being evaluated. The HPE is pointed out as the natural evolution of PMUs, which estimate the fundamental phasor in a synchronized way. There are several methods in the literature for estimating the harmonic phasor under time-varying conditions [23–26]. The vast majority of methods rely on sophisticated techniques that often have a high computational effort that limits their use in low-cost hardware. However, in [23], the authors show that the FFT can be still used to estimate the time-varying harmonic phasor as long as a good frequency estimator and a good interpolator are used.

Figure 8.8 shows the basic methodology proposed in [23] for estimating the HP. The block A represents the resampling method based on B-Spline interpolation. The B-Spline needs to be fed with the fundamental frequency estimation that is obtained by using a modified zero-crossing frequency estimator (Block B). In Block C, the signal is separated in the component fundamental $x_{60}[l]$ and the harmonic components $x_h[l]$. Then the FFT is used to estimate the harmonic components, every two cycles. The fundamental phasor is estimated using a single DFT.

Figure 8.9 shows the total vector error (TVE) for harmonics, up to 13th order, of the proposed B-spline method compared to the HPE [27], TFT [28], SIFE [29], and FT FIR [25] for off-nominal conditions. It can be noticed that the B-Spline and FT FIR methods perform much better than the other methods, mainly in higher harmonics, presenting a "œflatter" behavior. The SIFE and TFT methods overpass the 1% limit, while the HPE method touches the 1% limit for the 13th harmonic. As presented in [23], the B-Spline method presents the lowest computational effort and it performs well for dynamic signal variation, as required by the standard [30].

Figure 8.8   Dynamic phasor estimator using FFT and interpolated B-spline approach

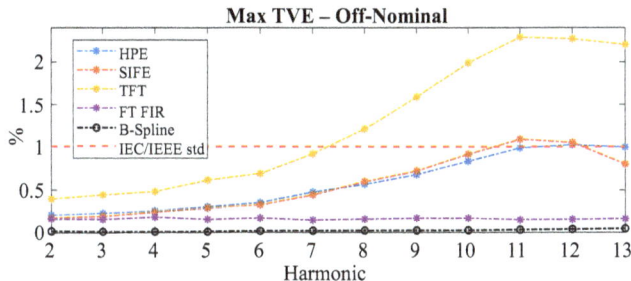

Figure 8.9 TVE for off-nominal frequency test

8.5.2 Supraharmonic distortion

Supraharmonic (SH) emissions have received increasing attention of the research community [31,32]. Recently, compatibility levels have been published by the IEC 61000-2-2 standard [33] and the standardization work on emission limits has started. The interesting for monitoring supraharmonic it is because several new component connected to the network, typically power electronic interface, has emitted frequency component in the frequency range between 2 kHz and 150 kHz. These components are in general not correlated with the fundamental frequency and for this has been named supraharmonics [34].

SH emissions are usually unintentional, caused by switching components of frequency converters or rectifiers, but there are also intentional ones, generated by the use of power line communication (PLC), which has become increasingly common with the adoption of smart-grids. Additionally, the SH emissions are time-varying and the spectral content can be broadband or narrowband. This means that the analysis considering only the frequency domain may be not enough to characterize the SH emission. For example, in [35], it is possible to observe an example of emission that when analyzed only in the frequency domain appears to have a broad band from 40 to 80 kHz; however, when the SH signal is analyzed in the time–frequency (TF) domain, it is possible to verify that it is a narrow-band emission with time-varying frequency and amplitude. This type of behavior is commonly observed in equipment with component switching frequency variation. Figure 8.10 shows the TF analysis of a voltage signal in a power grid contained a frequency inverter connecting a PV panels to the AC grid. From this figure, it is possible to note a packet of PLC communication around 0.01 s and in the frequency range of 50–90 kHz around 0.01 s. Also, it is possible to note a pulsing signal around 130 kHz.

This example shows that SH monitoring is challenging, as it requires high sampling rates and the use of methodologies that take into account the time-varying nature of emissions. To overcome some of these difficulties, some researchers have suggested to apply the methodology for electromagnetic compatibility tests that determine compliance with emission limits, which is discussed in the CISPR 16-2-1 standard [36]. For the SH frequency range, defined as band A (9–150 kHz), this standard suggests the use of the peak or quasi-peak (QP) detectors as illustrated in Figure 8.11(a).

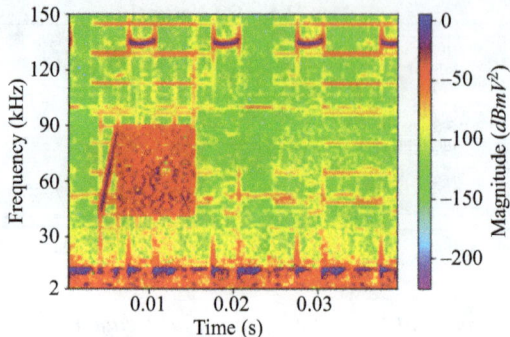

Figure 8.10   *Time–frequency SH analysis: PV inverter*

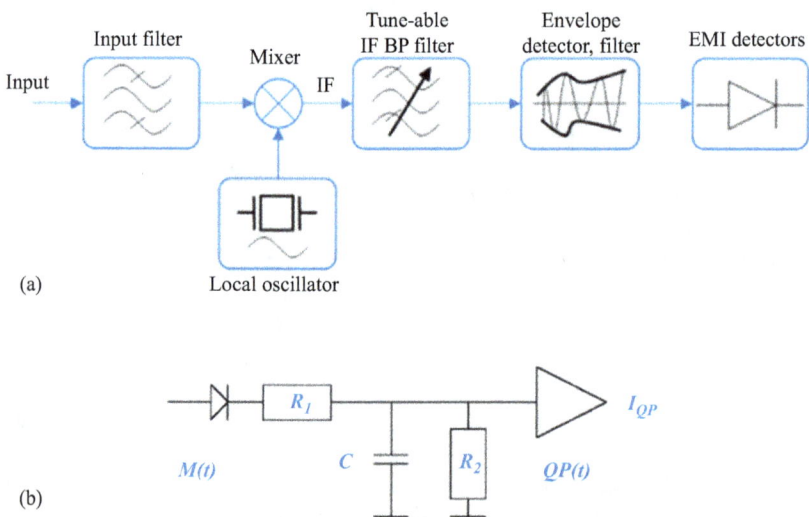

Figure 8.11   *(a) Block diagram of EMI-receiver. (b) An analog quasi-peak detector.*

In the block diagram of the EMI-receiver shown in Figure 8.11, oscillator must sweep the entire SH range at 200 Hz resolution. After the oscillator sets a specific frequency, the output of the mixer is filtered by a bandpass filter, whose specification is defined by the standard [36]. After this stage, the envelope of the filtered signal must be found and applied to the QP detector illustrated in Figure 8.11(b). The QP detector needs 1 sec to update the value for every frequency set in the oscillator, so the procedure to scan all frequencies in the SH range takes a long time if a single EMI-receiver is used. Of course, several parallel EMI-receiver can be employed at the expense of increasing hardware. Also, the procedure using STFT can be used to implement the QP detector as suggesting in [37]. However, the computational effort is still high.

If we take into account only the narrow band SH emission, it is possible to reduce the computational effort using the phase locked loop (PLL) technique instead of the frequency scanning method specified by the CISPR 16 standard, as suggested in [38]. Through the use of PLL units, the highest energy components of the signal are obtained and processed through the QP detector, enabling the analysis of the SH distortion levels. The block diagram of this approach is illustrated in Figure 8.12. The proposed PLL structure identifies the frequency of the highest supra-harmonic at same time that it extracts this component. Then, the SH component is used in the demodulator stage. The other processing stages are similar to the corresponding ones shown in Figure 8.11.

This methodology is useful for evaluating tonal SH emission such as generated by converter and inverter, commonly used in distributed generation and storage system energy. The main advantage of this methodology is instead of scan all frequencies in the SH band, the PLL tracks the main energy components and estimates the respective QP values. As SH is sparse in the frequency domain, a small number of PLL-based QP detectors are implemented in hardware, normally three are sufficient, reducing the computational cost of the detector.

Figure 8.13 shows a representation of the proposed methodology as suggested in [38]. Using this methodology with N parallel PLL, N quasi-peak component can be tracked and measured. The PLL is by nature an adaptive filter, so it is able to track time varying SH component.

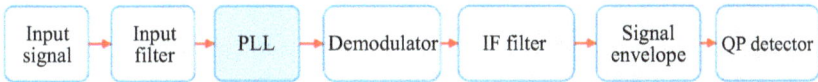

Figure 8.12   Modified CISPR 16 using a PLL for unintentional SH emission evaluation

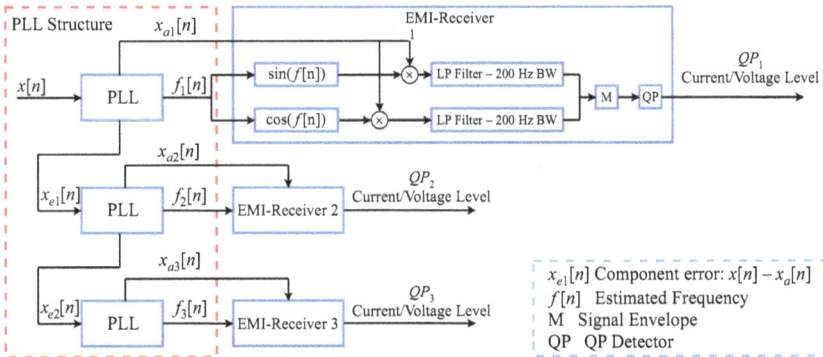

Figure 8.13   Representation of the cascade PLL methodology

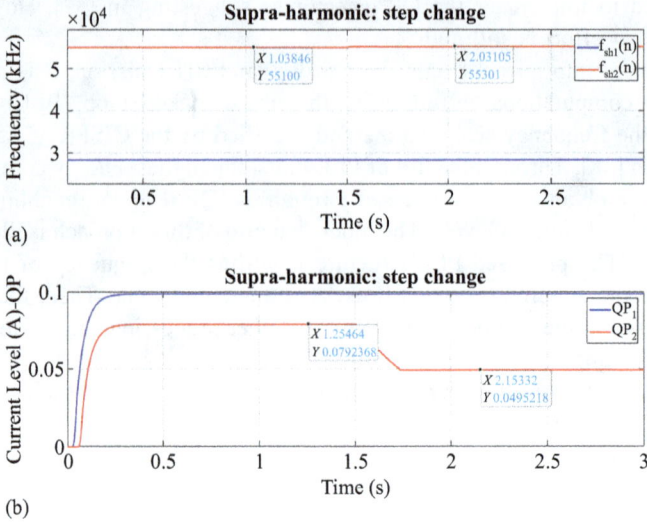

Figure 8.14   PLL-based QP estimation: (a) frequency of the components; (b) QP values

As an example of QP estimation based on PLL approach, consider a signal given by

$$x(t) = A_{sh1}cos(2\pi f_{sh1}t) + A_{sh2}\cos(2\pi f_{sh2}t) \tag{8.19}$$

where $A_{sh1}= 0.1, f_{sh1} = 28.5$ kHz are, respectively, the magnitude and frequency of the SH1 and A_{sh2} and f_{sh2} the magnitude and frequency of the SH2. The SH1 parameters are constant, while the SH2 parameters vary in steps, changed from $A_{sh2} = 0.08$ and $f_{sh2} = 55.1$ kHz to $A_{sh2} = 0.05$ and $f_{sh2} = 55.3$ kHz, respectively, at time 1.5 s.

Figure 8.14 shows the performance of the two PLLs and QP detectors activated for this case. As the supraharmonic component 1 does not change, the tracked frequency f_{sh1}, as well as the respective QP value QP_{sh1}, do not change. For component 2, it is possible to observe the new frequency tracking, as well as the new QP value.

8.6   A monitoring system based on the SED concept

To face the new challenges with the increasing penetration of large blocks of renewable energy, mainly wind and solar power plants, the power grid may face adverse conditions from those that usually occurred in the past, even in the transmission level. For example, Ref. [39] discusses the impacts of the integration of renewable sources in high voltage (HV) networks, with the growth in the harmonic distortion level and voltage fluctuation being two of the most affected parameters.

This degradation of the electrical PQ must be closely monitored as it may negatively impact the system operation. For example, there is limited information in the literature that makes a correlation between the decrease in PQ in the transmission systems with the increase of failures in the protection system or the early aging of equipment, such as the power transformer. This gap is due in part to the lack of continuous monitoring of the PQ in HV systems and the poor integration and interoperability between the various intelligent electronic devices (IED) at the substations. To face these challenges, an efficient monitoring system is an important element of the grid to keep the reliability and PQ under control.

In this sense, the concept of SED is of great importance in modern grids because it paves the way for open systems that embody supplier independent standard, interoperability, non-proprietary software and hardware and support easy upgrades [40,41]. The concept of edge computing emerged to overcome some difficulties inherent to cloud computing, such as latency. The term SED has been used in the specialized literature, especially in [42]. According to this philosophy, much of the processing is carried out close to the point where the information is produced, through an open architecture, such as an industrial computers, which receive different information through a local communication network.

There are several works in the literature that present substation automation solutions based on the IEC 61850 protocol. Some successful experiences of implementing systems based on IEC 61850 can be found in [43–49]. However, with regard to the philosophy of the substation of the future, as mentioned in [40,41], the IEC protocol is just one of the components to characterize a SED. The papers presented in [50,51] present a prototype of a multi-functional SED called PQMC (PQ monitoring center) which uses an open architecture with interoperability and scalability characteristics whose applications are implemented as software tasks on an industrial computer. The concepts used in these works are very important for monitoring modern networks, so some of them are highlighted below.

8.6.1 The SED

Figure 8.15 shows an illustrative diagram of SED. This figure shows that the process bus (PB) receives information from various devices installed in the substation. For example, the voltages and currents of the secondary of the measurement transformers are digitized in the merging unit (MU) and published in the PB, the information regarding the charging of the batteries, the power generation of the PV panels, etc., may be published in the PB. The most common format for publishing samples of voltage and current from the AC side is the IEC 61850-9-2 Sample Values (SV) format [52], but there are other formats established by the IEC61850 standard and others. The application layer has a plug and play characteristic, that is, a new application can be added to the device in a simple way, without this functionality interfering with the performance of applications already in operation. The entire logical and functional system is implemented on an industrial computer, revolutionizing the concept of IEDs as it is today. In this sense, it is possible to overcome the barrier of interoperability, vendor independence, and scalability.

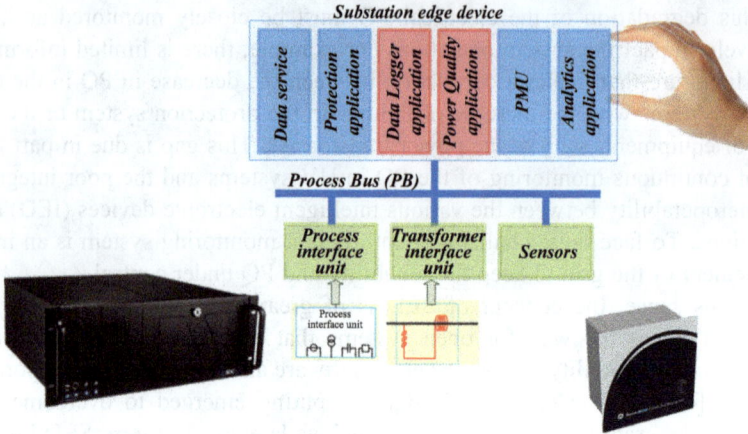

Figure 8.15   Illustration of an SED. The different functionalities of a substation are implemented in a computer and no longer in IEDs.

The PQMC presented in [51] has implemented as software tasks the functionality of PQ, data logger, harmonic phasor estimator, and data service. The system was installed in two substations that connect a 230 kV transmission line in the northern region of Brazil.

Additionally, to the general rules for PQ monitoring presented in

(a)   Be adherent to the concepts and standards associated with the SED.
(b)   Provide communication and sharing information with synchronized phasor measurement systems.
(c)   Allow systematic online monitoring and records of PQ with time synchronization.
(d)   Seek the methodological improvement and indicators employed in the processes related to verification/ evaluation of the PQ.

8.6.2   The system architecture of a real-time PQMC

The main components of the real-time PQMC, presented in [51], are shown in Figure 8.16. The voltage and current of the AC side are processed by the MU, which publishes the samples using the IEC-61850-9-2LE protocol. The SV published by the MU are processed by different applications running in the PQMC. For example, applications that compute the PQ parameters, the harmonic phasor, and detect disturbances (DDR – digital disturbance recorder).

Some useful PQ indicators (PQI), such as the total harmonic distortion (THD), voltage sags and swells, system frequency, among others are estimated and published to the data concentrator through the advanced message queueing protocol (AMQP) protocol (ISO/IEC 19464) [53]. Using the AMQP, it is possible to combine information that comes from different substations, providing a powerful tool for analyzing the behavior of the monitored power system. All published data has a

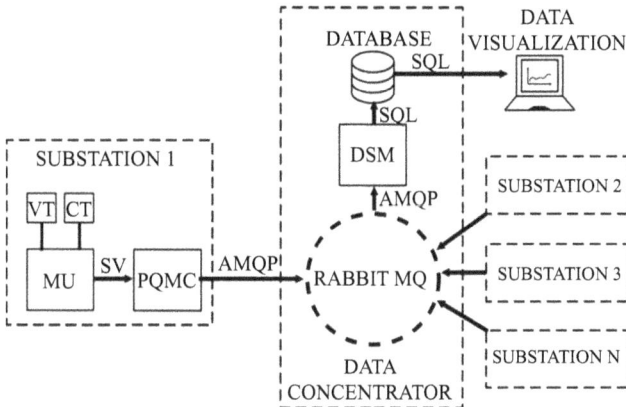

Figure 8.16   The proposed real-time PQ monitoring system

time stamp inserted by using a strategy involving a network time protocol (NTP) server and the sample count (SmpCnt) value of the arrived SV packets.

Figure 8.16 shows the message oriented middleware (MOM) used to inter-connect the system modules. The RabbitMQ was used as message broker which lies on the message queuing approach [53–55] and allows for the communication between the system nodes by means of the publisher–subscriber principle [55–57]. The Rabbit MQ routes the messages that can be located in different substations to the data consumers, such as the data storage module (DSM), fault localization module, etc. One of the main advantages of the message broker is the possibility to increase the numbers of producers and consumers without degrading significantly the system performance [56,57]. Furthermore, the use of a message broker makes the interoperability between the modules connected to it easier, so that it is simple to send data to or receive data from any component, regardless of its hardware, operating system, or even the programming language used to develop it.

Figure 8.17 illustrates the logical architecture of the PQMC. The main com-ponents in it are: the data acquisition module (DAM), the signal processing routines (SPR$_1$ $\cdots$ SPR$_N$), an event-triggered oscillography (ETO) and the data exportation module (DEM).

The DAM module receives the SV packets and extracts the samples of the current/voltage signals that will be processed by the SPRs. In addition, the timestamp is obtained to synchronize the data produced by the PQMC. This timestamp is obtained by using the internal clock of the PQMC and the SmpCnt value received in the SV packets. This strategy allows for the PQMC to compensate network delays, since the timestamp attached to the DAM output data is referred to the moment in which the MU sampled the voltages and currents from the VT and CT, respectively.

In the proposed structure, each voltage and current channel can be processed in parallel and the PQI are stored into an output data queue (ODQ). Additionally, the event-triggered oscillography (ETO) saves the samples when one or more

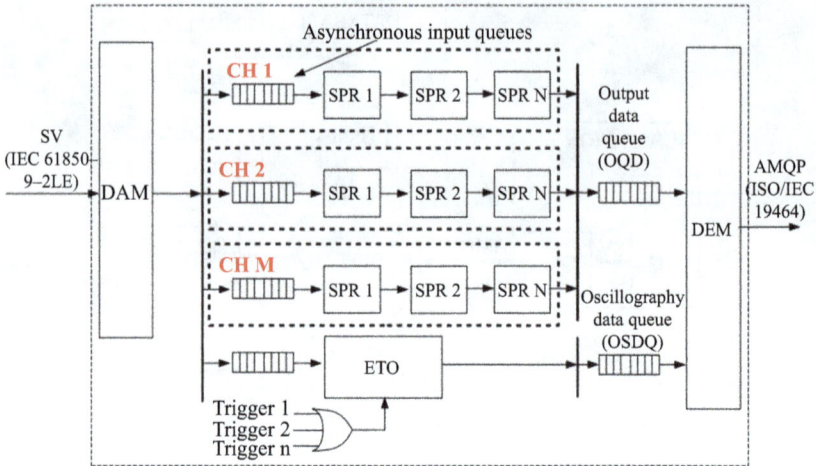

Figure 8.17   Logical architecture of the PQ monitoring center

monitored PQI, such as THD and voltage value, surpasses a predefined threshold value. The samples are stored in the oscillography data queue (OSDQ) together with the timestamps get from GPS precision clock.

The asynchronous queue approach gives the system great scalability due to the use of a single producer and multiple consumers design pattern. A single DAM supplies data to multiple signal processing routines, which can be independently augmented or improved. In summary, the proposed architecture allows for the modification of the PQMC features just by reconfiguring the software and without hardware changes, which is quite adherent to the SED concept.

8.6.3   Results from the PQMC

This section presents some results of the implemented system. The presented results were obtained from laboratory setup and field environment. The laboratory setup is presented in Figure 8.18(a) and (b) In Figure 8.18(a), a transmission power system was simulated in real-time digital simulator (RTDS) and a different disturbance was generated in the power system. The real-time voltage and current were amplified and sent to the MU. Then the MU published the samples in the process bus of the PQMC and the PQI were estimated.

The setup presented in Figure 8.18(a) was used to test the harmonic phasor estimator. The time varying harmonics are generated in the CMC 256 source and sent to MU that published the corresponding SV at the PB of the PQMC. Then the TVE was evaluated.

The proposed PQ monitoring system was also installed in a real 230-kV transmission system in Brazil. Figure 8.19 shows the single-line diagram of this subsystem. The PQMC was installed at substation SB1 and SB2. The transmission line is 130 km long. In this case, some estimated PQI were compared with the ones produced by commercial equipment.

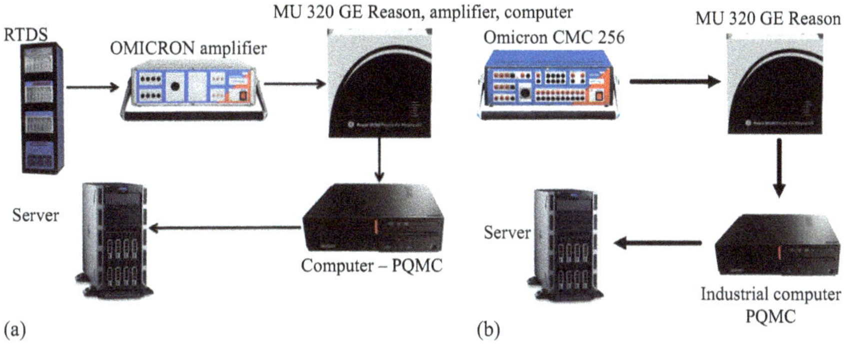

Figure 8.18   Experimental setups: (a) simulation of transmission power system in RTDS; (b) test of time varying harmonics

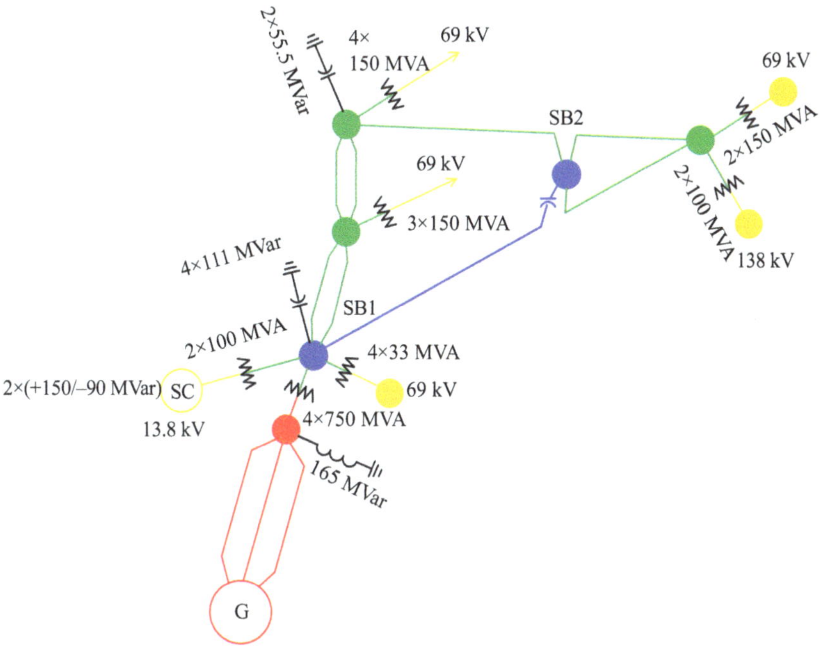

Figure 8.19   Diagram of the monitored power system

8.6.4   Results of PQI from laboratory setup

In Figure 8.20, a power system was simulated in the RTDS. The simulated system includes harmonic distortion and sag. A 9.64% of THD was generated connecting current harmonic sources in the network and a sag was generated by a short circuit in transmission line. In the upper figure, the voltage signal is presented together with the corresponding RMS values. The implemented methodology estimates separately the instantaneous RMS of the fundamental component and of the

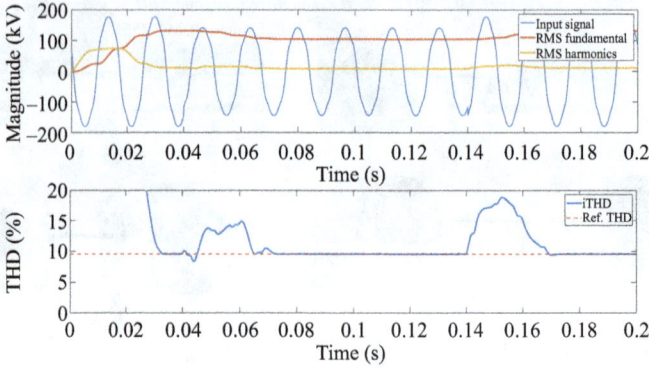

Figure 8.20   RMS and THD estimation in a signal with a dip

Table 8.1   RMS values and frequency

	RTDS	**PMU**	**PQMC**
VA (kV)	128.02	128.04	127.99
VB (kV)	128.02	127.90	127.98
VC (kV)	128.02	128.01	127.99
IA (kA)	1.030	1.030	1.031
IB (kA)	1.030	1.030	1.029
IC (kA)	1.030	1.030	1.030
Frequency (Hz)	60.00	60.00	60.00

harmonic components. Using this approach, it is possible to compute the instantaneous total harmonic distortion (iTHD), shown in the bottom figure together with the reference THD (9.64%). It can be noted when there is a sudden variation in the magnitude of the signal, there is also a significant variation in the iTHD. This variation is useful since it was used as a trigger to the data logger module.

To validate the PQMC results in the steady-state scenario, a comparison was made using the values calculated by the RTDS and also with the commercial PMU results. The PMU was configured to send the THD results along with the synchrophasors and frequency. A signal containing harmonics and no voltage variation was used in this test. Table 8.1 shows the results for the RMS and frequency and Table 8.2 the results for THD.

Observing Tables 8.1 and 8.2, it can be noted that the parameters calculated by the PQMC present very small errors when compared with the commercial PMU and the reference values of the RTDS. In fact, the error in the RMS estimation is less than 0.025% and 0.5% for the THD in the worst case.

8.6.5   Harmonic phasor estimation

To test the performance of the HPE, the signal presented in Table 8.3 was used. The column named C37.118.1 shows the maximum error defined by the standard [58]

Table 8.2   THD (%)

	RTDS	PMU	PQMC
VA	4.86	4.85	4.85
VB	4.86	4.85	4.85
VC	4.86	4.85	4.85
IA	4.44	4.42	4.43
IB	4.44	4.44	4.45
IC	4.44	4.48	4.46

Table 8.3   Limits established by the C37.118.1-2014

Test		C37.118.1		
Voltage signal		**TVE (%)**	**FE (mHz)**	**RFE (Hz/ s)**
Off-nominal Frequency	$\cos{(2\pi ft)}$ $55 \leq f \leq 65$	1	5	0.1
Harmonics	$2\cos{(2\pi f_o t)} + 0.1\cos{(2\pi i f_o t)}$ $2 \leq i \leq 50$	1	25	–
Frequency Ramp	$2\cos{(2\pi(55 + 1t)t)}$ $0 \leq t \leq 10$	1	10	0.2
Amplitude Modulation	$(1 + 0.1\cos{(2\pi f_m t)}) \times \cos{(2\pi f_o t)}$ $f_m = 1, 2, 3, 4, 5$	3	300	14
Phase Modulation	$\cos{(2\pi f_o t + 0.1\cos{\ (2\pi f_a t - \pi)})}$ $f_a = 1, 2, 3, 4, 5$	3	300	14

for the fundamental phasor. These limits were used as parameters for the harmonic phasor as well. Table 8.4 presents the error for the case where the harmonic amplitude is 5% of the fundamental component. As can be seen, the proposed HPE is complied with the standard limits. The column named $\delta(\%)$ is the standard deviation obtained in the test.

8.6.6   Field results

After the installation of both equipment in the field (SB1 and SB2 of Figure 8.19), the system is continuously monitored. An example of the acquired measurements from the PQMC and the PMU, for an interval of 15 min of the phase A voltage, is shown in Figure 8.21.

From Figure 8.21, it is possible to note that the parameters calculated by both equipment are very close, despite the parameters of the PMU presenting more oscillation, more evident in the THD. Examining the behavior of the parameters, it is possible to note that, for this interval, the system is operating in its normal condition, since the RMS of the voltage is close to its nominal value, the frequency is varying around 60 Hz and the THD level is close to 1%. The maximum relative difference between the PQMC and PMU results, considering the PMU as the

Table 8.4   TVE$_h$ values obtained by the implementation in the PQMC $A_h = 5\%$

Test signals	TVE$_{avg}$(%)	TVE$_{max}$(%)	δ(%)
Off-nominal frequency	0.226	0.819	0.132
Frequency ramp	0.235	0.797	0.127
Amplitude modulation	1.122	2.216	0.528
Phase modulation	0.778	1.547	0.347

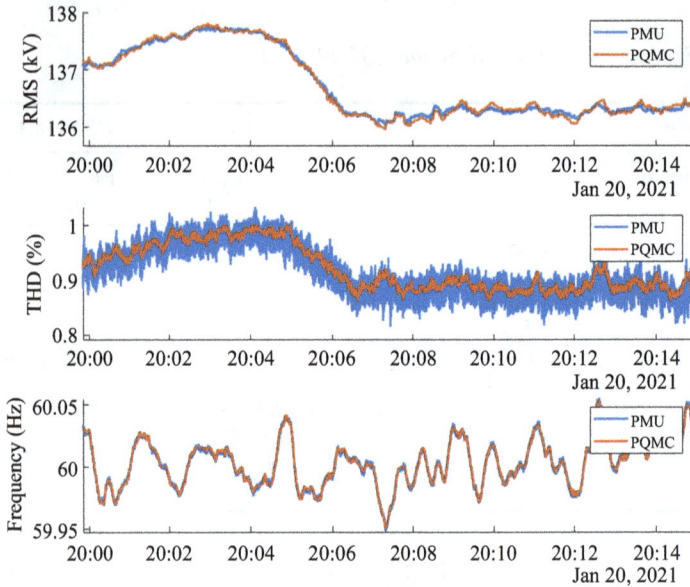

Figure 8.21   Field results from PMU and PQMC

reference, were: 0.2% for the RMS, 6.8% for the THD, and 0.1% for frequency. The high value for the THD difference is due to the oscillations present in the PMU result. The value calculated by the PQMC is smoother due to the use of the Hanning window, which attenuates the unwanted components in a more accentuated way than the rectangular window used by the PMU.

As explained before, the data logger module can produce oscillography signals based on a trigger generated by voltage disturbances or iTHD. An example of collected oscillography is shown in Figure 8.22.

Observing Figure 8.22, it is possible to note a severe sag in phase B, while the magnitude of the other phases increases slightly. This event did not cause the relay to trip and consequently was not recorded by conventional equipment. However, the information captured by the developed data logger can be of great importance to identify intermittent faults that have the potential to become permanent faults.

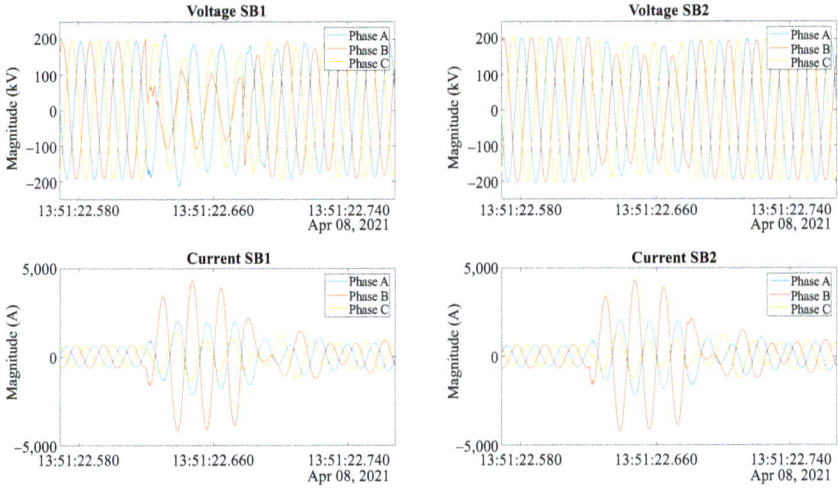

Figure 8.22   Oscillography signal collected in SB1

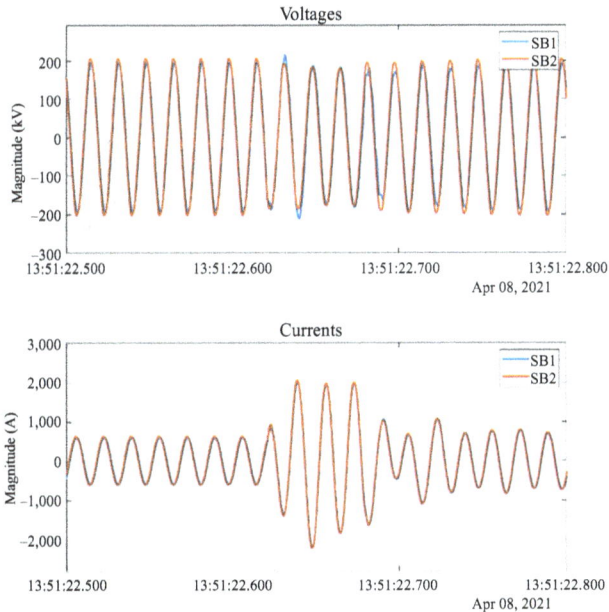

Figure 8.23   Synchronized oscillography signal from both SB1 and SB2

It is important to note that despite time difference between the trigger signal on two substations, the stored samples are synchronized with the GPS, so it is possible to analyze the oscillography of SB1 and SB2 together to observe the impact of some phenomena in both monitored points, as shown in Figure 8.23.

8.7 Conclusion

This chapter addressed some aspects of monitoring systems applied to current electrical networks in which distributed generation, ESSs and interconnection with AC networks are inserted. Initially, some PQ concepts for DC systems were presented, showing that this area deserves greater attention from the scientific and technical community. Then the chapter introduced the concept of detecting novelty in waveforms. The motivation to present this concept is that during the evolution of mixed systems in which there is the interconnection of DC and AC systems, new disturbances can occur and in this sense the waveform capture, using an intelligent detection is of great importance to understand the involved phenomena. The novelty concept guarantees the storage of waveforms that actually present information relevant to the analysis, not storing redundant information. Next, the chapter presented two important points in relation to the PQ in current networks: the concept of dynamic harmonic phasor and supraharmonics. These two phenomena deserve the attention of researchers given the new context of distributed generation and new loads introduced in the system, especially electric vehicles and ESSs. The estimation of the harmonic phasor under dynamic conditions is of great importance to identify and locate the harmonic sources, as well as to carry out a fair attribution of responsibility. Regarding supraharmonic distortions, the chapter highlighted the significant increase in these high-frequency distortions and the challenges to monitor and measure them through low-cost hardware. The methodology using PLL was presented. Finally, the chapter presented important aspects about monitoring systems using the concept of fully digitized substation and the substation edge computing (SED). It is suggested that the concept of SED be introduced in new monitoring systems as a guarantee of interoperability, independence and scalability. Some results of a monitoring system based on this concept were presented.

Acknowledgment

The authors are grateful to the various funding agencies that have contributed to the development of research over the years, in particular by the Coordination for the Improvement of Higher Education Personnel (CAPES) under Grant 001, National Council for Scientific and Technological Development (CNPq) under the grants 404068/2020-0, Fundação de Amparo à Pesquisa do Estado de Minas Gerais (FAPEMIG) under the grant APQ-03609-17, and National Institute of Electric Energy (INERGE).

References

[1] Rosewater D, Preger Y, Mueller J, *et al. Electrical Energy Storage Data Submission Guidelines, Version 2*. Livermore, CA: Sandia National Laboratories, 2021.

[2] International Electrotechnical Commission. Electromagnetic compatibility (EMC) – Part 4-30: Testing and measurement techniques – Power quality measurement methods. *IEC 61000-4-30*. 2003.

[3] Sanchez S, Molinas M, Degano M, *et al.* Stability evaluation of a DC micro-grid and future interconnection to an AC system. *Renewable Energy*. 2014;62:649–656.

[4] Wang H, Tian J, Yan J, *et al.* Definition and influencing factors of power quality in DC microgrids. In: *2021 IEEE 4th International Electrical and Energy Conference (CIEEC)*. IEEE, 2021. p. 1–5.

[5] Van den Broeck G, Stuyts J, and Driesen J. A critical review of power quality standards and definitions applied to DC microgrids. *Applied Energy*. 2018;229:281–288.

[6] Standard M. Aircraft electric power characteristics. Department of Defense Interface Standard (MIL-STD-704F). 2004.

[7] CENELEC. EN 50155. Railway applications—Rolling stock—Electronic Equipment. 2019.

[8] Silva LRM, Kapish EB, Duque CA, *et al.* The concept of novelty detection applied to power quality. In: *2016 IEEE Power and Energy Society General Meeting (PESGM)*. IEEE, 2016. p. 1–5.

[9] Chen Q, Hu G, Gu F, *et al.* Learning optimal warping window size of DTW for time series classification. In: *2012 11th International Conference on Information Science, Signal Processing and their Applications (ISSPA)*. IEEE, 2012. p. 1272–1277.

[10] Salvador S and Chan P. Toward accurate dynamic time warping in linear time and space. *Intelligent Data Analysis*. 2007;11(5):561–580.

[11] Chen L, Özsu MT, and Oria V. Robust and fast similarity search for moving object trajectories. In: *Proceedings of the 2005 ACM SIGMOD International Conference on Management of Data*, 2005. p. 491–502.

[12] Levenshtein VI, *et al.* Binary codes capable of correcting deletions, inser-tions, and reversals. *Soviet Physics Doklady*. 1966;10:707–710.

[13] Marteau PF. Time warp edit distance with stiffness adjustment for time series matching. *IEEE Transactions on Pattern Analysis and Machine Intelligence*. 2008;31(2):306–318.

[14] Serra J and Arcos JL. An empirical evaluation of similarity measures for time series classification. *Knowledge-Based Systems*. 2014;67:305–314.

[15] Tan PN, Steinbach M, and Kumar V. *Introduction to Data Mining*. India: Pearson Education, 2016.

[16] Deza MM and Deza E. *Encyclopedia of Distances*. New York, NY: Springer. 2009.

[17] GROUP JW. C4.24/cired power quality and EMC issues with future elec-tricity networks. *Technical Brochures*. 2018. p. 719.

[18] Bollen M, Milanoviä J, and Cukalevski N. CIGRE/CIRED JWGC4.112 – power quality monitoring. *Renewable Energy and Power Quality Journal*. 2014;12:1037–1045.

[19]   de Oliveira MM, Lima MA, Silva LR, *et al.* Independent component analysis for distortion estimation at different points of a network with multiple harmonic sources. In: *2022 20th International Conference on Harmonics & Quality of Power (ICHQP)*. IEEE, 2022. p. 1–6.

[20]   Shu Q, Wu Y, Xu F, *et al.* Estimate utility harmonic impedance via the correlation of harmonic measurements in different time intervals. *IEEE Transactions on Power Delivery.* 2019;35(4):2060–2067.

[21]   Papic I, Matvoz D, Špelko A, *et al.* A benchmark test system to evaluate methods of harmonic contribution determination. *IEEE Transactions on Power Delivery.* 2018;34(1):23–31.

[22]   Monteiro HL, Duque CA, Silva LR, *et al.* Harmonic impedance measurement based on short time current injections. *Electric Power Systems Research.* 2017;148:108–116.

[23]   Aleixo RR, Lomar TS, Silva LR, *et al.* Real-time B-spline interpolation for harmonic phasor estimation in power systems. *IEEE Transactions on Instrumentation and Measurement.* 2022;71:9004009.

[24]   Chen L, Zhao W, Xie X, *et al.* Harmonic phasor estimation based on frequency-domain sampling theorem. *IEEE Transactions on Instrumentation and Measurement.* 2020;70:1–10.

[25]   Duda K, Zieliński TP, BieÅ, A, *et al.* Harmonic phasor estimation with flat-top FIR filter. *IEEE Transactions on Instrumentation and Measurement.* 2020;69(5):2039–2047.

[26]   de Melo ID, Pereira JLR, Variz AM, *et al.* Harmonic state estimation for distribution systems based on synchrophasors. In: *2016 IEEE 16th International Conference on Environment and Electrical Engineering (EEEIC)*. IEEE, 2016. p. 1–6.

[27]   Chen L, Zhao W, Xie X, *et al.* Harmonic phasor estimation based on frequency-domain sampling theorem. *IEEE Transactions on Instrumentation and Measurement.* 2020;70(9001210):1–10.

[28]   Platas-Garza MA and de la O Serna JA. Dynamic harmonic analysis through Taylor-Fourier transform. *IEEE Transactions on Instrumentation and Measurement.* 2010;60(3):804–813.

[29]   Chen L, Zhao W, Wang Q, *et al.* Dynamic harmonic synchrophasor estimator based on sinc interpolation functions. *IEEE Transactions on Instrumentation and Measurement.* 2018;68(9):3054–3065.

[30]   IEEE. *IEEE Standard for Synchrophasor Measurements for Power Systems.* IEEE Standard C371181. 2011.

[31]   Espn-Delgado Á, Rönnberg S, Busatto T, *et al.* Summation law for supraharmonic currents (2–150 kHz) in low-voltage installations. *Electric Power Systems Research.* 2020;184:106325.

[32]   Alkahtani AA, Alfalahi ST, Athamneh AA, *et al.* Power quality in microgrids including supraharmonics: issues, standards, and mitigations. *IEEE Access.* 2020;8:127104–127122.

[33]   IEC-61000-2-2: Amendment 1 – Electromagnetic Compatibility (EMC) – Part 2-2: Environment – Compatibility Levels for Low-Frequency

Conducted Disturbances and Signalling in Public Low-Voltage Power Supply Systems, 2017.

[34] Bollen M, Olofsson M, Larsson A, *et al.* Standards for supraharmonics (2 to 150 kHz). *IEEE Electromagnetic Compatibility Magazine.* 2014; 3(1):114–119.

[35] Larsson A, Lundmark M, and Bollen M. Distortion of fluorescent lamps in the frequency range 2–150 kHz. In: *International Conference on Harmonics and Quality of Power*: 01/10/2006–05/10/2006; 2006.

[36] CODE P. Specification for Radio Disturbance and Immunity Measuring Apparatus and Methods – Part 3: CISPR Technical Reports. 2003.

[37] Schwenke M and Klingbeil D. Application aspects and measurement methods in the frequency range from 2 kHz to 150 kHz. In: *25th International Conference on Electricity Distribution*, 2019.

[38] Mendes TM, Ferreira DD, Silva LR, *et al.* PLL based method for supraharmonics emission assessment. *IEEE Transactions on Power Delivery.* 2021;37(4):2610–2620.

[39] Shafiullah G. Impacts of renewable energy integration into the high voltage (HV) networks. In: *4th International Conference on the Development in the in Renewable Energy Technology (ICDRET)*. ICDRET, 2016. p. 1–7.

[40] Hunt R, Flynn B, and Smith T. *The Substation of the Future: Moving Toward a Digital Solution*, vol. 17. IEEE, 2019. p. 47–55.

[41] Aftab MA, Hussain SS, Ali I, *et al.* IEC 61850 based substation automation system: a survey. *International Journal of Electrical Power & Energy Systems.* 2020;120:106008.

[42] Hunt R, Flynn B, and Smith T. The substation of the future: moving toward a digital solution. *IEEE Power and Energy Magazine.* 2019;17(4):47–55.

[43] Semjan A JNEs. Experience sharing – challenges and solutions on IEC 61850 substation commissioning and supervision in Thailand. In: *IEEE PES GTD Grand International Conference and Exposition Asia*. IEEE, 2019. p. 228–234.

[44] Vardhan H, Ramlachan R, Szela W, *et al.* Deploying digital substations: experience with a digital substation pilot in North America. In: *71st Annual Conference for Protective Relay Engineers (CPRE)*. IEEE, 2018. p. 1–9.

[45] Al Obaidli S, Subramaniam V, Alhuseini H, *et al.* IEC 61850 beyond compliance: a case study of modernizing automation systems in transmission power substations in Emirate of Dubai towards smart grid. In: *Saudi Arabia Smart Grid (SASG)*. IEEE, 2017. p. 1–9.

[46] Gaouda AM, Abdrabou A, Shaban KB, *et al. A Smart IEC 61850 Merging Unit for Impending Fault Detection in Transformers*, vol. 9. IEEE, 2016. p. 1812–1821.

[47] Ali NH EM. *Accelerating the Protection Schemes Through IEC 61850 Protocols*, vol. 102. New York, NY: Elsevier, 2018. p. 189–299.

[48] Wang Y and Wang J. Harmonic analysis of power quality monitoring device based on IEC 61850. In: *International Conference on Electrical and Control Engineering*. ICECE, 2010. p.1117–1120.

[49]   Kang Y and Li L. Modeling and implementation of power quality monitoring IED based on IEC 61850. In: *International Conference on Computational and Information Sciences*. ICCIS, 2011. p. 148–151.

[50]   Duque CA, Silva LMR, Soares GM, *et al.* Continuous power quality monitoring for power transmission systems. In: *IEEE Power & Energy Society General Meeting (PESGM)*. IEEE, 2020. p. 1–5.

[51]   Alberto M, Soares GM, Silva LR, *et al.* Newly implemented real-time PQ monitoring for transmission 4.0 substations. *Electric Power Systems Research*. 2022;204:107709.

[52]   IEC-61850-9-2:2011: Specific Communication Service Mapping (SCSM). Sample Value over ISO/IEC Requirements for Functions and Device Models, 2011.

[53]   Vinoski S. *Advanced Message Queuing Protocol*, vol. 10. New York, NY: IEEE, 2006. p. 87–89.

[54]   IEC-61588-2004: Precision Clock Synchronization Protocol for Networked Measurement and Control Systems, 2004.

[55]   Banavar G, Chandra T, Strom R, *et al.* A case for message oriented middleware. In: *International Symposium on Distributed Computing*. New York, NY: Springer, 1999. p. 1–17.

[56]   Blakeley B, Harris H, and Lewis R. *Messaging and Queueing Using the MQI*. New York, NY: McGraw-Hill, Inc., 1995.

[57]   Eugster PT, Felber PA, Guerraoui R, *et al. The Many Faces of Publish/ Subscribe*, vol. 35. New York, NY: ACM, 2003. p. 114–131.

[58]   IEEE. IEEE Standard for Synchrophasor Measurements for Power Systems – Amendment 1: Modification of Selected Performance Requirements. *IEEE Standard* C371181. 2014.

Chapter 9

Applications of battery energy storage systems for distribution systems

Vinicius C. Cunha[1], Ricardo Torquato[1], Tiago R. Ricciardi[1], Fernanda C.L. Trindade[1], Walmir Freitas[1], Victor Riboldi[2] and Tuo Ji[2]

Distribution energy systems have experienced significant changes over the past few years, with an increase in the monitoring level and the integration of new agents. Distributed energy resources, such as photovoltaic (PV) generators, electric vehicle charging stations, and energy storage systems are examples of these new agents. These devices have the potential to revolutionize the way that electric energy is produced, transported, distributed, and consumed. Additionally, there is continuous pressure to fully utilise the assets of electric power systems to maximise the profits related to the heavy investments made in this sector.

In this context, this chapter presents applications developed for battery energy storage systems of different sizes, which are: small, deployed mostly in residential and commercial customers; medium, deployed mostly in industrial customers and low voltage (LV) distribution systems; and large, deployed mostly in medium voltage (MV) distribution systems and distribution substations. The results presented in this chapter are based on field experiences from a Brazilian distribution utility.

9.1 Introduction

Applications of battery energy storage systems are classified according to the scale of these devices. The main characteristics that define the available scales of energy storage systems on the market are presented in this section.

Energy storage systems are generally specified by their rated power and energy capacity. Small-scale energy storage systems are primarily designed for residential and commercial customers due to their small energy capacity. They are usually specified as up to 10 kW rated power and up to 15 kWh energy capacity.

[1]School of Electrical and Computer Engineering, University of Campinas, Brazil
[2]CPFL Energia, Brazil

For medium-scale energy storage systems, the rated power varies from 25 to 75 kW, which can be sustained for up to 3 h. Then, the capacity of these devices is specified in the order of tens to hundreds of kWh (e.g., from 25 to 225 kWh). These energy storage systems tend to be installed in industrial customers connected to the MV systems or along the LV systems.

For large-scale energy storage systems, the rated power varies typically from 300 to 2,000 kW, which can be sustained for up to 3 h. Then, the capacity of these devices is specified in the order of hundreds to thousands of kWh (e.g., from 600 to 6,000 kWh). These energy storage systems tend to be installed in distribution substations (power transformers), as well as in MV feeders, as the typical loading of a feeder is in the same range of these power capacities.

As energy storage systems have the potential to benefit the operation and planning activities in distribution energy systems, utilities, and third-party companies have developed pilot projects to further study applications of these devices. For example, in Australia [1], a large-scale energy storage system specified with 30 MW of rated power and 8 MWh of energy capacity was deployed to assist in the integration of 90 MW of wind generation. The American utility ComEd developed a pilot project with energy storage systems of 25 kWh to improve the supply of customers that undergo constant interruptions due to weather conditions [2]. The American utility San Diego Gas & Electric installed two energy storage systems with specifications of 500 kW of rated power and 1,500 kWh of energy capacity in two substations [3].

In 2010, the USA Department of Energy started the Pacific Northwest Grid Demonstration Project (Smart Grid Demo), a USD 179 million program with a duration of 5 years [4]. This program was led by the Pacific Northwest Laboratory (PNNL) in partnership with other entities, including 11 utilities from five different states. The Portland General Electric (PGE) distribution utility, which supplies 830 thousand customers in 52 cities in Oregon state, took part in this program through the Salem Smart Power Centre (SSPC) project. In this centre, there is a large-scale energy storage system with specifications of 5 MW rated power and 1.25 MWh energy capacity that is still operating currently, although the project was completed in 2015. The main purpose of this energy storage system is to reduce the peak demand, avoid or mitigate energy interruptions, especially for critical customers, perform energy arbitrage, and assist the intermittent energy injection of renewable generation.

The applications of energy storage systems presented in this chapter are based on results obtained from R&D projects developed in Brazil. Due to this, not only the computational simulations but also the real-world system data and measurements are utilised in these studies. Computational simulations are executed using OpenDSS and Python. The OpenDSS used for the simulations runs on both Windows and Linux, depending on the study executed, through the DSS Python module [5], complemented by custom C++ code that uses the DSS C-API library [5] directly.

The rest of this chapter is divided as follows. In Section 9.2, the applications of small-scale, behind-the-meter battery energy storage systems are presented. In

Section 9.3, applications of medium-scale battery energy storage systems are shown. In Section 9.4, applications of large-scale battery energy storage systems are presented. In Section 9.5, the most common applications of each battery energy storage system scale are summarized.

9.2   Application of small behind-the-meter battery energy storage systems

In this section, the benefits of small-scale, behind-the-meter battery energy storage system deployment in distribution systems are investigated. These benefits are assessed from the perspectives of customers and distribution utilities considering scenarios with high penetration of distributed energy resources, such as PV generation. In Section 9.2.1, a detailed analysis is presented for three real-world low voltage systems. In Section 9.2.2, the previous analysis is expanded for a complete real-world feeder.

9.2.1   Detailed analysis for LV systems

In this section, three real-world LV systems from a Brazilian utility are considered as test systems. In these systems, residential customers already have installed PV generators, as detailed in Table 9.1. For the following study, the deployment of energy storage systems on these customers is evaluated.

In Table 9.1, the total capacity of PV generators in kWp, the rated power of distribution transformers (i.e., MV/LV distribution transformers), and the number of customers are shown. Two definitions of PV penetration are considered. The first one, Penetration I, is the ratio of PV generation capacity to the rated power of the distribution transformer. The second one, Penetration II, is the ratio of the number of PV generators to the number of customers in the system.

The individual capacity of the PV generators in the three LV systems is shown in Table 9.2. Note that most of the PV generators have small, rated power capacity, meaning that the PV generation is distributed in the LV system.

Table 9.1   Description of PV generator characteristics for the three LV systems evaluated

System	Number of PV generators	PV generator capacity (kWp)	MV/LV transformer (kVA)	Number of customers	Penetration I (%)	Penetration II (%)
System 1	14	41.5	45	36	92	39
System 2	5	13.0	15	14	87	36
System 3	7	23.0	30	18	77	39

Table 9.2   Rated power of PV generators located in the three LV systems

System	Number of PV generators	PV generators rated power (kWp)			
		2	3.5	5	8
System 1	14	8	5	0	1
System 2	5	4	0	1	0
System 3	7	3	2	2	0

The analysis of small-scale battery energy storage systems is made by assessing three cases, as described below:

1. LV systems without PV generators and energy storage systems (no PV and no BESS).
2. LV systems with PV generators and without energy storage systems (with PV and no BESS).
3. LV systems with PV generators and energy storage systems (with PV and BESS).

In case 3, the energy storage systems are located at the same customers that have PV generators. They are specified as 3.6 kWp rated power and 12 kWh energy capacity. The energy storage systems are operated to shift the energy production by the PV generator from the off-peak period to the peak period. This characteristic makes the customers less dependent on the distribution system.

The comparison among the proposed cases allows, first, to identify the technical impacts caused by the presence of PV generators in the low voltage systems ('No PV and no BESS' and 'With PV and no BESS' cases). Afterwards, these technical impacts are mitigated by including small-scale energy storage systems in customers ('With PV and BESS' case).

After defining the case studies, additional considerations for the simulations conducted are described. The simulations are executed for 20 weeks considering a time step of 15 min. This time step is chosen because it is the same one defined for the customer demand profiles provided by the distribution utility. The irradiance curve used is based on field measurements with 1-min resolution. The irradiance profile is aggregated to the 15-min resolution.

Due to the presence of PV generators, the main issue that occurs in the tested systems is the violation of the upper voltage limit defined by regulation. These voltage violations represent the main aspect that limits the increase in PV penetration in distribution systems [6]. Additionally, the increase in technical losses and tap operations of voltage regulators may become some of the other potential impacts caused by the increase in PV penetration.

The rise in voltage magnitude is basically due to the power injection in the circuit, and it becomes more significant when the PV generation exceeds the power required by customers. In this scenario, energy storage systems can be used to limit the power injected into the LV system, which consequently mitigates the voltage rise.

Table 9.3 Description of PV generator characteristics for the three LV systems evaluated

System	Number of customers	Penetration I (%)	Penetration II (%)	Customers affected by precarious voltage transgressions (%)		
				No PV and no BESS	With PV and no BESS	With PV and BESS
System 1	36	92	39	0	100	0
System 2	14	87	36	0	86	7
System 3	18	77	39	0	83	0

To exemplify the impact of PV generation on the voltage level on customers, the limits adopted in these studies are the ones established by the Brazilian regulation [7], i.e., the voltage magnitude has to be kept between 92% and 105% of its rated value in LV systems. This voltage range is an adapted version of the ANSI definition named as range A (i.e., between 95% and 105%) [8]. The Brazilian regulation defines that if the voltage magnitude violates these limits for more than 3% of a specified measurement period, a precarious voltage transgression is flagged in the assessed customer. In this case, the distribution utility must financially compensate the customers affected due to the improper supplied voltage quality and fix this situation by a given deadline. Although another voltage range is defined in the Brazilian regulation, called the critical range, only the precarious range is considered in this chapter to illustrate the application of energy storage systems.

After defining the voltage limits adopted, the number of customers impacted by precarious voltage transgressions is shown in Table 9.3. As expected, the case without PV generation does not present any customer with voltage transgression. However, the inclusion of PV generation leads to a high number of customers affected by voltage transgressions (100% of the customers with voltage violations for system 1). Note that the number of customers with voltage transgressions correlates better with the Penetration I definition than with the Penetration II definition. This means that the PV power capacity installed in LV systems tends to better describe the voltage transgression impacts than the number of PV generators.

For the case with PV generators and energy storage systems, i.e., 'With PV and BESS', the number of customers impacted is zero for systems 1 and 3. The total elimination of the voltage transgressions occurs even in the situation where all customers of the low voltage system are impacted. When complete elimination of the voltage transgressions is not possible, e.g., system 2, observe that it is highly mitigated, reducing from 86% to 7%.

After a general discussion about the number of customers impacted by voltage transgressions, system 1 is thoroughly evaluated regarding the energy storage system performance. This system is chosen because it has 100% of customers affected by voltage transgressions in the 'With PV and no BESS' case and this issue is eliminated after including energy storage systems in customers.

The PV generators located in System 1 are distributed in six with phase-to-phase connections between phases A and B, one between phases B and C, and seven between phases C and A. The rated line voltage of these PV generators is 220 V. The total power capacity is 19.0 kWp for phase A, 11.7 kWp for phase B, and 10.8 kWp for phase C. The capacity is not perfectly balanced among the phases, where the power capacity of phase A slightly exceeds the power capacities of phases B and C. In percentage, the PV capacity installed in each of the three phases A, B, and C is 45.8%, 28.1%, and 26.1%, respectively, considering the total PV capacity installed in System 1.

The behaviour of precarious voltage transgressions for System 1 throughout the studied period is shown in Figure 9.1. In Figure 9.1, each dot represents the phase with the highest number of voltage transgressions per customer in the studied week. The colour differentiation refers to the case considered (i.e., 'No PV and no BESS', 'With PV and no BESS', and 'With PV and BESS'). The dashed line corresponds to the 3% limit of total measurements, which defines the minimum number of measurements to characterise the precarious voltage transgression. Customers above this dashed line have precarious voltage.

Note that in Figure 9.1, the voltage transgression metric for the cases 'No PV and no BESS' and 'With PV and BESS' is below the 3% threshold throughout the 20 weeks of analysis, as already shown in Table 9.3. For the case with the energy storage, 'With PV and BESS', there are few voltage transgressions in 3 weeks, that is, the green dots are not located at zero per cent. However, even in these cases, the number of voltage transgressions is lower than 3% of the total number of measurements, meaning that the voltage supplied to customers remains adequate. On the other hand, for the 'With PV and no BESS' case, all customers present a higher number of measurements than 3% for all 20 weeks, transgressing the adequate voltage limits. This means that the utility must financially compensate them for the

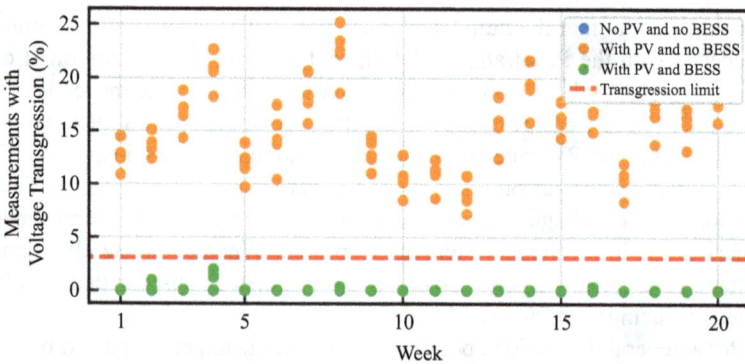

Figure 9.1   *Precarious voltage transgression metric for all customers in System 1 considering 20 weeks of analysis*

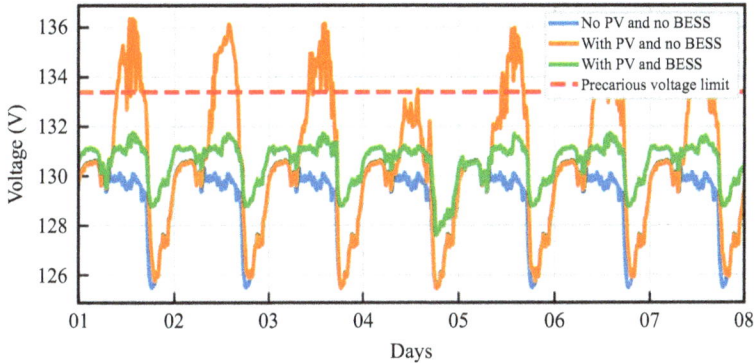

Figure 9.2   Voltage profile of phase A for a customer in System 1

precarious voltage supply and proceed with system reinforcements to fix this situation.

The voltage profile in phase A for a customer is shown in Figure 9.2. This phase is selected because it has the highest installed capacity of PV generation in this system. As expected, the 'With PV and BESS' case mitigates considerably the voltage rise compared to the 'With PV and no BESS' case. In the highest voltages peaks, the energy storage system prevents the voltage magnitude from increasing roughly 5 V (~4%). It represents a sufficient difference to classify the voltage supply as adequate for the case with an energy storage system and precarious for the case with no energy storage system.

Another point to highlight is that the energy storage system reduces the range between the maximum and minimum voltages supplied to the customer. In an operational perspective, a narrow range of voltages supplied to customers requires a smaller number of voltage regulator operations, reducing the maintenance costs of these devices. The range between the maximum and minimum voltages for the case 'With PV and no BESS' is 15 V whereas this range for the case 'With PV and BESS' is 8 V, which represents a reduction of 47%. Thus, energy storage systems tend to assist the operational aspects of distribution systems, reducing, for example, the operational degradation of voltage regulator devices.

9.2.2   General analysis for a complete feeder

In this subsection, the detailed analysis presented for three low voltage systems is generalized for a real-world feeder. The test system used in this study of applications for small-scale battery energy storage systems is shown in Figure 9.3. This feeder is a real Brazilian system, and its rated voltage is 11.4 kV for the MV system. It has three-phase transformers connected in delta on the MV side and wye grounded on the LV side.

The rated line voltage of the secondary systems is 220 V. The short-circuit level at the substation is 152 MVA. There are 2,069 customers connected in the LV

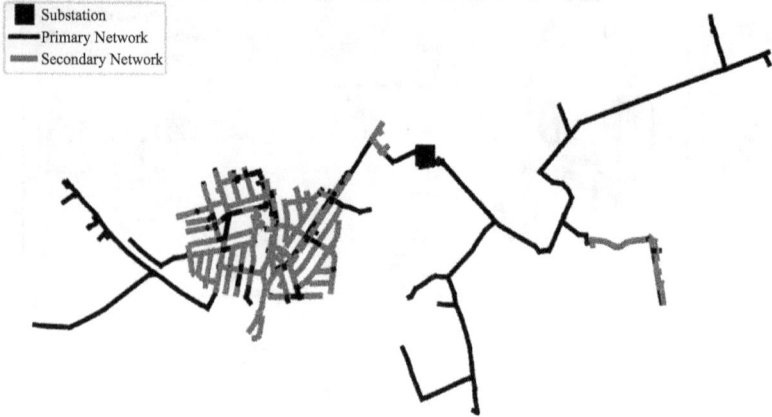

Figure 9.3 Single-line diagram of a real-world feeder used in the case study

systems, where 13% are single-phase (two-wires), 74% are two-phases (three-wires), and 13% are three-phases (four-wires).

In the study of this subsection, all customers are considered eligible to own a PV generator or an electric vehicle. The location of these emergent technologies is selected randomly, following a Monte Carlo method proposed in [9]. The penetration level of these technologies corresponds to the ratio between the number of customers with at least one of these emergent technologies and the total number of customers in the feeder. The penetration of PV generators and electric vehicles is increased from 5% to 50% in steps of 5%, where the following four cases are evaluated:

- **Base case:** There are no energy storage systems in the customers that own a PV generator, an electric vehicle, or both technologies.
- **30% of adoption:** Energy storage systems are included randomly in 30% of the customers who have a PV generator, an electric vehicle or both technologies.
- **50% of adoption:** Energy storage systems are included randomly in 50% of the customers who have a PV generator, an electric vehicle or both technologies.
- **100% of adoption:** Energy storage systems are included in 100% of the customers who have a PV generator, an electric vehicle or both technologies.

The technical impacts assessed in this subsection are voltage magnitude transgressions, as discussed in Section 9.2.1, and the following ones:

- **Voltage unbalance:** The ratio between negative and positive sequence voltages must remain 95% of the time below 2% on all three-phase buses.
- **Conductor overloading:** Line currents must remain below the corresponding conductor rating during at least 95% of the simulation time.

- **Distribution transformer overloading:** It must remain below the transformer-rated power for at least 95% of the simulation time. This is a conservative limit, as transformers are known to operate properly for loading levels up to 150% during reduced time intervals.
- **Technical losses:** Regulations do not establish technical limits for this factor. The technical losses are only compared with the base case.

Similar to the results presented in Section 9.2.1, the deployment of energy storage systems in LV customers considerably mitigates the voltage transgressions, as shown in Figure 9.4. Results are presented by means of bar charts where the bar represents the average result of the Monte Carlo simulation, and the upper whisker represents the 95th percentile of the results.

The voltage transgression impact is illustrated in Figure 9.4 because it has the most severe impact on the system. For example, roughly 50% of the customers have upper voltage transgressions considering the case with 50% penetration. This occurs because the substation voltage level is typically set close to the upper limit, i.e., 1.05 pu, as the traditional distribution systems used to have only LV transgressions (undervoltage). Due to this, even a small increase in the voltage level due to the presence of PV generators is sufficient to transgress the upper voltage limits.

In addition, the other technical impacts assessed are also mitigated with the increasing installation of energy storage systems, as summarized in Table 9.4. The results presented are the 95th percentile considering all scenarios run in the proposed Monte Carlo method. For the technical impacts related to voltage, i.e., voltage magnitude and unbalance, the percentage refers to the number of customers affected. The percentage related to the devices, e.g., transformers and lines, refers to the number of devices affected. The technical losses are shown in energy (kWh), which corresponds to the losses for one day of simulation.

Voltage unbalance and distribution transformer loading are also impacted by the presence of PV generators and electric vehicles, although not as much as the voltage magnitude. The voltage unbalance occurs mostly due to single- and two-phase devices (PV generator or electric vehicle charger) included in the distribution

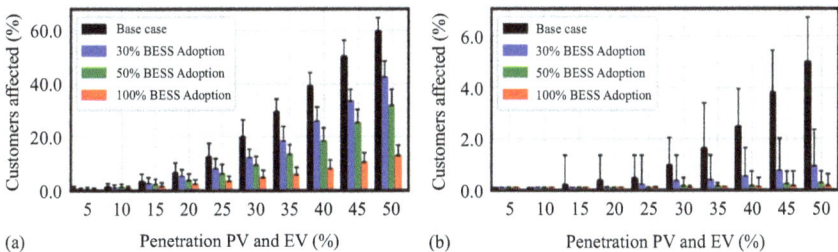

Figure 9.4   Customers affected by (a) upper and (b) lower precarious voltage transgressions

Table 9.4   Assessment of technical impacts considering two cases of PV generators and electric vehicles for four scenarios of battery energy storage system deployment

Technical impact	25% PV generators and electric vehicle penetration (95th percentile)				50% PV generators and electric vehicle penetration (95th percentile)			
	Base Case	30% BESS	50% BESS	100% BESS	Base Case	30% BESS	50% BESS	100% BESS
Upper precarious voltage transgression (%)	16.35	10.2	8.26	4.53	51.14	39.47	33.5	16.43
Lower precarious voltage transgression (%)	1.73	1.35	0.11	0.14	6.79	2.19	1.0	0.72
Voltage unbalance (%)	5.08	7.81	6.0	6.0	17.23	15.23	13.91	12.5
LV conductor overload (%)	2.83	1.83	1.5	1.42	6.39	3.88	2.55	1.97
Distribution transformer overloading (%)	2.63	1.32	1.32	1.32	14.47	5.26	1.32	1.32
Technical losses – MV system (kWh)	156.51	151.79	150.11	151.64	165.0	148.56	141.01	140.64
Technical losses – LV system (kWh)	450.69	406.39	403.15	373.05	603.07	498.8	449.33	383.21
Technical losses – distribution transformers (kWh)	255.44	233.59	226.05	221.0	334.69	267.42	234.45	217.53

system. The overloading of distribution transformers occurs mainly while matching electric vehicle charging and the peak load of the system. Line overload occurs only in the secondary lines, i.e., the primary lines do not experience overload. Technical losses decrease overall as the deployment of energy storage systems increases, with a larger benefit on the LV system and distribution transformers.

9.3   Application of medium battery energy storage systems in LV systems

In this section, the benefits of medium-scale battery energy storage systems deployment in distribution systems are investigated. These benefits are assessed for customers and distribution utility perspectives considering scenarios with high penetration of distributed energy resources, such as PV generation. In Section 9.3.1, a detailed analysis is presented for a real LV system. In Section 9.3.2, a general evaluation is made for a real complete feeder.

9.3.1   Detailed analysis for a LV system

In this case study of a medium-scale energy storage system, the objectives of the device installed at the LV side of an MV/LV distribution transformer are: (1) alleviate the transformer loading, especially at the generation peak (around noon) and load peak (around 6 pm); (2) reduce the solar energy surplus of the neighbourhood that is exported to the MV feeder to also reduce the energy that is imported from the MV feeder. This study is summarized in this subsection, and it is presented in detail in [10].

Figure 9.5 illustrates a schematic of the feeder, where the energy storage system is deployed in the LV side of a distribution transformer. The characteristics of the LV system, where the energy storage system is deployed, are presented in detail. The topology of the LV system is illustrated in Figure 9.6.

This system is connected to the MV primary feeder through a 11.4/0.22 kV Δ-Yg distribution transformer (75 kVA). The total number of customers connected to this LV system is 47, divided into 11 two-phase (three-wires) and 36 three-phase (four-wires) connected. As there are no consumption measurements for these customers, synthetic curves from [11] are used. The energy storage system is specified

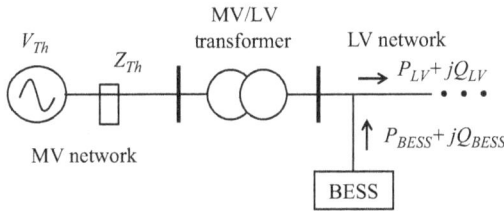

Figure 9.5   Schematic circuit to illustrate the location of the energy storage system

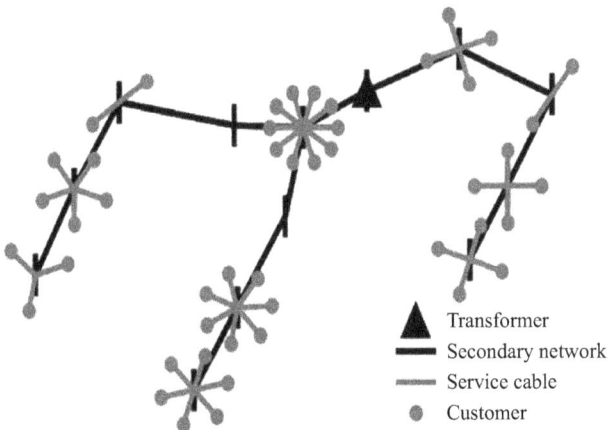

Figure 9.6   Single-line GIS diagram of the real LV system

as 100 kW rated power (either charging or discharging) and 255 kWh energy capacity. To prevent cell damage, the maximum depth of discharge (DoD) is 90%.

The number of customers with PV generators is 27 (i.e., roughly 57%). These customers are selected because they already have PV generators in practice. The nominal PV capacity installed in this system is 67.5 kWp (2.5 kWp per PV system), which corresponds to 90% of the transformer rated capacity. Therefore, this proportion expresses a high PV penetration scenario. The 27 PV generators are monitored in real-time and their power generation is recorded with 1-min resolution. These measurement data are aggregated to a 15-min resolution before their utilisation in the simulations.

9.3.1.1   Description of the energy storage system operation modes

To evaluate the performance of the energy storage system, three operation modes are studied. The first operation mode is the fixed time period. In this case, the energy storage system charges and discharges at pre-fixed periods of the day. This is the simplest operation mode of an energy storage system because it does not require any measurement from the LV system to determine when the energy storage system must operate and with which power. The energy storage system is charged from 9 am to 5 pm and discharged from 5 pm until when the state of charge (SoC) reaches its minimum allowable value. Both the charge and discharge active power settings are fixed. This mode is also known as the time-shift or load-shift.

The second operation mode is called fixed power thresholds. For this operation mode, the discharging operation occurs when the direct power flow through the transformer (i.e., from the MV to the LV system) exceeds a pre-set direct power threshold (P_{th_dir}). On the other hand, the charging operation occurs when the reverse power flow through the transformer (from the LV to the MV system) exceeds a pre-set reverse power threshold (P_{th_rev}). The power injected ($P_{BESS} > 0$) or consumed ($P_{BESS} < 0$) by the energy storage system during these operations is equal to the difference with respect to the threshold. It can be expressed mathematically as follows:

$$P_{BESS}(t) = \begin{cases} P_{LV}(t) - P_{th\_dir}, & if \ P_{LV}(t) > P_{th\_dir} \\ 0, & if \ P_{th\_rev} < P_{LV}(t) < P_{th\_dir} \\ P_{LV}(t) - P_{th\_rev}, & if \ P_{LV}(t) < P_{th\_rev} \end{cases} \quad (9.1)$$

This operation mode requires real-time measurements of active power flow at the BESS connection point ($P_{LV}(t)$). This mode is also known as peak-shaving.

The third operation mode is named as load following. In this operation mode, the active power flow on the MV/LV distribution transformer is supplied completely by the energy storage system, as shown below:

$$P_{BESS}(t) = -P_{LV}(t). \quad (9.2)$$

The energy storage system is supposed to charge whenever the LV system injects active power back to the main grid (PV generation exceeding the LV system load). Similarly, it has to discharge whenever the LV system consumes power from the main grid (LV system load exceeding the PV generation). To be possible, the active power profile of the LV system needs to be measured in real time, so that the energy storage system can accurately follow such a profile.

9.3.1.2 Performance of the energy storage system considering three operation modes

The objective of these studies is to investigate the performance of the three control modes in alleviating transformer peak loading, particularly the reverse power flows (peak shaving), and to prevent solar energy surplus from being injected back into the primary feeder as much as possible. First, without the energy storage system, based on a yearly simulation, the maximum direct peak observed was 81.6 kW and the maximum reverse peak was −60.2 kW. The average daily direct peak is 52.8 kW and the average daily reverse peak is −40.8 kW. Furthermore, 128.3 MWh was consumed (customer loads and system losses) while 90.7 MWh was produced by the 27 PV systems. This resulted in 96.1 MWh of energy being imported from the MV to the LV system, and 58.5 MWh being exported from the LV to the MV system. Notice that without the energy storage system, roughly two-third of the produced solar energy is not directly used within the LV system and it is injected into the MV system.

For the fixed time operation mode, the charging setting is defined as 20 kW flat from 9 am to 5 pm and the discharging setting is defined as 20 kW flat starting at 5 pm until the energy storage system SoC is 25%. In this mode, the yearly maximum direct peak was reduced from 81.6 to 61.2 kW (25%) and the reverse from −60.2 to −50.5 kW (16%). In terms of average daily values, the reduction was from 52.8 to 37.6 kW (29%) and from −40.8 to −23.2 kW (43%). In terms of energy balance, the values of energy consumed (customer loads and system losses), as well as the solar generation are roughly the same as the same profiles were used. However, the volume of energy imported was reduced from 96.1 to 72.5 MWh (25%), and the energy exported was reduced from 58.5 to 21.4 MWh (63%). Due to the 90% round-trip efficiency of the energy storage system and 1% of rated capacity for ancillary services power consumption, 13.5 MWh corresponds to energy storage system losses, equivalent to 15% of the solar energy produced.

For the fixed power threshold operation mode, the charging operation is set whenever the transformer reverse power flow exceeds −11 kW and the discharging operation is set whenever the direct power flow is above 17 kW. The simulation results have shown that only in 140 days (38%) the energy storage system operation successfully keeps the transformer loading within the range [−11, 17] kW. In 174 days (48%), due to low PV generation allied with a tolerance to partially export the solar energy surplus (represented by the −11 kW setpoint) and high peak consumption during the previous evening, the energy storage system did not fully charge. It means that it was not capable of supplying the peak consumption of the LV system, resulting in exceeding the 17-kW direct peak reference. Regarding the reverse peak shaving, in 51 days (14%), it was not possible to limit injections in the MV system below 11 kW, because the energy storage system did not completely discharge on the previous night and exhausted its storage capacity before the reverse power flow reduced below the threshold.

In this mode, the yearly maximum direct peak was slightly reduced to 74.0 kW (9%) and the reverse to −56.7 kW (6%). In terms of average daily values, the reduction was to 30.2 kW (43%) and to −15.1 kW (63%), all compared to the case without energy storage system. In terms of energy balance, the values of imported energy were reduced from 96.1 to 76.3 MWh (21%), and of energy exported were reduced from 58.5 to 27.8 MWh (52%). The energy storage system losses were 10.9 MWh. Note that this mode resulted in a better performance in terms of reducing the average daily values of both peaks, but in a worst performance in terms of reducing the maximum yearly values of both peaks and the volume of solar energy surplus exported, compared with the 'Fixed Time Periods'.

For the load-following mode, the energy storage system can perform better than the previously discussed modes in terms of energy exchanges with the MV system. It occurs because the energy storage system is reasonably oversized in terms of energy capacity for the evaluated LV system. The simulation results have shown that with this operation mode, compared to the case without an energy storage system, the exported energy surplus can be reduced from 58.5 MWh to only 3.3 MWh (i.e., it was reduced from 64.5 to 3.6% of the PV generation). On the other hand, the imported energy has reduced from 96.1 MWh to 54.1 MWh. The energy storage system losses were 13.2 MWh. Regarding the transformer peaks, with this operation mode, the yearly maximum direct peak was slightly reduced to 74.0 kW (9%) and the reverse peak to −48.4 kW (20%). In terms of average daily values, the reduction was to 40.5 kW (23%) and to −7.7 kW (81%), all compared to the case without an energy storage system. Although it has not resulted in a greater direct peak reduction (the previous methods showed better performance), this mode has proved to be the best approach to reduce the reverse peak over the year (20%), to reduce the average daily reverse peak (81%) and to reduce both the exported and imported energies between the LV system and the primary feeder.

9.3.2   General analysis for a complete feeder

Industrial customers that have installed a medium-scale battery energy storage system planned to obtain financial and operational benefits from this device. Due to that, the distribution utility has no control on the operation of these devices, meaning that the utility can be passively benefitted or impacted by their operation. In this context, this subsection investigates the technical impacts of medium-scale energy storage systems installed in customers connected to MV distribution systems.

For this study, 14 industrial or commercial customers connected to a real MV system are elected to have an energy storage system. The operation of the energy storage system is defined by the tariff contracted by the customer, that is, the energy storage will charge during the cheapest period of the day and discharge during the most expensive period of the day. Three types of tariffs that exist in the Brazilian regulation [12] are considered: (a) conventional: fixed tariff values throughout the day for demand and energy consumption; (b) green: fixed value for demand and two values for energy (a higher tariff for peak hours, which are between 6 pm and 9 pm, and a lower tariff for off-peak hours, which are the rest of the day); and (c) blue: two values for demand and energy.

The energy storage systems are specified based on the consumption characteristics of the customer to which it is connected. The nominal power of the energy storage system is defined as 25% of the contracted demand of the customer. The energy capacity of this device is designed so that it can maintain its rated power for 3 h. For example, a customer with contracted demand of 100 kW will have an energy storage system with a specification of 25 kW rated power and 75 kWh energy capacity.

The tariff definition of the 14 customers evaluated is summarized in Table 9.5. Note that roughly 71% of the customers have a green tariff, where there are two energy prices, one for peak hours and another for off-peak hours, and the demand price is fixed in only one value for these two periods.

Figure 9.7 illustrates the operation of an energy storage system in a customer with a green tariff. In this tariff, the cost of energy in the peak period is 3.8 times higher than in the off-peak period. Due to this, the energy storage system operates to charge during the off-peak period and discharge the total energy demanded by the customer during the peak period.

The analysis of the results for the feeder is based on the demand profiles and on metrics related to them, such as daily peak demand and load factor. The load factor (F_{LF}) is defined as the ratio of the mean active power (P_{mean}) to the maximum

Table 9.5 Distribution of customers according to the defined tariff

Tariff	Conventional	Green	Blue
Number of customers	4 (28.6%)	10 (71.4%)	0 (0.0%)

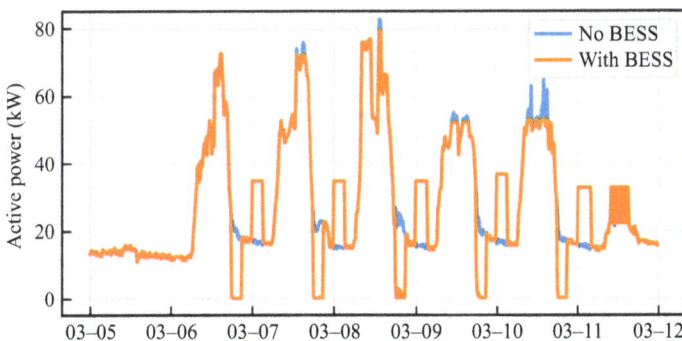

Figure 9.7 Demand profile of a customer with and without an energy storage system, considering the green tariff for the customer

demand ($P_{\max}$) in a given location of the feeder (e.g., transformer, customer, substation, etc.). It can also be understood as an expression of how much energy was used in a time period (E_t), versus how much energy would be used if the peak power demand was sustained during the same time period ($P_{\max}t$) [13]. It is mathematically formulated as follows:

$$F_{\mathrm{LF}} = \frac{P_{\mathrm{mean}}}{P_{\max}} = \frac{E_t}{P_{\max}t}. \tag{9.3}$$

Figure 9.8 presents the demand profile of the feeder considering the cases with and without energy storage systems in the 14 customers.

As shown in Figure 9.9, there is at least 1% peak demand reduction in 175 days. The average value of peak demand reduction is 4.6% and the maximum value is 12%. In 22 days, the feeder demand considering the case with energy storage systems in the customers increase by at least 1%. In these days, the peak demand

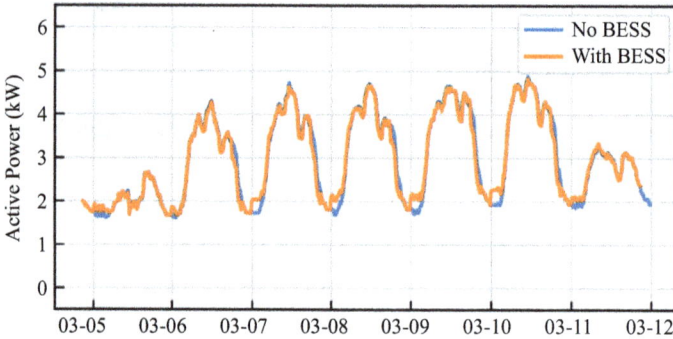

Figure 9.8   *Demand profile of the feeder with and without energy storage systems in 14 evaluated customers*

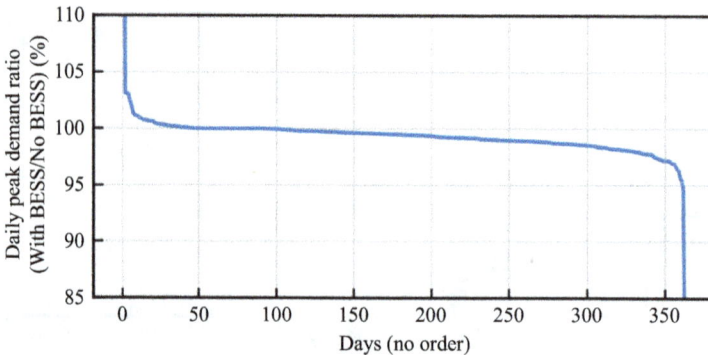

Figure 9.9   *Ratio of the daily peak demand of the feeder considering the cases with and without energy storage systems in 14 evaluated customers*

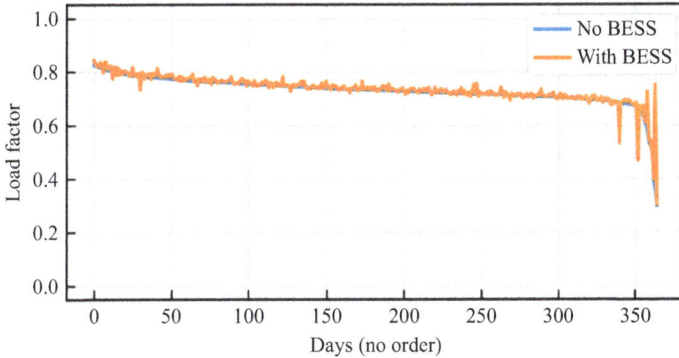

Figure 9.10   Daily load factor of the feeder considering the cases with and without energy storage systems in 14 evaluated customers

increases after the peak period, occurring from 9 pm to midnight due to the charging operation of the energy storage systems.

Figure 9.10 shows the impact of the energy storage systems operation on the daily load factor of the studied feeder. As on most of the days (297 days of the year), the daily peak demand reduces and the average demand increases due to the operation of the energy storage systems, the daily load factor overall increases. The average increase in the daily load factor is 3.2%, reaching a value as high as 20.5% for the daily evaluation.

9.4   Application of large battery energy storage systems in distribution feeders and substations

In this section, the benefits of large-scale battery energy storage system deployment in distribution systems are investigated. A large-scale battery energy storage system with a rated power of 1 MVA and energy capacity of 2 MWh is tested in a feeder to improve the demand profile (Section 9.4.1), backup power supply for off-peak demand (Section 9.4.2), and energy arbitrage (Section 9.4.3). Furthermore, a theoretical application of energy storage systems for voltage sag mitigation using simple charts is presented in Section 9.4.4.

9.4.1   Improvement of the demand profile of a feeder

To evaluate the performance of a large-scale battery energy storage system, the time operation mode is selected. In this operation mode, the energy storage system is configured to charge from 10 pm to 5 am with a constant charging power of 257 kW. It is assumed that the available energy capacity of the energy storage system for charging and discharging is 90% of its rated energy capacity (2 MWh), as 10% represents the emergency reserve. The discharging period is set from 6 am to 6 pm with a constant discharging power of 150 kW. The demand profile of the test feeder

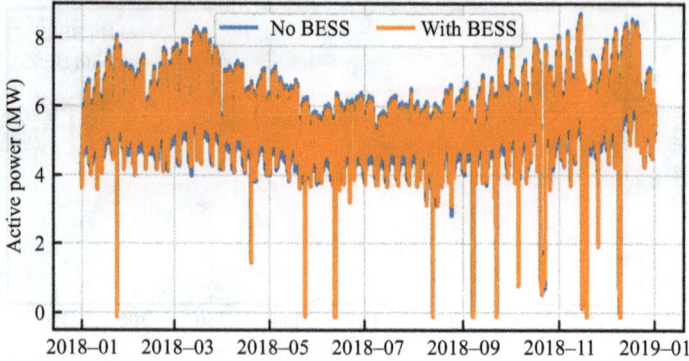

Figure 9.11   Yearly demand profile of the test feeder considering the cases with and without the energy storage system

with and without the operation of the energy storage system is shown in Figure 9.11. As expected, the daily peak demand is reduced throughout the year, more specifically in 327 days (89.6%) with a daily average reduction of 129 kW. The maximum peak demand of the feeder is reduced by 134 kW, from 8.7 MW to 8.57 MW (a reduction of 1.5%).

One advantage of the time operation mode is the constant peak reduction throughout the year. This is possible because the demand profile of a feeder has minor variations from day to day, making it less difficult to predict the best periods for charging and discharging. The demand profile is relatively stable even across different seasons. The peak demand reduces in 162 out of 212 days in summer/spring (76.4%) and in 109 out of 153 days (71.2%) in winter/autumn.

Two additional metrics are evaluated to verify the improvement in the demand profile of the feeder, that is, the load factor (10.3) and the dispersion factor. The dispersion factor is calculated by the difference between the maximum and minimum demand of the feeder, normalized by its average demand, as shown below:

$$F_{disp} = \frac{P_{max} - P_{min}}{P_{med}}. \tag{9.4}$$

The dispersion factor is expected to be low, ideally close to zero.

The operation of the energy storage system in the time operation mode increases the load factor and reduces the dispersion factor of the feeder. In this yearly evaluation, the feeder presented a large load factor, varying daily between 0.7818 and 0.8970 in 90% of the time, with an average value of 0.8328. As the energy storage system is included in the analysis, the daily average value of the load factor increases by 2.1%. This increase is practically constant for the statistical values between the 5th and 95th percentiles, roughly 2%.

For the dispersion factor, the feeder already had a set of values relatively small, between 0.2718 and 0.7921 in 90% of the time, with an average of 0.4531. The presence of the energy storage system reduced the dispersion factor by 11.8% considering the daily average value of this metric.

9.4.2 Backup supply for off-peak demand

If the active power demand of the feeder is lower than the rated power of the energy storage system (1 MW), the battery may be utilised to completely supply the feeder. In this scenario, this subsection evaluates whether the energy storage system can supply both the active and the reactive power demand of a feeder. For example, this application isolates the feeder from the utility in terms of active power demand, which can help the utility to execute a maintenance activity.

To completely supply the feeder for a certain period, the energy storage system is operated in a load following mode. In this operation mode, the energy storage discharges to totally supply the feeder during a pre-defined period. When not operating in the discharge mode, the storage system is charged at a fixed rate. Note that the charging of the storage system is also avoided during the peak demand period of the feeder.

For example, Figure 9.12 illustrates a feeder completely supplied for certain periods by an energy storage system. This energy storage system is operated in the load-following mode. The discharging period takes place during the light loading condition of the feeder, that is, when the active power demand of the feeder is lower than 750 kW. In this case, the energy storage system completely supplies the active power demand of the system between 1:30 am and 4:00 am. Due to this, the demand profile shows a period of zero demand in the case with the energy storage system. The energy storage system is charged in two periods, that is, before and after the discharge event. It occurs to avoid the energy storage system from charging during the peak load of the feeder. The total period in which the energy storage system is able to sustain the complete feeder with a 750-kW demand is 2.5 h.

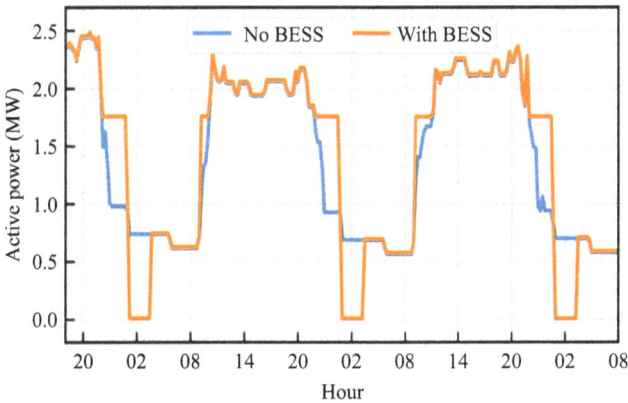

Figure 9.12   Demand profile of a feeder with an energy storage system operating in a load following mode

9.4.3   Energy arbitrage

The energy arbitrage consists of the operation mode of the energy storage system that is controlled by the price of the energy. It means that the energy storage system charges when the energy price is reduced and discharges when the energy price is increased. In this subsection, the tariff mode considered is the green tariff, as explained in Section 9.3.2, meaning that two values for the cost of energy are considered. The highest cost value is defined in the peak demand period, that is, between 6 pm and 9 pm and the lowest cost value is defined in the off-peak demand period, that is, between 9 pm and 6 pm.

To illustrate the energy arbitrage, Figure 9.13 shows the demand profile of a feeder with and without an energy storage system. Although the demand price is fixed in the green tariff, there are two contracted demands motivated by the peak and off-peak energy price (4.2 MW and 3.8 MW, respectively).

In Figure 9.13, observe that the charging period of the energy storage system does not affect the peak demand of the feeder. The charging period allows the utility to financially benefit from the energy storage system, as the energy costs are higher during the peak period between 6 pm and 9 pm. Note that the period with the highest energy cost does not necessarily match the peak demand of this analysed feeder as the definition of peak and off-peak periods for tariff purposes comprises a wide power system area.

In Table 9.6, energy consumption during the peak period is reduced by 69 MWh, which represents 11.3% of the energy consumed in the 77 days analysed. During the off-peak period, the energy consumed increases by 92 MWh. The energy increased during the off-peak period is higher than the energy reduced during the peak period mostly because of the losses in the energy storage system. In terms of costs, the percentage variations due to storage system installation have the same order of magnitude as the energy variations. The increase in costs during the off-peak period is 1.7% and the decrease in costs during the peak period is 11.5%. This leads to a total energy cost mitigation of 2.1%.

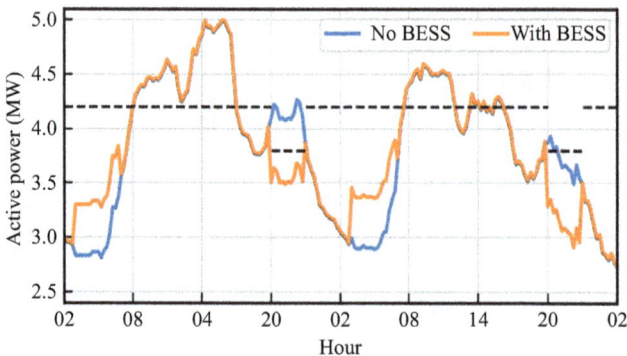

Figure 9.13   Demand profile of a feeder with an energy storage system operating in the energy arbitrage mode

Table 9.6   Summary of the energy consumed and energy costs during peak and off-peak periods

Case	Energy off-peak (MWh)	Energy peak (MWh)	Cost energy off-peak (USD)	Cost energy peak (USD)	Cost energy (USD)
No BESS	5,427	610	303,000	129,200	432,200
With BESS	5,519 (+1.7%)	541 (−11.3%)	308,000 (+1.7%)	115,000 (−11.5%)	423,000 (−2.1%)

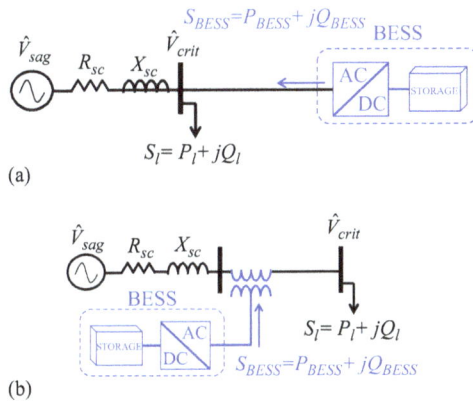

Figure 9.14   (a) Shunt and (b) series connection for an energy storage system

9.4.4   Voltage sag mitigation

This subsection presents a practical approach to determine whether a shunt/series energy storage system can mitigate voltage sags, assuming system information, such as short-circuit impedance and historical voltage sags, and load information is available. This is done by analysing a chart obtained analytically, which can be easily constructed based on simple information from the system and load, without resorting to computer simulations. This complete practical approach is developed in [14].

Figure 9.14 presents an equivalent circuit with a critical bus in evidence. This circuit consists of a system bus, the system short-circuit impedance and a load. The system bus voltage is $\hat{V}_{sag}$ and it represents the upstream system. The load represents a critical process that cannot be interrupted during voltage sags, so that its terminal voltage must be maintained above pre-defined limits when the upstream system experiences voltage sags. The installation of an energy storage device, connected either in parallel or in series with the load, is investigated as a means of providing sufficient power, so that the voltage at the critical bus ($\hat{V}_{crit}$) can ride through voltage sag events without causing interruption in the load process. The idea consists of proposing a practical guide where utility and industry engineers can easily determine whether a given energy storage system has sufficient capacity to meet the voltage requirements of the critical load. This guide is based on the following equations for shunt and series connections of the energy storage system [14], respectively,

$$S_{BESS}(\theta) = V_{crit}^2 \frac{(R_{th}\cos\theta + X_{th}\sin\theta) - \sqrt{(R_{th}\cos\theta + X_{th}\sin\theta)^2 - (R_{th}^2 + X_{th}^2)\left(1 - \frac{V_{th}^2}{V_{crit}^2}\right)}}{(R_{th}^2 + X_{th}^2)},$$

(9.5)

$$S_{BESS}(\theta) = |I_l|^2 \left\{ [(R_{sc} + R_l)\cos\theta + (X_{sc} + X_l)\sin\theta] - \sqrt{[(R_{sc} + R_l)\cos\theta + (X_{sc} + X_l)\sin\theta]^2 - A} \right\},$$

$$A = 4\frac{V_{crit}^2}{V_{th}^2}\frac{1}{|I_l|^4}\left\{\frac{V_{crit}^2}{V_{th}^2}\left[(R_{sc} + R_l)^2 + (X_{sc} + X_l)^2\right] - (R_l^2 + X_l^2)\right\}$$

(9.6)

In the equations above, R_{th} and X_{th} are the Thévenin equivalent resistance and reactance, respectively, obtained from the parallel connection between the short-circuit impedance ($R_{sc} + jX_{sc}$) and the load equivalent impedance ($R_l + jX_l$). The load rated current is I_l. V_{crit} is the minimum voltage that must be sustained so that the load is not disconnected from the circuit.

Four case studies (two with shunt connection and two with series connection) are analysed to verify the accuracy and applicability of the proposed charts and formulas. To validate the proposed formulation, these curves are also obtained through repetitive power flow simulations in the DSS Extensions implementation of the OpenDSS software [5]. The four case studies are listed below:

- Case 1: S_{sc} = 500 MVA, S_l = 1.0 MVA, and shunt BESS with capacity up to S_{ESS} = 35 MVA is considered viable.
- Case 2: S_{sc} = 100 MVA, S_l = 1.0 MVA, and shunt BESS with capacity up to S_{ESS} = 10 MVA is considered viable.
- Case 3: S_{sc} = 100 MVA, S_l = 1.0 MVA, and series BESS with capacity up to S_{ESS} = 0.3 MVA is considered viable.
- Case 4: S_{sc} = 100 MVA, S_l = 4.0 MVA, and series BESS with capacity up to S_{ESS} = 1.2 MVA is considered viable.

In the four cases, the data collected for the first step of the practical guide correspond to V_{base} = 27.6 kV, X_{sc}/R_{sc} = 1.0, pf_l = 0.8 inductive (X_l/R_l = 0.75), V_{sag} = 0.58 pu and V_{crit} = 0.70 pu.

The power capacity considered for each energy storage system candidate is first verified to check if it satisfies the minimum capacity requirement indicated by the proposed approach, i.e., if their rated power is higher than the minimum required. The minimum rated power required in Case 1 is 42 MVA. Thus, the 35 MVA energy storage system is not a viable solution for voltage sag mitigation. On the other hand, the minimum rated power required in Case 2 is 8.4 MVA and the minimum rated power required in Cases 3 and 4 are 0.17 MVA and 0.69 MVA,

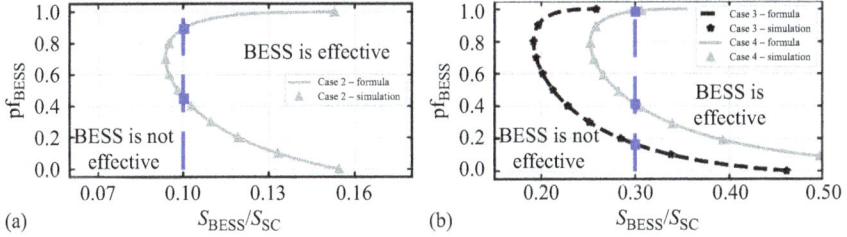

Figure 9.15 Proposed charts to determine the characteristics of energy storage systems that mitigate voltage sags

respectively. Therefore, the energy storage systems available in these three cases satisfy the minimum requirement and, as such, should be studied in more detail.

The final step is to analyse the Power Factor practical chart, which can quickly indicate to utility engineers what the power factor range is in which the storage should be operated so that the system can ride through the voltage sag under study. These, curves shown in Figure 9.15, are obtained analytically by using (10.5) (for Case 2 – shunt) and (10.6) (for Cases 3 and 4 – series). Observe that Power Factor curves obtained from the proposed formulas are similar to the results obtained in the simulations. This confirms that the proposed method enables engineers to assess if storage can ride through voltage sags only by using a few formulas and charts, without needing to run any computer simulation.

9.5 Summary of applications

The small-scale energy storage systems have the potential to be deployed in residential and commercial customers, as these devices have a small energy capacity. However, the applications of these small-scale devices can benefit not only their owners but also the distribution utility indirectly. The utility tends to have a passive behaviour considering the small-scale energy storage systems because the customers are the owners of these devices and operate them to maximize their own interests. The most common applications and benefits of small-scale energy storage systems are listed below:

- Energy arbitrage (shift the consumption considering the energy cost).
- Loading reduction of equipment and conductors.
- Reduction in technical losses.
- Mitigation of rise/drop of voltage magnitude in a steady-state condition.
- Mitigation of voltage fluctuations due to the presence of PV generation.
- Peak shaving, that is, reduction of the peak loading level.
- Shift in demand time.
- Increase in the load factor of the equipment of the distribution system.

As opposed to the scenario presented for small-scale energy storage systems, utilities tend to have a more active behaviour in acquiring and operating medium-scale energy storage systems. Third-party agents may also have a share of these devices if the economic and regulatory aspects allow. The most common applications and benefits of medium-scale energy storage systems are listed below:

- Energy arbitrage (shift the consumption considering the energy cost).
- Voltage support.
- Reduction in technical losses.
- Mitigation of rise/drop of voltage magnitude in a steady-state condition.
- Backup energy for non-programmed interruptions.
- Peak shaving, that is, reduction of the peak loading level.
- Postponement of investments in the system.
- Increase in the load factor of the equipment of the distribution system.

In the case of large-scale energy storage systems, the installation and operation are mostly done by the utilities. It occurs because these systems tend to be deployed near the distribution substation due to their size. The most common applications and benefits of large-scale energy storage systems are listed below:

- Energy arbitrage (shift the consumption considering the energy cost).
- Voltage support.
- Reduction in technical losses.
- Mitigation of rise/drop of voltage magnitude in steady-state condition.
- Backup energy for non-programmed interruptions.
- Feeder/substation peak shaving, that is, reduction of the peak loading level.
- Postponement of investments in the system.
- Increase in the feeder/substation transformer load factor.
- Postponement or reduction in the electric energy supply contracts.
- Ramping support for loads or renewable energy generation.
- Mitigation of voltage sag.

Acknowledgements

This study was financed in part by the Coordenação de Aperfeiçoamento de Pessoal de Nível Superior – Brasil (CAPES) – Finance Code 001, by the Brazilian National Council for Scientific and Technological Development (CNPq) (Process 304373/2020-6), by São Paulo Research Foundation (FAPESP) (Processes 2017/10476-3, 2020/10523-4) and by CPFL Energia under the grant ANEEL R&D program PD-00063-3047/2018.

References

[1] ElectraNet. *Boosting Reliability on Lower Yorke Peninsula*, 2022. https://www.electranet.com.au/electranets-battery-storage-project/ [Accessed 17 Feb 2022].

[2]   Business Wire. *ComEd Conducting Illinois First Community Energy Storage Pilot*, 2017. https://www.businesswire.com/news/home/20170316005291/en/ComEd-Conducting-Illinois%E2%80%99-Community-Energy-Storage-Pilot [Accessed 17 Feb 2022].

[3]   SDG&E. *Energy Storage Implementation*, 2014. https://energy.gov/sites/prod/files/2014/06/f17/EACJune2014-4Bialek.pdf [Accessed 17 Feb 2022].

[4]   Pacific Northwest. *Smart Grid Technology Performance*, 2015. https://www.smartgrid.gov/document/Pacific_Northwest_Smart_Grid_Technology_Performance.html [Accessed 17 Feb 2022].

[5]   P. Meira and D. Krishnamurthy. *DSS Extensions: Multi-platform OpenDSS Extensions*, 2022. https://dss-extensions.org/ [Accessed 17 Feb 2022].

[6]   R. Torquato, D. Salles, C. O. Pereira, P. C. M. Meira, and W. Freitas, A comprehensive assessment of PV hosting capacity on low-voltage distribution systems, *IEEE Transactions on Power Delivery*, vol. 33, no. 2, pp. 1002–1012, 2018.

[7]   ANEEL. *Proceedings of Distribution Energy Systems in the National Power Systems – PRODIST – Module 8*, 2022. https://www.aneel.gov.br/modulo-8 [Accessed 24 Feb 2022] (Portuguese)

[8]   ANSI. *ANSI C84.1*, 2006. https://webstore.ansi.org/ [Accessed 24 Feb 2022].

[9]   V. C. Cunha, R. Torquato, T. R. Ricciardi, W. Freitas, and B. Venkatesh, Assessing energy storage potential to facilitate the increased penetration of photovoltaic generators and electric vehicles in distribution systems, in *2017 IEEE Power & Energy Society General Meeting*, 2017, pp. 1–5.

[10] C. O. Pereira, V. C. da Cunha, T. Ricciardi, *et al.*, Pre-installation studies of a BESS in a real LV system with high PV penetration, in *2019 IEEE PES Innovative Smart Grid Technologies Conference – Latin America (ISGT Latin America)*, 2019, pp. 1–6.

[11] R. Torquato, Q. Shi, W. Xu, and W. Freitas, A Monte Carlo simulation platform for studying low voltage residential systems, *IEEE Transactions on Smart Grid*, vol. 5, no. 6, pp. 2766–2776, 2014.

[12] ANEEL. *Proceedings of Tariff Regulation*, 2022. https://www.aneel.gov.br/procedimentos-de-regulacao-tarifaria-proret [Accessed 24 Feb 2022] (Portuguese).

[13] Austin Energy. *Understanding Load Factor*, 2022. https://austinenergy.com/wcm/connect/8fe76160-0f73-4c44-a735-529e5c7bee61/understanding LoadFactor.pdf?MOD=AJPERES&CVID=kC1aR-I#:~:text=Load%20factor%20is%20an%20expression,over%20a%20period%20of%20time [Accessed 24 Feb 2022]

[14] V. C. Cunha, F. C. L. Trindade, and B. Venkatesh, A practical method to assess the potential of energy storage systems to mitigate voltage sags, *Electric Power Systems Research*, vol. 201, 107525, 2021.

Chapter 10

An extended hosting capacity approach including energy storage

Sicheng Gong[1], Vladimir Ćuk[1], Tiago Castelo de Oliveira[1] and J.F.G.(Sjef) Cobben[1]

This chapter proposes an evolved concept of "hosting capacity" using the term of "feasible region" for installing additional loads or generations. Through converting the grid model into a more compact one, "hosting capacity region" not only is promising to further exploit the grid potential for power delivery but also benefits grid operation feasibility investigation with concise formulas. Facing the derived hosting capacity, originally complicated energy storage optimization problems can be represented algebraically, which is more efficient and friendly for computer processing. Case study based on a 10.5 kV Dutch grid has been implemented, eventually demonstrating the validity of relevant assessment and optimization methods.

10.1 Concept of hosting capacity

Referring to [1,2], "hosting capacity" is a term to quantify the acceptable capacity of loads or generations in certain profiles, for integration to specific points of connection (POCs) in an existing grid. It is an essential index in industrial practices to help distribution system operator (DSO) allocate a suitable integration capacity for an energy unit, which is in the waiting list for installation. As illustrated in Figure 10.1, originally designed for distributed generations or loads, "hosting capacity" is treated as a maximal capacity according to some performance index thresholds.

Unfortunately, not all grid performance indices can keep monotonic to the unit integration amount. Instead, those indices are typically not in linear relation to the power of the connected installation [3]. Moreover, with a rising penetration rate of intermittent energy units, energy storage systems (ESSs) and prosumers are needed for grid regulation, which can play roles of generation and load simultaneously, asking for a span over the whole real number field [4,5]. A single cut-off value can

[1]Electrical Energy Systems Group, Eindhoven University of Technology, The Netherlands

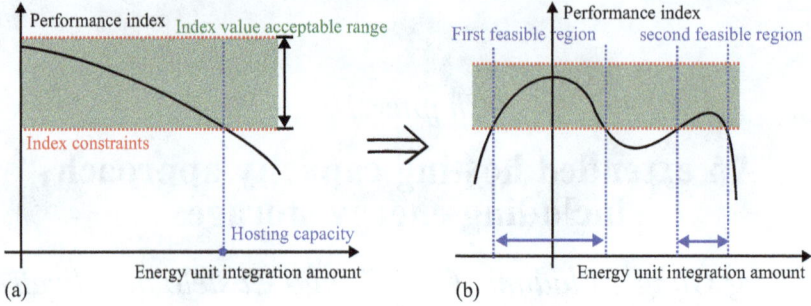

Figure 10.1   *Conceptual evolution of hosting capacity. (a) Monotomic-index-based concept. (b) Generic-index-based concept.*

no longer explicitly describe capacity constraints in these scenarios, naturally stimulating the conceptual evolution of "hosting capacity."

As illustrated in Figure 10.1, facing a non-monotonic vertical-axis-crossing plot, "hosting capacity" can be represented by two disjoint segments. Therefore, intuitively, we take the "feasible region" term as a suitable tool to boost the conceptual evolution of "hosting capacity." Considering more POCs and performance indices investigated, the essence of "hosting capacity" seems more apparent for us, which should be seen as a combinational feasible region of integration capacity, indicating that one connection becomes dependent on another connection (or more than one).

10.1.1   Performance index clarification and simplification

In advance of detailed discussions on how such concept evolves, we need turn to its original format, figuring out respective performance indices that deserve our attentions. Technically, hosting capacity determination should involve in voltage level, cable (or transformer) capacity, harmonic distortion, voltage dip, etc. [6,7]. These indices seem complicated and mutually influenced, while we can simply categorize them into two aspects: power quality and facility limit. The former one is defined by grid codes and the latter one refers to facility datasheets.

For the convenience to further exploit the essence of hosting capacity, we mainly investigate the impacts of voltage level and cable capacity on hosting capacity. Such simplification makes sense as this book is talking about energy storage, while harmonic distortions and voltage dips will involve in filters and controllers, eventually overextending our scope. Moreover, instead of considering stochastic unit profiles or temperature-dependent cable parameters, a restrained concept of "hosting capacity" is investigated, which is derived based on a deterministic scenario [8].

10.1.2   Hosting capacity region definition

So far, we still have a basic question: how to represent the so-called "hosting capacity," through several scalar value thresholds or combinational constraining cuts? Even for only one POC, its integration power can be divided into two aspects: active power and reactive power, and there definitively exists mutual relationship between their thresholds. With a raised number of involved POCs, which is denoted as n, the hosting capacity needs to be represented by $2n$ variables, including each POC's integrated active and reactive power. Therefore, we let "hosting capacity" evolve into "hosting capacity region," which is defined to be the combinational feasible region of these $2n$ variables. Definitively, in practices, we can use less dimensions if part of variables are assumed constant.

Unlike "dynamic hosting capacity" in [8,9] to expand hosting capacity in a time-series domain, "hosting capacity region" extends the original concept to a higher-dimension domain. This evolved concept can bring several benefits for DSO, not only extending the grid operation state space but also converting the grid model into a more computer-friendly one. More illustrations are provided as follows.

10.1.2.1   Grid power delivery potential exploitation

Starting from the 2-dimensional (2D) perspective, we can look into a combinational feasible region of active power unit integration over two POCs. As shown in Fig. 10.2 (a), the orange parallelogram denotes their respective feasible region boundary, which is derived in Section 10.2.2.4. Here we use the results in advance to illustrate how evolved "hosting capacity" exploits the grid potential to deliver more power.

Based on the previous "hosting capacity" concept, the DSO will set separate upper and lower thresholds to both installations, and the corresponding feasible region will become a rectangular denoted by blue lines. It can be seen that some state space is dropped accordingly. With Bus 1 injecting more active power, Bus 2 can absorb more active power without violating any voltage or current limits. The original concept covers up such possibility. Instead, the combinational "hosting capacity region" will cover all possible scenarios, allowing the grid to deliver more power in some extreme scenarios.

In Figure 10.2(b), a similar phenomenon can be observed when focusing on a single POC. By using the results at Section 10.2.1.2 in advance, the combinational

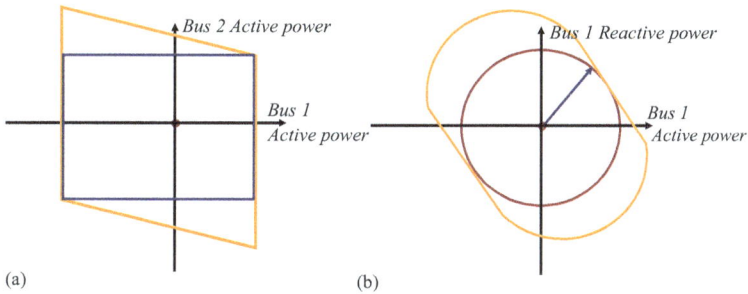

Figure 10.2   Hosting capacity region cases. (a) Active power over 2 POCs. (b) Active and reactive power over single POC.

feasible regions are denoted by the orange boundary. Still using the original concept, the DSO will set a scalar value constrain on apparent power over such POC, leading to a round feasible region and operation space loss. Both cases in Figure 10.2 provide a vivid demonstration on how such new "hosting capacity region" exploits the grid power delivery potential.

10.1.2.2   Grid problem format transformation

In conventional grid research works, the grid model itself commonly suffers a high data volume, as it needs to include nominal voltage/capacity value, cable/transformer parameters, grid topology, and certain energy unit profiles. It may cause inefficiency for model processing. Even for a simple operation scenario, the computer has to run iterative power flow calculation codes and check relevant constraints. When we use such model for further investigations, frequent power flow calculation and quantities of operation scenarios will cause a high hurdle to power engineers to realize quick and efficient solving.

Meanwhile, through derived hosting capacity region, the original grid model can be put aside when we investigate operation feasibility problems, for instance minimal energy storage size. Since such region can be easily processed by advanced computational algebra technology, we successfully transform the previous complicated problems into less computing-intensive ones. Moreover, the hosting capacity region contains less information to hinder retrieving grid parameters and existing loads. The DSO can share this region to relevant POC operators, encouraging their combinational operation while without extra grid and user data privacy concerns.

Although we can use power flow tools to solve as many scenarios as possible in advance in a brute-force way, intended for a resolution-limited feasible region, an analytical solution to the region directly is still preferred. Such solution can be efficient and easy to repeat, especially when we change some grid parameters or installations for a new consideration of the same grid.

10.1.3   *Hosting capacity region with ESS*

Referring to a derived hosting capacity region, it unlocks flexibility-based concepts (e.g. market-based one) for the DSO and aggregators, through setting combinational connection capacity limits to POCs. Meanwhile, in practices, some users are willing to legitimately request a larger connection capacity even with higher connection expenses. The DSO cannot fudge such request. Instead of physically upgrading grid cables or transformers, energy storage equipment installation is a promising alternative to satisfy such request, especially facing high power intermittent energy units, for instance emerging ultra-fast electric-vehicle (EV) charging infrastructure systems in urban areas.

Geometrically, with ESS integrated, the relevant hosting capacity region is equivalently reshaped. As illustrated in Figure 10.3, still looking into a 2D hosting capacity region, with installing bidirectional active power regulation equipment (maximum regulation power is denoted as P_r) over the same POC, the region is reshaped accordingly. The operation point is covered by the updated hosting

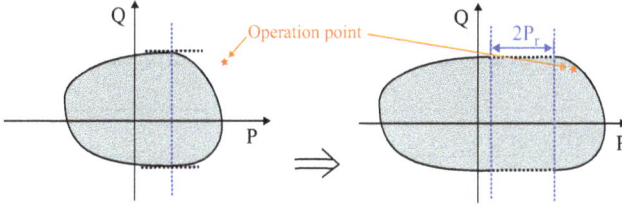

Figure 10.3   Hosting capacity region reshaping with ESS

capacity region, indicating such operation status becomes acceptable after ESS integration. The derived minimal value of P_r will be a solid reference for DSO when deploying relevant ESS.

In later parts of this chapter, we will discuss on how to assess the hosting capacity region, and how ESS explicitly imposes impacts on such region. It can be seen from results that the evolved concept of "hosting capacity" will further benefit DSO to optimally design grid integration and regulation capacity.

10.2   Hosting capacity region assessment

10.2.1   Distribution grid model

In this chapter, we start with a limitation on radial distribution systems as shown in Figure 10.4. According to [10], DistFlow model is used to mathematically describe the distribution grid model as given in (10.1). Corresponding variable definitions in (10.1) are provided in Table 10.1. For further illustration, the coupling between bus injection power and line transmission power is denoted by (10.1a) and (10.1b). The relationship among bus voltage level, line current level and line transmission power are written in (10.1c) and (10.1d). Equations (10.1e) and (10.1f) are variable constraints defined by unit profiles, grid codes, and cable datasheets:

$$P_{ij} = \sum_{k:(j,k)\in E} P_{jk} + r_{ij}l_{ij} - P_j \tag{10.1a}$$

$$Q_{ij} = \sum_{k:(j,k)\in E} Q_{jk} + x_{ij}l_{ij} - Q_j \tag{10.1b}$$

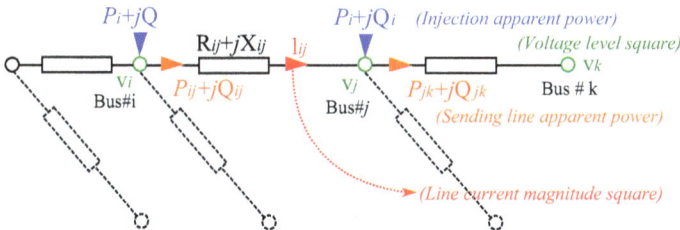

Figure 10.4   Distribution grid schematic

Table 10.1   General notations in DistFlow model

Index	Meaning
Constant parameters	
$E/(j,k)$	Grid graph/connection between node j and node k
r_{ij}/x_{ij}	Line resistance/reactance between node i and node j
P_i^{min}/P_i^{max}	Minimum/maximum of active power in node i
Q_i^{min}/Q_i^{max}	Minimum/maximum of reactive power in node i
v_i^{min}/v_i^{max}	Minimum/maximum of voltage level square on node i
l_{ij}^{max}	Maximum of current magnitude square from i to j
Decision variables	
P_j	Equivalent injection active power in node j
Q_j	Equivalent injection reactive power in node j
v_j	Square of voltage level over node j
l_{ij}	Square of current magnitude from node i to node j
P_{ij}	Injected active power from node i to node j
Q_{ij}	Injected reactive power from node i to node j

$$v_j = v_i - 2(r_{ij}P_{ij} + x_{ij}Q_{ij}) + (r_{ij}^2 + x_{ij}^2)l_{ij} \tag{10.1c}$$

$$l_{ij} = \frac{P_{ij}^2 + Q_{ij}^2}{v_i} \tag{10.1d}$$

$$v_i^{min} \le v_i \le v_i^{max} \tag{10.1e}$$

$$l_{ij} \le l_{ij}^{max} \tag{10.1f}$$

$$P_i^{min} \le P_i \le P_i^{max} \tag{10.1g}$$

$$Q_i^{min} \le Q_i \le Q_i^{max} \tag{10.1h}$$

Although there are a quantity of decision variables to be solved in a deterministic operation scenario, we only care about the feasibility region of P_i and Q_i, since only these variables are concerned when deriving the hosting capacity region. In advance of introducing technical details for hosting capacity region assessment, preliminary analysis on DistFlow model will be implemented.

However, you may still own a basic question, regarding the cable capacitance ignorance in DistFlow model. There are two ways to reconsider cable capacitance when Π-model is adopted. The first one is to figure out an equivalent constant reactive power appliance over each POC, as each POC voltage level is constrained and we can ignore such deviations. The second one is to create more POCs with zero voltage level and without any energy unit integrated. Those POCs are connected to original POC with pure capacitive lines. It will definitely increase our computation burden when deriving the hosting capacity region, while no accuracy sacrifice is caused.

10.2.1.1 Ergodic testing

Instead of mathematical processing for hosting capacity region assessment with high-performance computers, we can still get a rough view through ergodic testing. Although an unconstrained quantity of testing scenarios may not be very elegant for an engineer to answer this question, they still deserve in the initial stage to help find some clues.

Normally, when discussing about hosting capacity, P_i and Q_i are assumed determined, then numerical methods are adopted to solve (10.1a)–(10.1h). Through checking whether l_{ij} and v_i meet (10.1e) and (10.1f), the respective operation point will be evaluated. However, in this part, we try to look into the same problem reversely, and implement ergodic testing with v_i and l_{ij} varying. With known v_i and l_{ij}, P_i and Q_i can be calculated directly as shown in (10.2), where power flow calculation can be avoided with a direct solution. Through solving quadratic equations, we can derive several feasible operation points at one time:

$$v_j = v_i - 2(r_{ij}P_{ij} + x_{ij}Q_{ij}) + (r_{ij}^2 + x_{ij}^2)l_{ij} \tag{10.2a}$$

$$\Rightarrow Q_{ij} = aP_{ij} + b, \, a \leftarrow \frac{-r_{ij}}{x_{ij}}, \, b \leftarrow \frac{v_i - v_j + (r_{ij}^2 + x_{ij}^2)l_{ij}}{2x_{ij}} \tag{10.2b}$$

$$\Rightarrow P_{ij}^2 + (aP_{ij} + b)^2 - v_i l_{ij} = 0 \, \left(\text{considering } l_{ij} = \frac{P_{ij}^2 + Q_{ij}^2}{v_i}\right) \tag{10.2c}$$

$$\Rightarrow P_{ij}, Q_{ij} \text{ are solved by root formula} \tag{10.2d}$$

$$\Rightarrow P_j \leftarrow \sum_{k:(j,k)\in E} P_{jk} + r_{ij}l_{ij} - P_{ij}, \, Q_j \leftarrow \sum_{k:(j,k)\in E} Q_{jk} + x_{ij}l_{ij} - Q_{ij} \tag{10.2e}$$

Meanwhile, even keeping the same sampling resolution on v_i and l_{ij}, the total testing scenario number will increase exponentially with grid scale raising. In some practices, it is unnecessary for us to derive the combinational hosting capacity region of all POCs in the grid. Therefore, instead of (10.2), we can keep the conventional method of ergodic P_j and Q_j testing, especially when the expected hosting capacity region is in less dimensions.

10.2.1.2 Non-convex hosting capacity region

The introduced ergodic testing method can be employed in a simple grid case as shown in Figure 10.5. Such grid is composed of a slack bus, a receiver bus and a piece of power cable. According to grid codes, the receiver bus owns its voltage level flexibility of 10%.

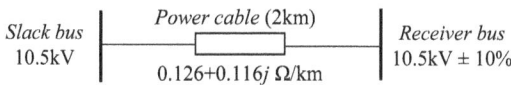

Figure 10.5   Schematic of a simple grid case

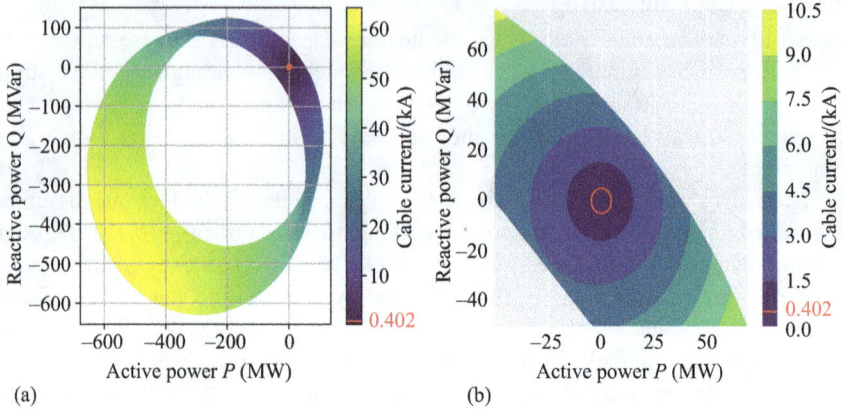

Figure 10.6   Hosting capacity region of a simple grid case. (a) Overall region. (b) Zoomed region.

Without power cable current limits, taking the receiver bus as a POC, its respective hosting capacity region is given in Figure 10.6(a), where horizontal and vertical axis indicate active and reactive power separately. The colorbar illustrates corresponding cable current level distribution in such region. Geometrically, the hosting capacity region is non-convex. That sounds not as good news, as non-convexity will commonly cause extra obstacles when we try to numerically solve some geometrical problems. Fortunately, current level constraints are inevitable in industrial practices. In this example, according to cable datasheet, it current level threshold is equal to 402 A. We highlight practical feasible region by red contours in Figure 10.6(a), which is seen as a small red dot.

Moreover, we can zoom the low-current part in Figure 10.6(a) to be Figure 10.6(b). Through drawing contouring lines, we can find the hosting capacity region can shrink into a convex one with current level threshold decreasing. Regarding a practical 402 A threshold, it is low enough to ensure a convex respective hosting capacity region.

Now, you may be confused why we so desire the convexity. To answer it, we will give some brief explanations here, and more details will be provided later in coming sections. As shown in Figure 10.7, the region benefits from convexity due to the feasibility guarantee for the connection line between two existing feasible points. Based on three feasible points, we can derive a feasible planar triangle region accordingly. With emerging known feasible points, the derived feasible region comes more and more close to the original one, and the derived one can be easily identified by a limited number of linear constraints, which is more friendly for numerical processing. That explains why we so desire the convexity.

In summary, the hosting capacity region defined by DistFlow cannot be ensured convex, while such convexity can still be met in some specific scenarios.

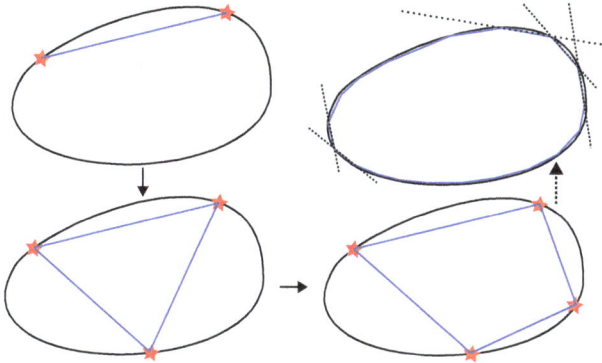

Figure 10.7 Convex region linearization

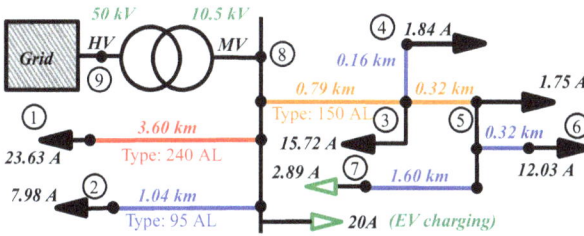

Figure 10.8 Testing case schematic

The convexity can benefit our future hosting capacity region exploitation, and we should take it seriously.

10.2.1.3 Testing case explanation

In this work, we will use a branch of standard Dutch MV grid case as a common testing case in coming sections. The grid schematic has been given in Figure 10.8, where original loads share the same power factor of 0.98. The power flow tool is based on Pandapower library [11]. Gurobi solver is also employed to help solve optimization problems in this chapter [12]. All computation is finished on a computer with an Intel i7-9750H processor running at 2.60 GHz using 15.8 GB of RAM, running Windows 10 Enterprise version.

Bus 9 is assumed a slack bus. The EV charging station is connected to Bus 8 and its nominal charging current is set to 20 A. Meanwhile, considering bidirectional charging converter technology and "Vehicle to Grid" policy, through taking Bus 8 as a specific POC, we will try to exploit its hosting capacity region. Bus 7 is selected as well to assess their combinational feasible region. Such region will provide an important reference for converter capacity redesigning and policy adjustment. Cable and transformer impedance are listed in Table 10.2, which considers high-temperature operation for a conservative result.

Table 10.2   Power cable and transformer parameters

Object	Resistance (70 °C)	Reactance (50 Hz)
240 AL cable	163 mΩ/km	83 mΩ/km
150 AL cable	265 mΩ/km	99 mΩ/km
95 AL cable	412 mΩ/km	113 mΩkm
36 MVA transformer (50 kV/10.5 kV)	0.0022 pu	0.065 pu

10.2.2   Linearized DistFlow model

10.2.2.1   Model explanation

Our intuitive to deal with such imperfect reality is rational simplification. In accordance with [10], a linearized DistFlow model has been proposed as shown in (10.3). Compared to (10.1), we use some cancelling symbols to represent how such linearized model is derived. Especially, since l_{ij}-related items have been ignored in (10.3a)–(10.3c), (10.3d) can be rephrased as an inequality constraint directly. Equation (10.3d) is a nonlinear while convex constraint. The reason why we still name such model "Linearized DistFlow Model" is that all equality constraints become linear now.

Since all constraints are convex, regarding (10.3), its respective feasible region is theoretically convex. Such convexity can powerfully support our exploitation on hosting capacity region as stated above. Moreover, such linearization makes senses when l_{ij} is comparatively low. As shown in Figure 10.6(b), the region is naturally convex with low l_{ij} in a simple grid case.

$$P_{ij} \simeq \sum_{k:(j,k)\in E} P_{jk} + r_{ij}l_{ij} - P_j \tag{10.3a}$$

$$Q_{ij} \simeq \sum_{k:(j,k)\in E} Q_{jk} + x_{ij}l_{ij} - Q_j \tag{10.3b}$$

$$v_j \simeq v_i - 2(r_{ij}P_{ij} + x_{ij}Q_{ij}) + (r_{ij}^2 + x_{ij}^2)l_{ij} \tag{10.3c}$$

$$l_{ij} \leq l_{ij}^{max}, \quad l_{ij} = \frac{P_{ij}^2 + Q_{ij}^2}{v_i} \leq l_{ij}^{max} \tag{10.3d}$$

10.2.2.2   Heuristic convex hull algorithm

The basic idea of convex region exploitation has been explained above. Meanwhile, it still has not explained how we can define those boundary points efficiently. There are several classic convex hull determination algorithms, and many researchers have investigated a lot in this topic [13,14]. However, it seems irrational to employ these algorithms directly, as they are designed for a group of known points. In our case, instead of randomly choosing testing points, we can manage how testing points are generated, and we can use such capability to realize heuristic region assessment.

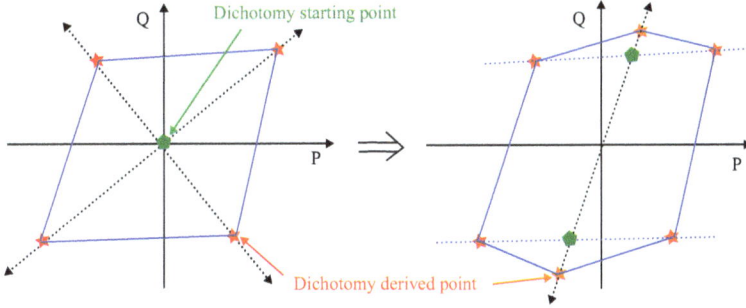

Figure 10.9   2D hosting capacity region exploitation

Regarding the scalar value boundary exploitation, Dichotomy method in Algorithm 10.1 is intuitive and efficient. Based on such algorithm, we can continue from a simple case, where the hosting capacity region is limited in 2D. The corresponding heuristic exploitation flow chart has been given in Figure 10.9. Through selecting two initial linear trajectories, we can determine four boundary points by employing Algorithm 10.1. In the initial stage, all lower-bound starting points are set zero point, as an initial status without any extra energy unit integration is assumed acceptable. Later on, we will select another direction, while the lower-bound starting point in Algorithm 10.1 can be updated as the conjunction point between the connection line and this new linear trajectory. By iteratively running Algorithm 10.2, the region will expand continuously. Through converting these connection lines as linear constraints, the hosting capacity region will be algebraically represented.

When a boundary pair in one direction is derived, it still needs to figure out how to connect these lines. How we can heuristically calculate cutting lines determines our assessment efficiency. As shown in Figure 10.9, with two more

---

Algorithm 10.1: Dichotomy algorithm

Data: Lower-bound starting point S_l, upper-bound starting point S_u
Result: Derived boundary point S

1 $S_a \leftarrow S_l$, $S_b \leftarrow S_u$;
2 Set stopping criteria ε;
3 **repeat**
4    $S \leftarrow \frac{S_a + S_b}{2}$;
5   Run grid simulation programs for testing;
6   **if** *S is acceptable* **then**
7       $S_a \leftarrow \frac{S_a + S_b}{2}$;
8   **else**
9       $S_b \leftarrow \frac{S_a + S_b}{2}$;
10 **until** $|S_b - S_a| \leq \varepsilon$;

Algorithm 10.2: Linearized Hosting Capacity Region Exploitation in $\mathbf{R}^2$

Data: Orthogonal vector basis set $\mathbf{V} = \{\mathbf{v}_1, \mathbf{v}_2\}$
Result: Half space set $\mathbf{H}$

1 Set stopping criteria ε;
2 Run Dichotomy to find bounds $\alpha_{1+} \cdot \mathbf{v}_1$ and $\alpha_{1-} \cdot \mathbf{v}_1$;
3 $\mathbf{v}_{1+} \leftarrow \alpha_{1+} \cdot \mathbf{v}_1$, $\mathbf{v}_{1-} \leftarrow \alpha_{1-} \cdot \mathbf{v}_1$;
4 Run Dichotomy to find bounds $\alpha_{2+} \cdot \mathbf{v}_2$ and $\alpha_{2-} \cdot \mathbf{v}_2$;
5 $\mathbf{v}_{2+} \leftarrow \alpha_{2+} \cdot \mathbf{v}_2$, $\mathbf{v}_{2-} \leftarrow \alpha_{2-} \cdot \mathbf{v}_1$;
6 Initialize Sequence $\mathbf{Q} \leftarrow [\, \mathbf{v}_{1+}, \mathbf{v}_{2+}, \mathbf{v}_{1-}, \mathbf{v}_{2-} \,]$;
7 **repeat**
8   Initialize integer number $i \mathbin{\text{¡}} \text{size}(\mathbf{Q})$;
9   Select the ith and $(i+1)$th element in $\mathbf{Q}$, denoted by $\mathbf{v}_i$, $\mathbf{v}_{i+1}$;
10   (if $i+1 > \text{size}(\mathbf{Q})$, $\mathbf{v}_{i+1} \leftarrow \mathbf{v}_1$)
11   Initialize $\mathbf{v}_0$ as a average of $\mathbf{v}_i$ and $\mathbf{v}_{i+1}$;
12   Run Dichotomy to find bounds $\alpha_{0+} \cdot \mathbf{v}_0$ and $\alpha_{0-} \cdot \mathbf{v}_0$;
13      $\mathbf{v}_{0+} \leftarrow \alpha_{0+} \cdot \mathbf{v}_0$, $\mathbf{v}_{0-} \leftarrow \alpha_{0-} \cdot \mathbf{v}_0$;
14   **if** $i > \text{size}(\mathbf{Q})/2$ **then** j$\leftarrow$ i-size($\mathbf{Q}$)/2+1;
15   **else** j$\leftarrow$ i+size($\mathbf{Q}$)/2+1;
16   Insert $\mathbf{v}_{0+}$ between $\mathbf{v}_i$ and $\mathbf{v}_{i+1}$ in $\mathbf{Q}$;
17   Insert $\mathbf{v}_{0-}$ between $\mathbf{v}_j$ and $\mathbf{v}_{j+1}$ in $\mathbf{Q}$;
18 **until** $\text{size}(\mathbf{Q}) \geq \varepsilon$;
19 Initialize half space set $\mathbf{H} \leftarrow \emptyset$;
20 **for** $\mathbf{v}_i$ *in* $\mathbf{Q}$ **do**
21   Use $\mathbf{v}_i$ and $\mathbf{v}_{i+1}$ to generate half space h_i;
22   (if $i+1 > \text{size}(\mathbf{Q})$, $\mathbf{v}_{i+1} \leftarrow \mathbf{v}_1$)
23   Add h_i to $\mathbf{H}$;

---

boundary points (orange) generated, top and bottom lines in the left quadrangle should be replaced by four other lines in the right hexagon. So we can use a list to denote the connection sequence of all points. When a new pair is derived, we can insert them appropriately to the previous list. Inspired by such principle, Algorithm 10.2 is written down to illustrate the whole procedure of 2D hosting capacity region assessment.

10.2.2.3   Region correction

Since the linearized DistFlow model acquires convexity with the sacrifice of its accuracy through Algorithm 10.2, it is logical to understand and even correct such region, to ensure that all working points would not exceed any limits. In other words, we should avoid letting the linearized and accurate region overlap, and the linearized one must keep totally covered by the accurate one.

From the perspective of active power, (10.3a) is recycled and we repeat it as (10.4) for better reading experience. Compared to the accurate model, the linearized one provides an under-estimator of injection power P_j:

$$P_{ij} \simeq \sum_{k:(j,k)\in E} P_{jk} + r_{ij}l_{ij} - P_j \tag{10.4}$$

Through using $\widehat{P}_j$ to denote the linearized one, it can be rewritten as

$$\widehat{P}_j = P_j - r_{ij}l_{ij} \le P_j \tag{10.5}$$

Although l_{ij} is dynamic in various scenarios, considering l_{ij} is naturally constrained, we can still derive

$$\widehat{P}_j \le P_j \le \widehat{P}_j + r_{ij}l_{ij}^{\max} \tag{10.6}$$

Therefore, if C and C are used to denote the exact and linearized hosting capacity region, we can conclude that

$$\bigcap_{(i,j)\in E} C(C + r_{ij}l_{ij}^{\max})(C + x_{ij}l_{ij}^{\max}) \subseteq R \tag{10.7}$$

As soon as P_j is selected as a dimension of C, $C + r_{ij}l_{ij}^{\max}$ denotes a new region generated by shifting C along the axis of P_j by $r_{ij}l_{ij}^{\max}$, otherwise it can be taken as a copy of C directly. $C + x_{ij}l_{ij}^{\max}$ shares a similar definition. Therefore, we can employ region shifting and intersection following the principle of (10.7).

Moreover, referring to (10.3), we can write (10.8), where P_l, Q_l, P_n and Q_n denote vectors of line/node active/reactive power. M is a matrix whose entity is defined to be 1 or -1, as long as its column index is the parent or child node of its row index in the grid. M can be proved nonsingular in a radial network through mathematical induction methods, which is saved in this chapter. Therefore, the mapping from the feasible region of P_n and Q_n, to that of P_l and Q_l is a bijection. In plain words, C can be assessed through the feasible region of P_l and Q_l:

$$MP_l = P_n , MQ_l = Q_n \tag{10.8}$$

Since the slack bus voltage is assumed constant, using (10.9b), we can calculate line current level between such slack bus and its child bus, based on determined P_{ij} and Q_{ij}. Afterwards, using (10.9a), child bus voltage level can be derived through known l_{ij}, P_{ij}, and Q_{ij}. Through induction rules, we can conclude that fixed P_l and Q_l will determine values of all l_{ij} and v_j eventually. Combined with the conclusion in last paragraph, although aiming for feasible region of P_n and Q_n, that of P_l and Q_l should be more concerned when evaluating assessment performance:

$$v_j = v_i - 2(r_{ij}P_{ij} + x_{ij}Q_{ij}) + (r_{ij}^2 + x_{ij}^2)l_{ij} \tag{10.9a}$$

$$l_{ij} = \frac{P_{ij}^2 + Q_{ij}^2}{v_i} \tag{10.9b}$$

In (10.9), the term of $(r_{ij}^2 + x_{ij}^2)l_{ij}$ is still ignored, with an underestimation of the voltage level. The impact should be considered. Let $\mathcal{D}$ denote feasible region of $\mathbf{P}_l$ and $\mathbf{Q}_l$. It is fortunate that any point in $\mathcal{D}$, can ensure derived v_j of each node keeps lower than $v_i^{\min}$. Meanwhile, v_i may exceed $v_i^{\max}$. Moreover, l_{ij} is overestimated, indicating the corresponding l_{ij} based on the accurate model keeps lower than $l_{ij}^{\max}$.

Therefore, in advance of utilizing (10.7) for region shifting and intersection, we need to also use another simplified convex model (10.10) and run Algorithm 10.2, which provides over-estimated v_i and ensures the corresponding v_i from the accurate model keeps lower than v_i^{max}. Intuitively, to ensure all voltage and current level well constrained, we can intersect $\mathcal{D}_1$ and $\mathcal{D}_2$, which denote $\mathbf{P}_l$ and $\mathbf{Q}_l$ feasible region derived from (10.3) and (10.10) separately:

$$P_{ij} = \sum_{k:(j,k)\in E} P_{jk} - P_j \tag{10.10a}$$

$$Q_{ij} = \sum_{k:(j,k)\in E} Q_{jk} - Q_j \tag{10.10b}$$

$$v_j = v_i - 2(r_{ij}P_{ij} + x_{ij}Q_{ij}) + (r_{ij}^2 + x_{ij}^2)l_{ij}^{max} \tag{10.10c}$$

$$P_{ij}^2 + Q_{ij}^2 \le v_i l_{ij}^{max} \tag{10.10d}$$

$$v_j^{min} \le v_j \le v_j^{max} \tag{10.10e}$$

Considering bijection relationship between $\mathcal{C}$ and $\mathcal{D}$, let $\mathcal{C}_1$ and $\mathcal{C}_2$ denote the hosting capacity region derived by (10.3) and (10.10) separately. It is concluded that constrained voltage and current levels are guaranteed for any $\mathbf{P}_n$ and $\mathbf{Q}_n$ point in $\mathcal{C}$, which is calculated as

$$\mathcal{C} = \mathcal{C}_1 \mathcal{C}_2 \tag{10.11}$$

Eventually, with recycling original constraints as (10.12), the corrected region is ensured totally covered by the accurate one, eventually becoming acceptable as stated above. Figure 10.10 is given to summarize and illustrate the whole region

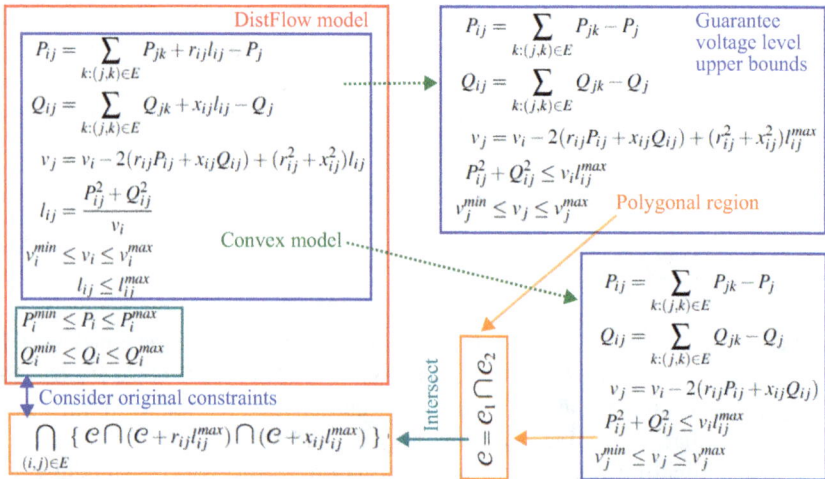

Figure 10.10 Hosting capacity region correction flow chart

correction progress through a flow chart:

$$P_j^{\min} \leq P_j \leq P_j^{\max}$$
$$Q_j^{\min} \leq Q_j \leq Q_j^{\max}$$
(10.12)

10.2.2.4   Case study

Using the testing case in Figure 10.8 at Section 10.2.1.3, with only active power inspected, we can derive the hosting capacity region regarding Buses 7 and 8. Through running Algorithm 10.2, Figure 10.11(a) is derived, where purple ones are boundary points derived from (10.3), and red ones are from original load flow model. Without region correction, these two regions overlap. Similarly, referring to (10.10), the purple boundary points are derived as shown in Figure 10.11(a) with over-estimated voltage level results. It still can be seen that these two regions overlap.

As shown in Figure 10.12(a), the light blue and light red polygons are hosting capacity regions with under-estimated and over-estimated voltage levels. The grey part is the intersection of both regions, which is adopted as the updated hosting capacity region referring to (10.11). The good news is that this updated one is almost totally covered by the accurate one (red points). Moreover, through zooming the boundary part, we can still figure out such covering relationship. Eventually, through employing (10.7) afterwards, the final hosting capacity region, denoted by bronze part in Figure 10.12(a), is zoomed and confirmed covered by accurate boundary points (red). Moreover, through computing the ratio of final assessed region to the accurate one, which is up to 92.07% in this case, the performance of proposed hosting capacity region assessment technology has been further illustrated.

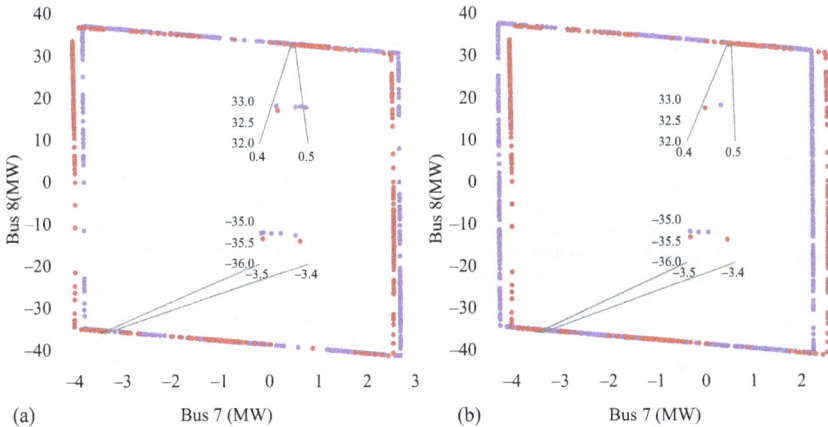

Figure 10.11   Hosting capacity region boundary points. (a) With under-estimated voltage levels. (b) With over-estimated voltage levels.

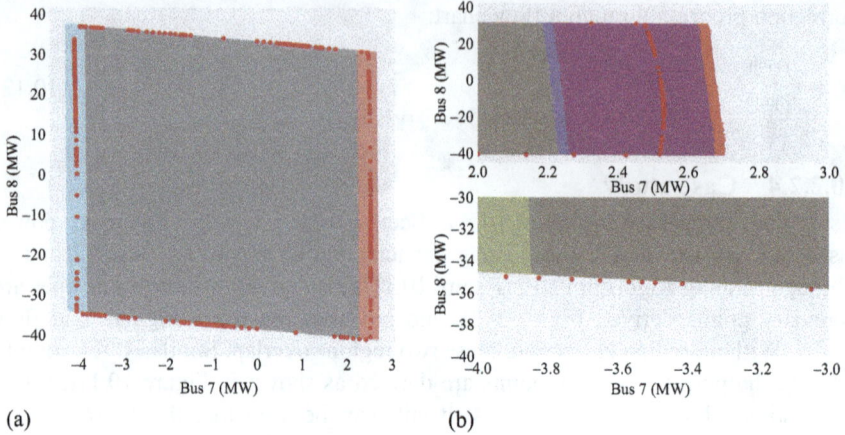

(a)   (b)

Figure 10.12   *Hosting capacity region correction. (a) Updated hosting capacity*
region. (b) Zoomed hosting capacity region.

10.2.3   Relaxed DistFlow model

10.2.3.1   Model description

As explained above, the linearized DistFlow model can contribute to a valid hosting capacity region, while with potential region space sacrifice. Meanwhile, referring to [15], there seems another route to utilize model convexity for hosting capacity region assessment. In advance of further discussions, for convenience of model description, we can denote an exact solution by S and the exact feasible region by $\mathcal{R}$ constrained by (10.1) as given in (10.13). Thus, we can provide an alternative expression about hosting capacity region $\mathcal{C}$ as illustrated in (10.14), which is derived from hosting power values in $\mathcal{R}$. Each element in C responds to a specific S, while only bus injection power is concerned:

$$
\begin{aligned}
S &\equiv \{P_{ij}, Q_{ij}, P_j, Q_j, v_j, l_{ij} \mid (i,j) \in E, (1)\text{ismet}\} \\
\mathcal{R} &\equiv \{S \mid S \text{ meetsDistFlowModel}\}
\end{aligned}
\tag{10.13}
$$

$$
\mathcal{C} \equiv \{\{P_j, Q_j \mid (i,j) \in E\} \mid R \neq \varnothing\}
\tag{10.14}
$$

Convex relaxation on DistFlow model has been deeply investigated in the topic of optimal power flow calculation, which relax (10.1d) to be (10.15). Instead, in this chapter, we focus on its corresponding feasible region. After relaxation, the new feasible region $\mathcal{R}^*$ becomes convex where $\mathcal{R} \subseteq \mathcal{R}^*$:

$$
l_{ij} \geq \frac{P_{ij}^2 + Q_{ij}^2}{v_i}
\tag{10.15}
$$

Most new generated points will meet all constraints in (10.1) except (10.1d), which means $S^* \notin \mathcal{R}$. Equation (10.16) is given for further illustration, where the subscript $*$ to distinguish elements in S^*:

$$l_{ij}^* > \frac{P_{ij}^*2 + Q_{ij}^*2}{v_i^*} \tag{10.16}$$

As mentioned before, S^* is still infeasible, thus we would like to map a certain S^* to another feasible solution $\widetilde{S}$, indicating $\widetilde{S} \in \mathcal{R}$. Through shifting S^*, we can regenerate another point defined as (10.17), where the hat symbol distinguishes elements of the regenerated solution $\widetilde{S}$. ε is an independent scalar variable. With a suitable value for ε, S can be confirmed to meet all constraints in (10.1), indicating a new exact feasible solution while without repetitive powerflow calculation:

$$\widetilde{v}_i = v_i^*, \widetilde{v}_j = v_j^* \tag{10.17a}$$

$$\widetilde{l}_{ij} = l_{ij}^* - \varepsilon \tag{10.17b}$$

$$\widetilde{P}_{ij} = P_{ij}^* - r_{ij}\varepsilon/2 \tag{10.17c}$$

$$\widetilde{Q}_{ij} = Q_{ij}^* - x_{ij}\varepsilon/2 \tag{10.17d}$$

$$\widetilde{P}_i = P_i^* - r_{ij}\varepsilon/2 \tag{10.17e}$$

$$\widetilde{P}_j = P_j^* - r_{ij}\varepsilon/2 \tag{10.17f}$$

$$\widetilde{Q}_i = Q_i^* - x_{ij}\varepsilon/2 \tag{10.17g}$$

$$\widetilde{Q}_j = Q_j^* - x_{ij}\varepsilon/2 \tag{10.17h}$$

10.2.3.2 Point-wise region assessment

The proper value determination of ε is given in Algorithm 10.3, where Gauss–Seidel method is adopted to ensure a stable solution. The grid bus number is denoted by N, then Algorithm 10.3 will run $N - 1$ times due to $N - 1$ links. In a

---

Algorithm 10.3 : ε value determination

Result: ε

1 Initialize $\widetilde{v}_i = v_i^*$, $\widetilde{P}_{ij}^{(0)} = P_{ij}^*$, $\widetilde{Q}_{ij}^{(0)} = Q_{ij}^*$, $\widetilde{l}_{ij}^{(0)} = l_{ij}^*$;

2 Initialize $\varepsilon^{(0)} = \widetilde{l}_{ij}^{(0)} - [(\widetilde{P}_{ij}^{(0)})^2 + (\widetilde{Q}_{ij}^{(0)})^2]/\widetilde{v}_i$, $n = 0$;

3 **while** $|\varepsilon^{(i)}|$ *is larger than threshold* **do**

4      $n \leftarrow n + 1, \widetilde{l}_{ij}^{(n)} \leftarrow \widetilde{l}_{ij}^{(n-1)} - \varepsilon^{(n-1)}$;

5      $\widetilde{P}_{ij}^{(n)} \leftarrow \widetilde{P}_{ij}^{(n-1)} - r_{ij}\varepsilon^{(n-1)}/2$;

6      $\widetilde{Q}_{ij}^{(n)} \leftarrow \widetilde{Q}_{ij}^{(n-1)} - x_{ij}\varepsilon^{(n-1)}/2$;

7      $\varepsilon^{(n)} \leftarrow \widetilde{l}_{ij}^{(n)} - [(\widetilde{P}_{ij}^{(n)})^2 + (\widetilde{Q}_{ij}^{(n)})^2]/\widetilde{v}_i$;

8 **end**

9 $\varepsilon \leftarrow l_{ij}^* - \widetilde{l}_{ij}^{(n)}$;

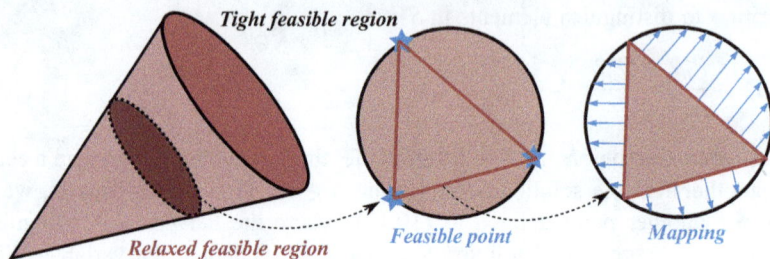

Figure 10.13   Hosting capacity region exploitation principle

distributive computation scheme, at most $N - 1$ processor can run Algorithm 10.3 simultaneously as each ε will respond to a certain link independently.

The principle of capacity reconfiguration is illustrated in Figure 10.13. Equation (10.4d) is a second-order conic constraint, whose respective feasible region is a cone as represented in Figure 10.13. The surface of this cone represents the exact feasible region defined by tight constraint (10.1d). If we look into its crossing section, all feasible solutions S calculated by powerflow calculation tools are only possible to be located upon the bound of this circle, which is denoted by the circular bold black line. Through connecting these feasible points, we can derive several lines. Except endpoints, these lines are out of the tight feasible region while still belong to the relaxed one. In accordance with (10.17), we will calculate ε and map infeasible points to the circular bold line, eventually deriving a new set of feasible points and realizing hosting capacity region exploitation.

Now, you may also notice that such method is point-wise, indicating we cannot directly figure out extra linear boundaries to help formulate a polygon, which is essential to represent the hosting capacity region. Instead, we are using its convexity to generate a point very close to its respective true solution. Moreover, with distributive computation capability of such method, repetitive power flow calculation can be avoided, replaced by parallel ε computation in (10.17) using Algorithm 10.3. Therefore, after running Algorithm 10.2 with several boundary points derived, we can take it as a fast-computing interpolation method when we want to derive a cloud of boundary points to represent a more precise hosting capacity feasible region.

1.2.3.3 Case study

Still focusing on the testing case and results in Section 10.2.2.4, we can simplify the hosting capacity region as a quadrilateral defined by four vertices as listed in Table 10.3. Each point is on the edge of grid constraint violation, indicated by relatively low or high Bus 7 voltage, especially in the context of full transformer loading. Therefore, combining with previous geometrical results in Figure 1.12, we select these four vertices to realize interpolation as illustrated above.

However, only four vertices derived from powerflow calculation may not be sufficient to get the whole region due to natural accompanying reactive power

Table 10.3   Table of vertices in hosting capacity region

No.	Bus 7/8 load (MW)	Bus 7 voltage (pu)	Transformer load (%)
1	2.486/30.450	0.900	99.995
2	−4.104/36.998	1.100	99.998
3	−4.069/−34.904	1.100	100.001
4	2.520/−41.477	0.900	100.004

Table 10.4   Table of interpolation time expenses

Total points	PowerFlow point	Interpolation point	Computation time(s)
400	400	0	2.343
400	40	360	1.504
400	4	396	1.289

correction in Algorithm 10.3. When ε is derived, it imposes changes on P_i and Q_i simultaneously. However, in this case, we aim for a combinational feasible region only related to active power, and extra reactive power load is expected to keep zero. If the reactive power correction is huge, even we derive another feasible point successfully, it cannot be assumed an acceptable interpolated point.

As shown in Table 10.4, based on such four vertices, we compare three scenarios sharing the same boundary point number of 400. The first scenario derives all interpolations still based on powerflow calculation. The last one totally uses the proposed interpolation method only based on 4 known vertices. The middle one just compensates extra 9 points using powerflow calculation over each edge, and later utilizes the proposed method to get left points. Through comparing computation time cost in the same testing platform, the interpolation method is confirmed faster-speed than powerflow calculation. We also cite the results of exact region boundary points in Section 10.2.2.4, and compare them with interpolated points in the middle scenario in Table 10.4. The good matching between these two points groups in Figure 10.14(a) confirms the validity of the proposed interpolation method.

The reason why we still adopt the middle scenario instead of the last one, even in higher computation time cost, it ensures limited reactive power correction volume. As shown in Figure 10.14(b), during interpolation, the middle scenario owns maximal reactive power correction volume lower than 0.03 Mvar, while the worst scenario with only four powerflow points owns over 2 Mvar reactive power correction. In practices, it pays off to sacrifice computational efficiency to compensate for time expenses. It is still open for discussions on how to rationally select a proper powerflow point density, achieving a trade-off between computation time cost and reactive power correction volume in this case.

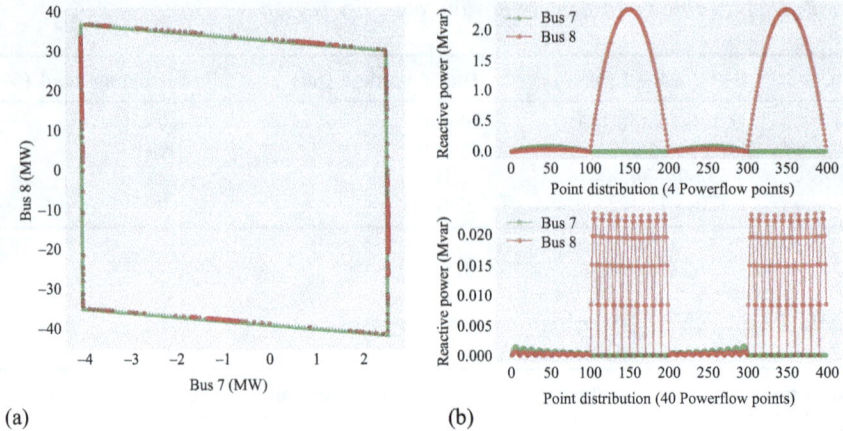

Figure 10.14   *Hosting point interpolation results. (a) Boundary points distribution. (b) Interpolation error distribution.*

10.3   Hosting capacity region reshaping with storage

The hosting capacity region assessment technology has been deeply discussed in the previous section. We have paid lots of efforts to derive a hosting capacity region, and it may satisfy the DSO by providing an essential reference to help set respective capacity limits on each relevant POC. Meanwhile, the results seem potential to create more values than that.

As mentioned above, in practices, although DSO sets relevant connection capacity limits to POCs, some users still need a larger connection capacity even with the willingness of higher connection expenses. Instead of physically upgrading grid cables, energy storage equipment installation is a promising alternative approach, especially considering most loads do not draw the maximal current in most of the time. With energy storage equipment integrated, the relevant hosting capacity region is equivalently reshaped. How to utilize such reshaping to help solve EES deployment problems is mainly discussed in this chapter.

10.3.1   Minimal energy storage capacity quantification

When ESS is integrated to a certain POC, its impacts on original hosting capacity region depends on its own output characteristics. Intended to quantify minimal ESS capacity to ensure an operation point, which was originally infeasible, become feasible again, we need to clarify the ESS characteristics in advance.

10.3.1.1   Considering apparent power constraints

Benefiting from a bidirectional inverter interface, the ESS can be flexible in its output power factor. In this scenario, we aim to calculate the minimal distance between the hosting capacity region and the operation point, which eventually quantify the minimal ESS capacity. There are many mature numerical methods investigated to

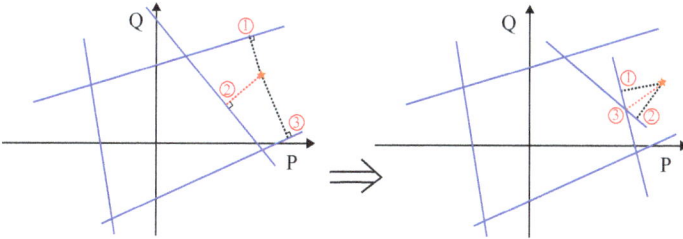

Figure 10.15 Minimal apparent power determination

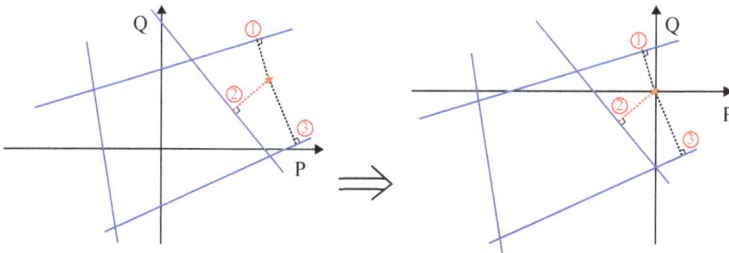

Figure 10.16 Coordinate shifting of hosting capacity region

calculate the distance from a point to a polygon in a planar space. However, in the beginning, we still choose to explain it geometrically for deeper understanding.

As shown by the left picture in Figure 10.15, when the polygon is not complicated and the operation point is well-posed, we can just calculate its distance to each boundary line as soon as this point meet this polygon. Unfortunately, as illustrated by the right picture, with polygon growing, the distance to one vertex can also be a solution, finally confusing and complicating our calculation progress.

Meanwhile, from an algebraic perspective, this problem can be solved efficiently. Through coordinate shifting shown in Figure 10.16, this problem can be represented as

$$\textbf{min}\ x^2 + y^2 \qquad \textbf{s.t.}\,(x, y) \text{ in shifted hosting capacity region} \qquad (10.18)$$

where x, y are horizontal and vertical coordinate values in the shifted coordinate system. Since the original hosting capacity region is defined by several linear bounds, the shifted one still keeps convex. The whole problem can be confirmed convex, which can be solved quickly while still ensuring the global optimum. The corresponding objective function value is equal to the expected minimal distance square. Eventually, such concise formulas successfully answer the same questions on minimal EES capacity, even in the absence of original grid model.

10.3.1.2 Considering constant power factor

In practices, ESS output power factor may be constrained constant, especially for full-power-mode converter-interfaced storage. In such scenario, we can still use

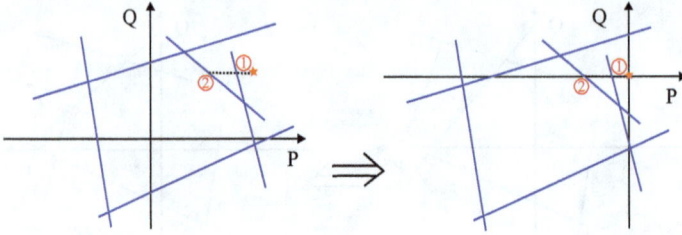

Figure 10.17   Coordinate shifting of hosting capacity region

coordinate shifting method as shown in Figure 10.17. Instead of geometrically computing intersection points between a slope-constant plot and boundary lines, if the power factor angle is set to α, through substituting y by $\tan(\alpha)x$, (10.18) is even simplified as there only exists only one variable. The final objective value can help quantify relevant ESS minimal capacity as well.

10.3.2   General optimal storage capacity design

In Section 10.2.2.4, we have assessed the hosting capacity region regarding pure active power over Buses 7 and 8. If we want to install two ESS over these two POCs separately, we still need to figure out an optimal deployment scheme. Caused by geographical differences, there may be various installation expenses for ESS. If such installation fees still keep linear to it capacity, while Buses 7 and 8 just shares various sloping rate β and γ. Following the same principle in (10.18), the planning problem can be represented as

$$\textbf{min } \beta|x| + \gamma|y| \qquad \textbf{s.t.} (x, y) \text{ in shifted hosting capacity region} \qquad (10.19)$$

The whole problem can be decomposed into four convex problems, where the symbol of x and y is fixed in each scenario. Meanwhile, there exists another path to solve (10.19). Through equivalently converting (10.19) as below, we import two non-negative auxiliary variables m and n while keeping the same optimal objective function value:

$$\min \ \beta m + \gamma n \qquad (10.20a)$$

$$\text{s.t. } m, n \geq 0 \qquad (10.20b)$$

$$-m \leq x \leq m \qquad (10.20c)$$

$$-n \leq y \leq n \qquad (10.20d)$$

$$(x, y) \text{ in shifted hosting capacity region} \qquad (10.20e)$$

10.3.3   Case study

Similar to case studies in Section 10.2.3.3, we still use four vertices to represent the feasible region. Each edge of such quadrilateral can be taken as a linear constraint,

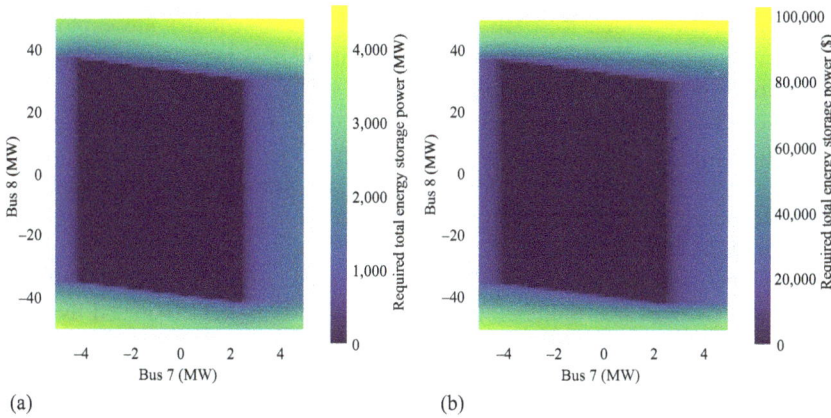

Figure 10.18   Energy storage capacity and cost optimization. (a) Optimal total energy storage capacity. (b) Optimal total energy storage cost.

defining a feasible half space by a cutting line. Looking into the storage capacity minimization problem, we can using the proposed algebraic optimization method to derive a conclusion. In order to demonstrate the validity of such method, we have tested all operation points, where Bus 7 active power is from −5 MW to 5 MW and Bus 8 is from −50 MW to 50 MW. The corresponding optimal total energy storage capacity is denoted by color distribution (Figure 10.18(a)). The dark blue part is equal to the feasible region, where the objective function value is equal to 0, indicating those operation points need no extra energy storage regulation.

Moreover, assuming the operation period set to 1 h, based on battery cost data in [16], we allocate 300$/kWh for Bus 7 and 650$/kWh for Bus 8. A general energy storage cost optimization problem is discussed. As illustrated in Figure 10.18(b), the cost distribution in various operation points has been provided. It can be seen from the results that the dark blue region keeps equivalent to the original hosting capacity region.

Both testing cases above have demonstrated the effectiveness of the proposed optimization method, which is mainly based on the derived hosting capacity region. Answering the same question, unlike conventional scenario-based testing methods, our proposed one is independent from the original grid model, indicating lower memory requirement and higher computation efficiency.

10.4   Summary

Based on the term of "feasible region," this chapter extends the concept of "hosting capacity." Through converting the grid model into a more compact one, "hosting capacity region" is confirmed promising to further exploit the grid potential for power delivery. Such evolved concept also benefits energy storage deployment investigation in a more compact format, which is independent from the conventional grid model. Respective region assessment schemes are exploited as well, including heuristic region growing and point interpolation, whose validity has been demonstrated by corresponding case study.

However, the proposed region assessed method is limited in 2D space. The results accuracy is high but still harmed by approximation. In future, the authors will put more efforts into the high-dimensional assessment technology and more accurate capacity region correction methods.

Funding

This work was supported by NEON (New Energy and mobility Outlook for the Netherlands, with project number 17628), a cross-over project financed by NWO (the Dutch Research Council).

References

[1]   Bollen M and Hassan F. *Integration of Distributed Generation in the Power System. Wiley Online Library*. New York, NY: IEEE Press, 2011.

[2]   Mulenga E, Bollen MH, and Etherden N. A review of hosting capacity quanti-fication methods for photovoltaics in low-voltage distribution grids. *International Journal of Electrical Power & Energy Systems*. 2020;115:105445.

[3]   Lehmann K, Grastien A, and Van Hentenryck P. AC-feasibility on tree networks is NP-hard. *IEEE Transactions on Power Systems*. 2015;31(1):798–801.

[4]   Zafar R, Mahmood A, Razzaq S, *et al.* Prosumer based energy management and sharing in smart grid. *Renewable and Sustainable Energy Reviews*. 2018;82:1675–1684.

[5]   Inês C, Guilherme PL, Esther MG, *et al.* Regulatory challenges and oppor-tunities for collective renewable energy prosumers in the EU. *Energy Policy*. 2020;138:111212.

[6]   Milanović J, Meyer J, Ball R, *et al.* International industry practice on power-quality monitoring. *IEEE Transactions on Power Delivery*. 2013;29(2):934–941.

[7]   Lamedica R, Geri A, Gatta FM, *et al.* Integrating electric vehicles in microgrids: overview on hosting capacity and new controls. *IEEE Transactions on Industry Applications*. 2019;55(6):7338–7346.

[8]   Oliveira TECD. *The Concept of Dynamic Hosting Capacity of Distributed Renewable Generation Considering Voltage Regulation and Harmonic Distortion*. Federal University of Itajubá, 2018.

[9]   Jain AK, Horowitz K, Ding F, *et al.* Dynamic hosting capacity analysis for distributed photovoltaic resources–framework and case study. *Applied Energy*. 2020;280:115633.

[10] Baran ME and Wu FF. Network reconfiguration in distribution systems for loss reduction and load balancing. *IEEE Power Engineering Review*. 1989;9 (4):101–102.

[11] Thurner L, Scheidler A, Schäfer F, *et al.* Pandapower – an open-source python tool for convenient modeling, analysis, and optimization of electric power sys-tems. *IEEE Transactions on Power Systems*. 2018;33(6):6510–6521.

[12] Gurobi Optimization, LLC. Gurobi Optimizer Reference Manual, 2022. https://www.gurobi.com.

[13] Avis D, Bremner D, and Seidel R. How good are convex hull algorithms? *Computational Geometry*. 1997;7(5–6):265–301.

[14] Berg Md, Kreveld Mv, Overmars M, *et al.* Computational geometry. In: *Computational Geometry*. New York, NY: Springer; 1997. p. 1–17.

[15] Low SH. Convex relaxation of optimal power flow – Part II: exactness. *IEEE Transactions on Control of Network Systems*. 2014;1(2):177–189.

[16] Cole W, Frazier AW, and Augustine C. *Cost Projections for Utility-Scale Battery Storage: 2021 Update*. Golden, CO: National Renewable Energy Lab. (NREL), 2021.

Chapter 11

Urban grid resilience in the context of new infrastructure

Trung Thai Tran[1], Minh Quan Tran[1], Thien-An Nguyen Huu[1] and Phuong Hong Nguyen[1]

Urban power grids are characterized by a bulky concentration of different load zones (commercial, industrial, and residential) and strict requirements on the grid resilience and reliability of power supply from distribution systems. The increasing power demand, due to the population and economic growth, is pushing the operation of urban power grids to its capacity limits. The installation of distributed energy resources (DER) like photovoltaic (PV), wind power, and energy storage (ESs) with proper control and coordination mechanisms can offer a possibility to improve grid resilience. In this chapter, emerging coordination utilizes dispatchable sources to enhance the restoration capability under different disruptive events such as grid malfunctioning, severe weather, malicious attacks, and operation missteps. These coordinated control methods are based on a multiple-time-scale hierarchical framework that can effectively manage a complex, multi-target requirement of grid resilience.

11.1 Introduction

Urban power grids are continuously upgrading to meet the rapid growth in energy demand and the desire to use greener energy resources. This rapid development results in many expansion and operation challenges [1]. Consequently, resilience becomes a critical aspect that needs to be considered for the development and modernization of urban power grids. Grid resilience refers to the capability of the system to anticipate, prepare, respond, mitigate disruptive events, and recover to the normal operational stage after these events.

Enhancing the resilience capability of urban power grids is based on several key factors, such as expansion and strengthening basic infrastructure [2] (e.g.,

[1]Eindhoven University of Technology, The Netherlands

install additional transmission facilities, upgrade existing transformers), meshed or ring operation [3], integration of DER units [4], islanding operation, and smart grid technologies [5]. In recent years, several advance transmission technologies have been implemented to optimize the expansion, and utilization of the existing infrastructure. However, the tight urban land resource and majority of the power lines being put underground are the main constraints for the infrastructure expansion. The meshed operation that allows the loads to be supplied by more than one feeder can be seen as a supportive solution. The smart grid technologies provide a two-way communication between entities in the grid. Thus, it gives the grid operator ability to observe the system, remotely control the local devices, allowing a self-healing of the grid when a contingency occurs. Several smart grid architectures have been proposed, such as holistic, multiple-dimensions smart grid architecture model (SGAM) [6], the BSI model for the security module of a smart meter gateway [7], ENISA smart grid security with security recommendations, and knowledge inventory [8].

Numerous benefits of DERs and ESs have been reported in the literature. The installation of diverse DERs and ESs close to the load centers helps reduce the stress of network components, enhance the continuity of supply for the critical loads and areas during emergency islanding. It increases the resilience of the grid and provide the black-start capability that is necessary to restore a system to normal operation after a disruptive event. ESs facilitate the generation of intermittent renewable energies and improve profits in normal operating conditions. Also, in some extreme events, ESs can provide ancillary services, e.g., dynamic voltage or frequency support and emergency supply, to prevent the urban power grid from the risk of voltage collapse or even blackout.

On the other hand, the increasing penetration of DERs with power electronic interfaces into the generation, transmission, and distribution areas has great impacts. It exposes many new technical challenges and risks on urban power grid control, operation, and stability. More sophisticated coordination, energy, and information management are required. Developing innovative technologies and control strategies becomes essential for the resilience enhancement of the next generation urban power grid.

This chapter aims to capture such development of emerging control technologies to coordinate and optimize the operation of DERs and to support urban power grids in term of automated actions, self-healing functionalities and long-term voltage stability support.

11.2   Resilience in urban power grids

According to [9], resilience is a function of time, showing the grid's ability to degrade gradually under a disturbance, then recovers to its pre-disturbance state. It can be divided into *short-term* and *long-term* resilience, as illustrated in Figure 11.1. *Long-term* resilience in energy systems refers to the ability to learn from previous events and adapt to changing conditions and future threats.

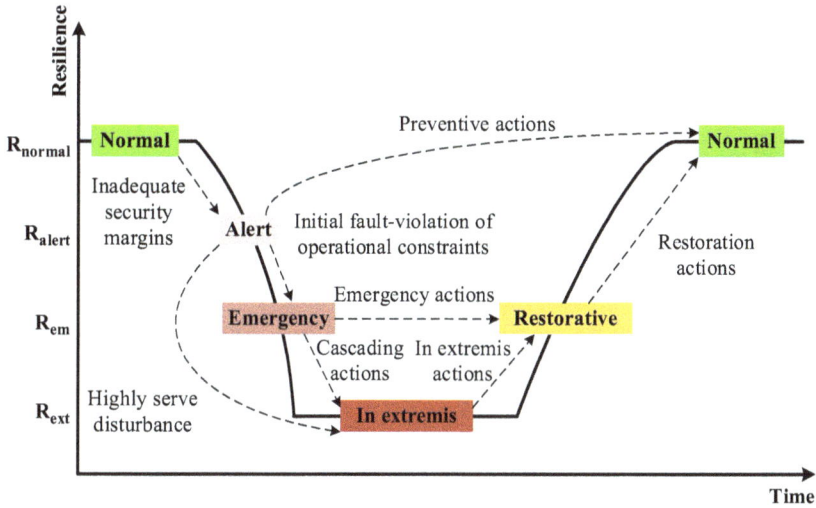

Figure 11.1   Resilience curve concept [9]

It is more about the planning process to upgrade the system to counteract with future predictable and unpredict disturbances. Whereas the *short-term* resilience of energy systems is the ability to prevent (precondition), effectively and rapidly react to (during), and restore the service after a disturbance or critical events.

The formal definition of resilience types has been evolving and varied depending on system ability to cope with disturbances. This chapter refers a *short-term resilience* as: (1) the ability to detect the faults and disconnect the most critical parts of the urban power grid to operate in off-grid mode in which the loads are continuously supplied by local generations and EES units; and (2) ability to identify the abnormal voltage dynamic and implement countermeasures [e.g., coordinating grid assets such as DERs and load tap changer (LTC) of transformer] to bring the voltage back to pre-fault condition after emergency conditions. Meanwhile, the ability to provide ancillary services from dispatchable sources to support optimal, reliable grid operation, and maximize the benefit from these sources is referred as *long-term resilience*. Table 11.1 details the services and application according with the resilience type.

11.3   Fault-initiated islanding for resilience enhancement

This section introduces typical hierarchical control structures and coordinated mechanisms for power electronic interface of DERs to provide functionalities that is necessary to enhance the resilience of urban power grids. The basis of hierarchical control is described. Then, features of fault-initiated islanding (FII), dynamic voltage support, and service provision are also presented in detail.

Table 11.1   Applications of ESs in resilience enhancement

Resilience	Service	Application
Long-term	Economic efficiency	• Renewable energy curtailment prevention • Loss reduction • Unit commitment
	Frequency restoration reserve (FRR)	• Frequency containment reserve (FCR) • Automatic FRR • Manual FRR
	Power balancing service	• Load following, load peak shaving, load valley filling • Generation/demand balance
Short-term	FII	• Provide backup power for critical, vulnerable loads, areas during a loss of a main supply
	Frequency regulation	• Support the system operator to control the frequency in a short period
	Voltage support	• Provide voltage support by modulating active and reactive power
	Small-signal stability	• Stabilize the power system in response to small perturbations (e.g., inter-area oscillations)
	Renewable energy capacity ramp rate	• Smooth the power generation from intermittent renewable energy (i.e., ramp rate control)

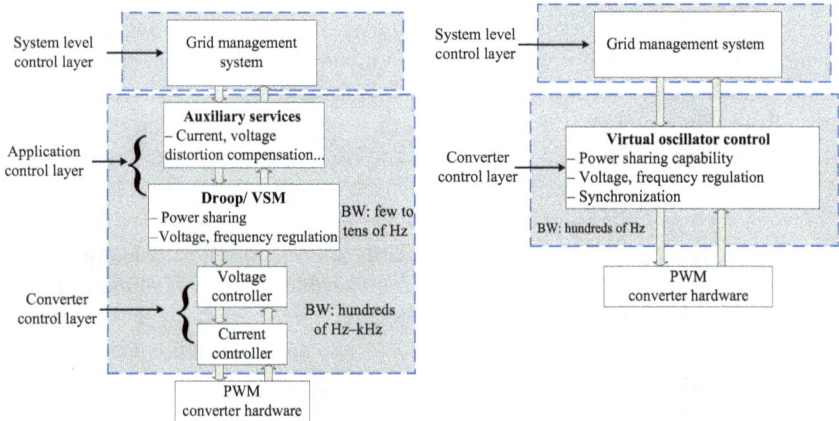

Figure 11.2   Typical hierarchical control structures for DERs and EESs

Different control methods have been described in the existing literature to provide a complex, multiple target coordination for diverse DERs in urban power grid, and hierarchical control structure is among the most effective solutions.

Typical hierarchical control structures for DERs are presented in Figure 11.2 which consists of multiple control layers with proper time-scale separation property

[10]. The overall control objective is to enable (a part) urban power grid to operate in both grid-connected and off-grid, i.e. islanded, mode, as well as seamless transition between two modes. The lowest control layer (*Converter Control Layer*) can use multi-stage controller combining of current, voltage and droop control loops [11]. The key control function of the layer is to maintain system stability during both steady-state and transient periods. The *Application Control layer* is the main control mechanism for voltage and frequency regulation, allowing DERs share the load powers in off-grid mode and follow the power setpoints in grid-connected mode. Advanced control technologies applied to this layer enable additional functions and auxiliary services such as voltage, current distortion compensation, disturbance rejection, and synchronization. The existing typical decentralized control method for this layer is the droop control method. A simple example of *P–V, Q–f* type droop control is presented as follows:

$$\omega_i - \omega_i^* = n_i Q_i \qquad (11.1)$$

$$\overline{V}_i - \overline{V}_i^* = m_i P_i \qquad (11.2)$$

where n_i and m_i are coefficients representing droop relations between angular frequency, reactive power (Q_i), and voltage magnitude ($\overline{V}_i$), active power (P_i), respectively.

Alternatively, single-stage controllers such as virtual oscillator control (VOC) [12], or self-synchronized synchronverters [13] approach can be used to ensure functionalities of both control layers. It means the complexity of the overall control structure is reduced, making it easier to implement in real devices.

The *System Level Control Layer* uses either a centralized control method relying on a full communication system, or a distributed control method with a spare communication system, or a fully decentralized method. The system-level control layer keeps the system voltage and frequency at the nominal values, guarantees the synchronization process, and controls power flow. The following sub-sections explain a control solution for resilience enhancement of urban power grid based on VOC-based hierarchical control structure, in which: (1) the FII is designed to disconnect critical parts of urban power grid from the main utility grid in contingency cases, (2) synchronization algorithm that allows seamless reconnection of the isolated parts, and (3) different control modes with dedicated switching mechanism to ensure a stable operation of the whole system, as presented in Figure 11.3 [14].

In this structure, DERs communicate with others through a communication network, either centralized or distributed based on different applications. The Andronov-Hopf Oscillator (AHO) is used for converter control layer. The dynamics of AHO is realized by two differential equations as shown in block AHO in Figure 11.3. The power tracking control and power sharing control adjust the internal parameters of AHO (i.e., V_n and ω_n) to regulate the active (P_i) and reactive power (Q_i) outputs of DER$_i$ following the commands from upper control levels (P_{ref} and Q_{ref}). The FII detection (FII-D) changes the status of switch S_1 and S_2 to change the operation mode of DER. The detailed configuration parameters of each control block can be found in [14,15].

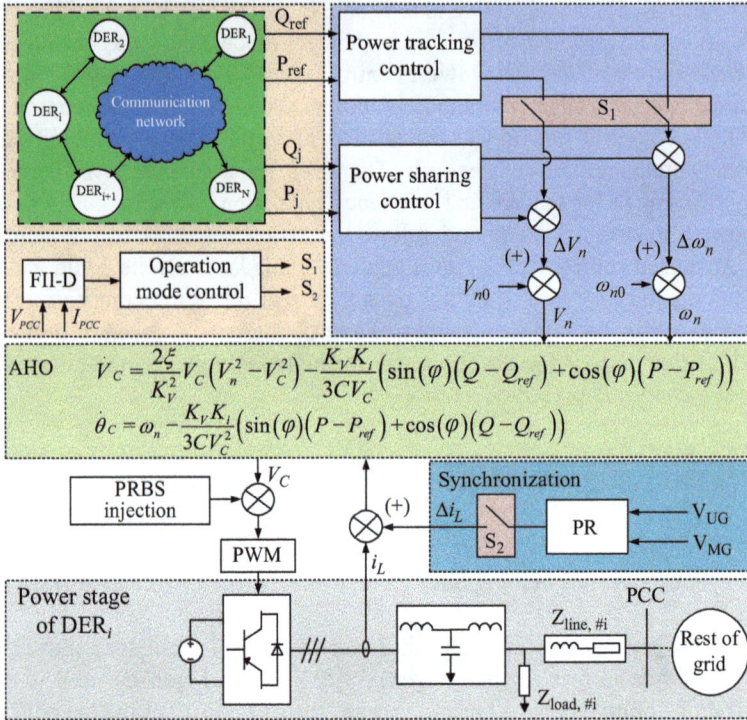

Figure 11.3   VOC-based hierarchical control structures for DERs and ESs

11.3.1   Fault-initiated islanding detection

FII is a method to disconnect (a part of) the urban power grid from the main utility grid and operate as autonomous islanded grids when a fault occurs [16]. In this occasion, the disconnected area is supplied by local generations such as PV and ES units, until the fault is cleared and this part is reconnected to the rest of the grid. In recent years, FII has received increasing attention as it allows network operator to reduce the risk of supply losses to critical loads, areas without huge investment in network components [17]. FII can be done through a sequence of actions, including fault detection, islanding and control-mode switch.

The FII detection can be generally categorized into communication and local methods, as shown in Table 11.2. The communication-based methods rely on high-speed two-way communication to detect the abnormal operation of remoted assets, then provides status checking information as an input for fault detection algorithm installed in a central controller. These methods can be applied for every type of dispatchable generations with reliable detection performance. The main disadvantages of these methods are the availability and high cost of communication system, and low flexibility to topology changes [18]. The local methods can be further divided into active (i.e., introducing disturbance to the grid and observe the

Table 11.2   Islanding detection solutions

Category	Technology	Characteristics
Centralized	Communication based [19,20]	- Require high-speed communication system - High detection performance with low non-detection zone - Vulnerable to network topology changes
Local	Active methods [21,22]	- Intentionally inject a disturbance to analyze the behavior of affected systems - Reduced size of non-detection zone - Trade-off between detection performance and quality of power supply
	Passive methods [23,24]	- Directly obtain system variables (voltage, frequency, power) to detect islanding - Have no impact on power quality, fast detection speed - Large non-detection zone
	Hybrid methods [25,26]	- Employs both active and passive detection techniques - Improve performance of individual methods - Effective in complex systems

response) and passive methods (i.e., observing locally the variation in voltage, frequency or power to make the correct tripping decision).

Figure 11.3 presents an active method based on the estimation of the impedance change is used to detect the disconnection of the main utility grid. The method consists of two steps. First, it injects a wideband pseudo random binary sequence (PRBS) to the control input of a dispatchable source and then a wideband non-parametric impedance at the PCC (Z_{PCC}) in frequency domain $Z_{PCC}(\omega) = V_{PCC}(\omega)/I_{PCC}(\omega)$ is estimated from the voltage (V_{PCC}) and current measurement (I_{PCC}), based on the discrete Fourier transform (DFT). Second, the parametric impedance estimation is obtained based on a complex curve fitting technique with a first-order polynomial function, as follows:

$$Z_{PCC}^{par}(s) = R_{PCC} + L_{PCC}s \tag{11.3}$$

It is known that the impedance of the utility grid is smaller than the impedance of the local load. Then, the estimated impedance output seen from a dispatchable source will be much smaller in grid-connected mode than in off-grid mode. Therefore, by continuously estimating the impedance, it is allowed the detection of islanding situation. In this case, a notification signal is sent to other dispatchable sources via communication network to change the operation mode accordingly.

11.3.2   Control modes of DERs

Based on the output of FII detection algorithm, the tripping action will be made. Islanding operation can only be successful when there is enough capacity from local generations, and the power electronic interfaces of DER units are switched to proper modes. In grid-connected mode, DER (e.g., PV and wind generation) control system regulates its power outputs to follow the reference signals according to optimal economical operation, and EES units are in stand-by mode or provide flexibility services following the requests from the network operator. Meanwhile, in the off-grid mode, the control objective is to provide continuous, reliable and stable power supply from DERs and/or ES units for the critical loads and areas, e.g., power-sharing control.

Power tracking control in grid-connected mode, network operators are responsible to distribute the active (P_{ref}) and reactive (Q_{ref}) power setpoints from an upper-level dispatch center to dispatchable sources. The control objective of power tracking control located in the *Application Control Layer* is to compensate for the mismatch between DER power outputs and the required setpoints, simply using a proportional-integral (PI) controller of the form (11.4) and (11.5):

$$\Delta V_n = \left(K_P^V + \frac{K_I^V}{s} \right)(P_i - P_{ref\_i}) \tag{11.4}$$

$$\Delta \omega_n = \left(K_P^\omega + \frac{K_I^\omega}{s} \right)(Q_i - Q_{ref\_i}) \tag{11.5}$$

where K_P^V, K_I^V, K_P^ω, and K_I^ω are the controller coefficients and are chosen by trial-error technique to simplify the design process. ΔV_n and $\Delta \omega_n$ are internal control variables of VOC associated with active and reactive power control, respectively.

Power sharing control in off-grid mode: it is necessary to maintain the power balance between generation and load. Power sharing control among dispatchable sources like ES and DER units are the most convenient method for power balance. This control mode can be realized by either only decentralized approach or with improved approach based on distributed average consensus control methods. An example of the consensus control is presented in [14] in which each DER and ES units iteratively exchange and update the information of their neighbors to recalculate internal control variables to force power outputs to reach a global consensus average value $[P_{ave}, Q_{ave}]$ as follows:

$$\Delta P_{ave} = \frac{1}{N} \sum_i^N P_i \tag{11.6}$$

$$Q_{ave} = \frac{1}{N} \sum_i^N Q_j \tag{11.7}$$

where N is the total number of dispatchable units, P_i and Q_i are the active and the reactive power output of each unit, respectively.

Synchronization framework: switching control mode in time is essential to maintain effective and stable operation of disconnected areas [15]. When the fault is resolved and the power supply from the utility grid is available again, the off-grid area can be reconnected again with a proper synchronization mechanism. To enable seamless transition between operation modes of the isolated area and prevent severe consequences, at the moment of closing the circuit breaker at the point of common coupling (PCC), the voltages at two sides of circuit breaker have to stay within a strict limit for a required time period [15]. The basic idea of the synchronization framework presented in Figure 11.3 is to adjust power setpoints of all dispatchable units in a distributed cooperative manner so that the voltage mismatch between two sides of circuit breaker is closely reduced to 0 using proportional resonant (PR) controller, given as follows:

$$G_{PR}(s) = K_P + K_I \frac{\omega_c s}{s^2 + 2\omega_c s + \omega_n^2} \tag{11.8}$$

where K_P and K_I are the proportional and resonant coefficients, respectively; ω_c and ω_n are the cutoff and the system nominal frequency, respectively.

11.3.3 Performance evaluation

This section validates the operation and performance of the discussed hierarchical control structure in supporting resilience enhancement of urban power grids by providing FII functionality. The parameters of the control schemes are presented in detail. Selected waveforms from several case studies are reported for discussion.

11.3.3.1 System under test

A European LV distribution network benchmark is modified to represent the operation of a part of an urban power grid, as shown in Figure 11.4. It consists of three LV subnetworks supplied by 20 kV–MV distribution system through a 20 kV/0.4 kV distribution transformer. The underground cables are used in the residential and industrial subnetworks, while the overhead lines are used in the commercial subnetwork. For simplification, the loads, including residential, industrial and commercial types, are modeled as constant impedance loads. Four 15 kVA-DERs (including 3 PV and 1 ES units) are appended into the tested system with the locations are shown in Figure 11.4. They exchange active and reactive power output information to their neighbors through spare communication network (noted with red dotted line). The circuit breaker CB is used to connect or disconnect the test system from the main utility grid. The controller of the main circuit breaker communicates with only DER$_1$ for distributed synchronization purpose. An individual DER is interfaced with the test system via an LCL filter and a DC–AC inverter and is controlled by the hierarchical control strategy. The system parameters are briefly described as follows:

- The system nominal voltage: 3-phase, 400 V (line to line), 50 Hz.
- DER capacities: 15 kVA; DC voltage: 800 V; the sharing ratio is 1:1:1:1; LCL filter parameters is 1.6 mH: 0.15 μF: 1.6 mH.
- The parameters of AHO are chosen following the design process presented in [27].
- Power tracking control: (K_P^V, K_I^V) 0.0025, 0.0002.
- Synchronization controller: (K_P, K_I, ω_c) 0.0098, 3, 0.63.

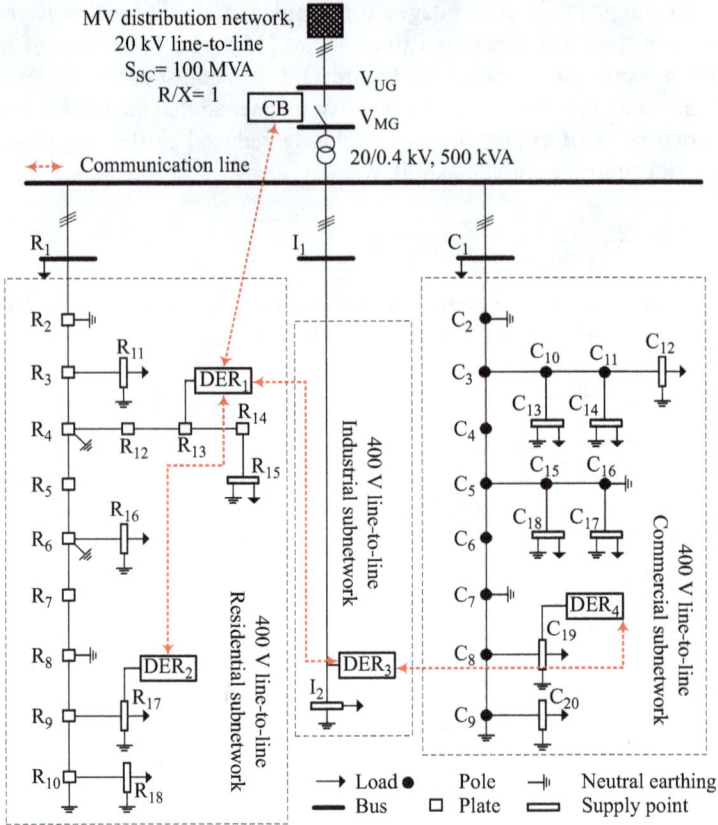

Figure 11.4   Topology of the tested LV urban power grid (modified from the CIGRE benchmark test network [28])

11.3.3.2 Transition from grid-connected to off-grid mode

This scenario aims at demonstrating the performance of the FII detection method to isolate the tested system from the main distribution system. The tested system is assumed to operate in grid-connected mode. Under the control of hierarchical controller, four DERs precisely track the active and reactive power setpoints from the dispatch center (see Figure 11.4(a) and (b) at $t = 0$–4 sec).

The impedance estimation algorithm is activated at $t = 2$ sec and is run continuously every 1 sec. The whole estimation process takes around 0.4 sec to finish. In grid-connected mode, the estimated magnitude of the impedance seen from the PCC of DER$_1$ is $|Z_{PCC}| \approx 0.6\ \Omega$ (see Figure 11.5(c)). At $t = 4$ sec, a fault is assumed to happen that causes the circuit breaker to open to disconnect the tested system from the main distribution system. From $t = 4$–4.4 sec, as the FII detection algorithm has not yet detected the disconnection, controllers of DERs are still on power tracking control mode. It causes a large transient of active and reactive power. If this situation remains for a long time period, system instability or severe

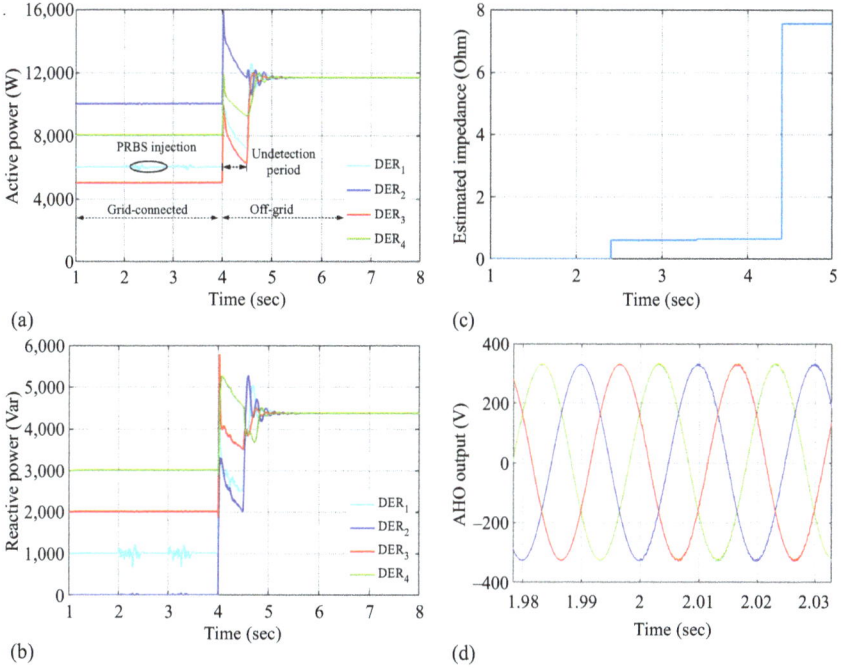

Figure 11.5   Transition from grid-connected mode to off-grid mode

consequences may happen. At $t = 4.4$ sec, the FII-detection sees a significant change in estimated impedance, and sends a control signal S_1 to change the operation mode of DERs to power sharing mode. After a short transient period, the isolated grid reaches a new equilibrium point where four DERs share the load powers equally.

11.3.3.3   Transition from off-grid to grid-connected mode

In this scenario, the isolated grid is assumed to operate in off-grid mode as a result of a fault occurred near the point of common coupling. Then, after a certain period of time, the fault is cleared and the isolated grid is allowed to reconnect to the distribution system. The following steps are used to verify the performance of the hierarchical controller.

- At $t = 0$–2 sec, the isolated grid is operated in off-grid mode. DERs, including ES units share the load power equally, governed by power sharing control based on distributed consensus control. The total load required is 47 kW + 17.6 kVar.
- At $t = 2$ sec, the isolated grid is allowed to reconnect to the distribution system.
- At $t = 10$ sec, the circuit breaker is closed.
- At $t = 15$ sec, the operation mode of DERs is changed to power tracking mode and their power outputs follow the setpoints from upper control level.

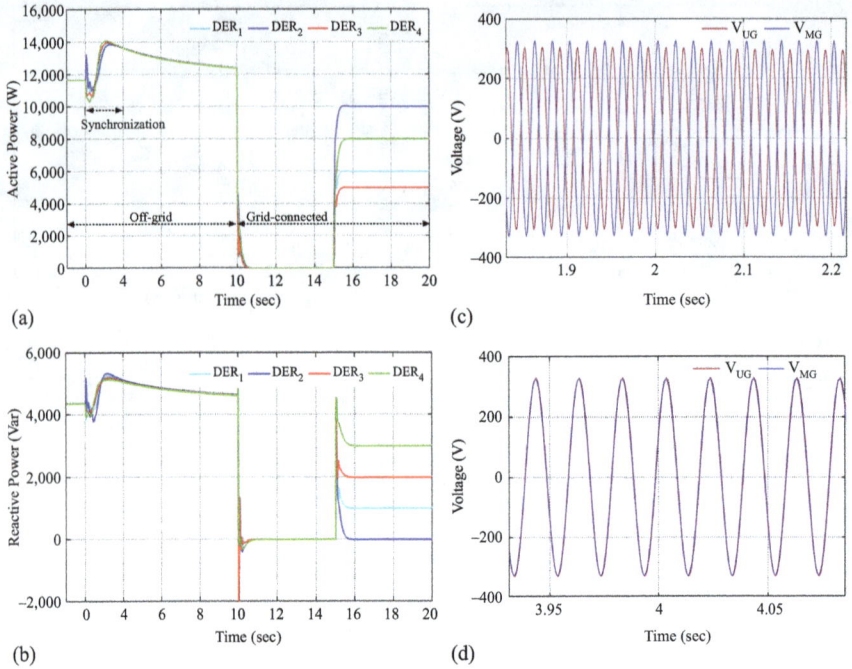

Figure 11.6   Transition from off-grid mode to grid-connected mode

As can be seen from Figure 11.6, at $t = 0 - 2$ sec, four DERs share the load power without power-sharing inaccuracy, as the effect of the power sharing control. When the synchronization control is activated at $t = 2$ sec, the voltage at the off-grid side V_{MG} is gradually synchronized with the voltage at the distribution side V_{UG} in about 2 sec (see Figure 11.6(c)), indicating that isolated grid can seamlessly connect to the main distribution system without causing severe consequences to the whole system. As the synchronization controller sends an additional control signal to the input of the controller of DER$_1$, its power outputs change accordingly. Both active and reactive power of the remaining DERs are forced to follow the change of DER$_1$ in a distributed manner with unidentical transient behaviors, as the effect of the distributed average consensus algorithm. The whole isolated grid will then converge to a new equilibrium point at which the synchronization criteria is satisfied (i.e., V_{MG} and V_{UG} are synchronized). At $t = 10$ sec, the isolated grid is connected smoothly to the distribution system with a small transient in a short period of time. As Figure 11.6(a) and (b) shows, both active and reactive power-sharing among DERs are not affected by the synchronization process.

After closing the circuit breaker, the control signal S_2 is disabled, and the S_1 is activated to change the operation mode of DERs from power sharing control mode to power tracking control mode with the setpoints are set to 0. The active and reactive power output of DERs follows exactly the reference values. At $t = 15$ sec, the dispatch center calls for different step changes in P_{ref} and Q_{ref} to each DER.

There is a short transient but it vanishes quickly, and all DERs response to the setpoint changes in about 0.4 sec.

11.4 Dynamic voltage support

In this section, an instability problem of the voltage in urban power grid in emergency conditions is discussed in detail. Then, a mechanism is presented in which DER units and existing grid controllers (e.g., LTC) are coordinated to improve the dynamic voltage stability.

11.4.1 Long-term voltage instability problem

To better understand the voltage instability problem that can occur in urban power grid during emergency cases, a European MV distribution network benchmark is used, as shown in Figure 11.7. The MV urban power grid is assumed to connect to the utility grid via two parallel transmission lines. A simulation is performed in

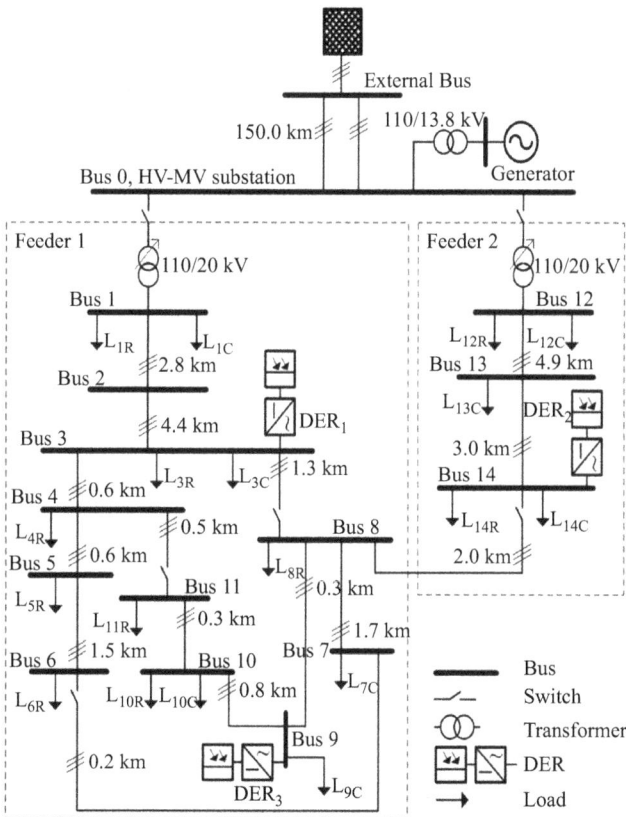

Figure 11.7 Topology of the tested MV urban power grid

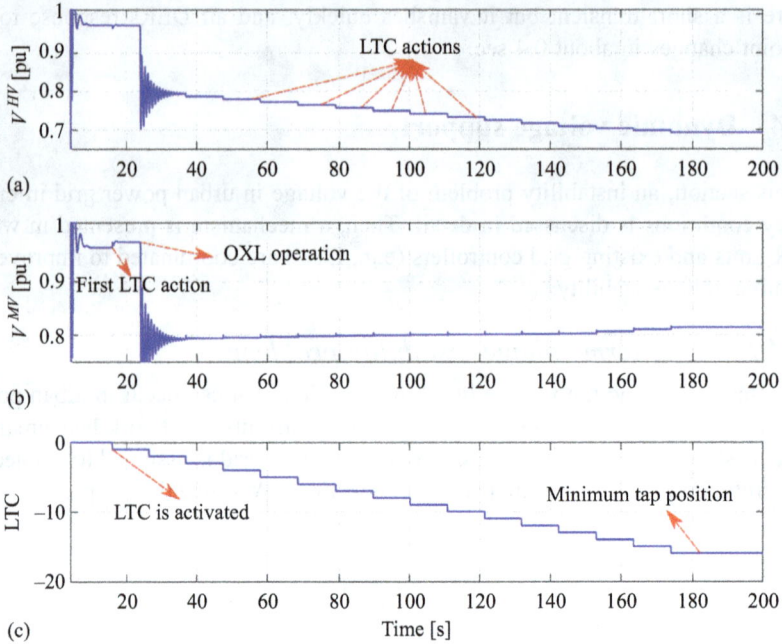

Figure 11.8 *Long-term voltage instability. VHV, VMV, and LTC are the voltage measured at Bus 0 and 1, and the tap position of the transformer connecting these buses, respectively.*

which a three-phase fault is considered at one of the two parallel transmission lines between external bus and Bus 0. This transmission line trips at time $t = 0$ to isolate the fault. After a short-term period with dynamics assumed to be stable, the system enters a long-term period where voltage stability is of concern. In general, this period involves slow-acting equipment such as LTC, over excitation limiter (OXL) of a synchronous generator and controlled loads.

As seen from Figure 11.8, a short transient period occurs following the tripping of the transmission line. After that, the voltage at two sides of the transformer V^{HV}, V^{MV} are stable at 0.973 pu and 0.928 pu, respectively. As a result, the LTC controller changes the tap position to increase V^{MV}, aiming to bring this voltage back to nominal value, i.e., 1 pu. It should be noted that the LTC action is activated every fixed period of 15 sec and the first activation is at $t = 15$ s.

At $t = 24$ s, the voltages V^{HV}, V^{MV} are dropped further to 0.815 pu and 0.824 pu, as the OXL circuit is activated to protect the generator winding from the fault. Consequently, the LTC controller continues to change the tap position of the transformer to increase the voltage V^{MV}. The increase of V^{MV} along with the LTC actions results in the increase of active and reactive power consumption from voltage-dependent loads, causing the increase in the total load consumption in the urban power grid. However, the limited power transfer capacity is reached due to the tripping of a transmission line. Consequently, the voltage V^{HV} is continuously decreased below the acceptable operating range. This may activate the low voltage protection system which can lead to cascading tripping of other transmission lines and even causing the whole system to be collapse.

11.4.2 Coordination mechanism for dynamic voltage support

In recent years, the coordination between DERs and LTC for voltage stability support has received increasing attention from researcher. In [29], a voltage-constrained centralized management was developed based on a physics-based model in the form of sensitive matrix. A centralized model predictive control (MPC) is developed in [30] to regulate the distribution network voltage using the steady-state voltage sensitivity analysis, which is extracted from an offline power flow calculation. In [31], the concept of distributed MPC was developed for long-term voltage coordination in multi-area power system. This method assumes that voltage of all buses can be observed locally via the phasor measurement units (PMUs). Then a distributed MPC is implemented using spare communication network.

Beside the physics-based methods, data-driven methods are also promising solutions to improve voltage monitoring and control. In [32], an artificial neural network (ANN) is applied to identify the voltage sensitivity matrix which then used for voltage prediction in the centralized MPC model. Recently, in [33], the authors use ANN to build a control knowledge between system dynamics and optimal control actions, thus supporting the controller to find optimal solutions faster.

This section presents a coordination method combining centralized MPC method and a voltage prediction model to support long-term voltage stability. The voltage evolution is predicted online based on system Jacobian matrix. Then, it is used as the predictor for the MPC model, where the optimal control action for LTC and DERs are determined. The primary goal of the method is to use available power supplies from DERs, especially ES units to effectively restore the voltage at the grid edges back to the pre-fault conditions with a minimum number of LTC control actions.

MPC is an advanced control method that uses a discrete-time model of a system to predict the future behavior of the desired control variables and compute a set of future control actions by optimizing an objective function with predefined constraints. The MPC can solve a single or multiple objectives together with discrete and continuous control variables. Here, the overall objective is to identify and call for available resources (LTC action and DER generation) to support restoring voltage after emergency conditions, expressed as follows:

$$\min \sum_{i=0}^{N_c-1} ||\Delta V(k+i)||_{R_v}^2 + ||\Delta V^T(k+i)||_{R_T}^2 \tag{11.9}$$

Subject to

$$\begin{cases} u^{min} \leq u(k+i) \leq u^{max} \\ \Delta u^{min} \leq \Delta u(k+i) \leq \Delta u^{max} \quad \text{for } i = 0, \; 1, ...Nc-1, \\ 0.9\text{pu} \leq V_k^{MV} \leq 1.1\text{pu} \end{cases}$$

where $\Delta u(k) = \left[\Delta P_k^{DER}, \Delta Q_k^{DER}, \Delta V_k^T \right]$ is the change of the control variables at time step k compares to step $k-1$. R_v and R_T are weight matrices for voltage regulation and LTC actions used to determine the priority of the control variables.

The voltage prediction model for MPC is expressed as follows:

$$V_{k+1}^{HV} = V_k^{HV} + \frac{\partial V_k^{HV}}{\partial P_k^{DER_j}} \Delta P_k^{DER_j} + \frac{\partial V_k^{HV}}{\partial Q_k^{DER_j}} \Delta Q_k^{DER_j} + \frac{\partial V_k^{HV}}{\partial V_k^T} \Delta V_k^T \qquad (11.10)$$

where V_k^{HV} is the voltage measurement at the time step k. $\partial V_k^{HV} / \partial V_k^T$ is the voltage sensitivity matrix with respect to an LTC position. $\partial V_k^{HV} / \partial P_k^{DER_j}$ and $\partial V_k^{HV} / \partial Q_k^{DER_j}$ are the voltage sensitivity matrices corresponding to the change of the reactive and active power, respectively. These terms can be obtained using the inverse of the system Jacobian matrix $\mathbf{J}$, as follows:

$$\begin{bmatrix} \Delta \delta_2 \\ \vdots \\ \Delta \delta_n \\ \Delta V_2 \\ \vdots \\ \Delta V_n \end{bmatrix} = \mathbf{J}^{-1} \begin{bmatrix} \Delta P_2 \\ \vdots \\ \Delta P_n \\ \Delta Q_2 \\ \vdots \\ \Delta Q_n \end{bmatrix} \qquad (11.11)$$

where $\Delta \delta_i$ and ΔV_i are the absolute change in voltage angle and voltage magnitude at bus i which corresponding to the change in active (ΔP) and reactive ($\Delta \delta_2$) power.

11.4.3 Performance evaluation

In this section, a simulation has been implemented using MATLAB®/Simulink® to evaluate the effectiveness of the coordination mechanism. The numerical results were obtained on a ThinkPad Laptop with an Intel Core (TM) i7-8750 central processing unit (CPU), 2.20 GHz processing speed and 16 GB random access memory (RAM).

The modified MV European benchmark distribution system is used as the tested system with the topology shown in Figure 11.6 and parameters are explained in [34]. The main system parameters are presented as follows:

- The MV distribution system is supplied by an external grid via two parallel transmission lines, represented by a 110 kV/50 Hz three-phase voltage source, with a short-circuit power of 500 MVA and R/X ratio of 0.1.
- The system nominal voltage: 3-phase, 20 kV (line to line), 50 Hz.
- The transformer with an LTC controller is installed between Buses 0 and 1, which is designed to keep the voltage at Bus 1 within a range from 0.985 pu to 1.015 pu.
- A synchronous generator with OXL contributing to the long-term voltage issue is installed at the HV/MV substation.

The coordination mechanism is used to support the voltage dynamics by optimizing the LTC action and mobilizing the available power supply from DERs.

The MPC algorithm is installed at the *Application Control Layer* of the hierarchical controller. The performance of the method is validated by comparing the voltage stability in the case of with and without the coordination.

Figure 11.8 presents a time sequence of the control actions. After a short stable transient period, the voltage V^{HV}, V^{MV} enter a long-term voltage period. The recorded pre-fault primary voltage is $V_0^{HV} = 0.996$ pu. As the voltage V^{HV} is dropped in contingency case, the coordination mechanism is activated to bring it back to the pre-fault value.

The simulation results show that, V^{HV} is smoothly brought back to the pre-fault value while keeping V^{MV} in a predefined limit (i.e., [0.9, 1.1] pu). In this case, the prediction horizon of MPC is two steps ahead. As can be seen from Figure 11.9, in the period from $t = 22$ sec to $t = 80$ sec, the voltage is slightly increased while the LTC is kept unchanged. This is the advantage of multi-objective based control. In this period, the power from DERs is still available. Thus, the MPC keeps the LTC position unchanged and use only power from DERs to support the voltage. In the next period of time $t = 80$ sec to $t = 140$ sec, the powers from DERs reaches their limits. Thus, the coordination mechanism must use the support from the LTC operation to keep increasing V^{HV} back close to the nominal value.

Figure 11.10 compares the performance of the coordination mechanism with different prediction horizon of the MPC controller. As shown, with $N = 2$, the MPC controller starts to change the LTC position at around $t = 80$ sec, while the time with $N = 5$ and $N = 8$ is around $t = 90$ sec and $t = 160$ sec, respectively. As the result,

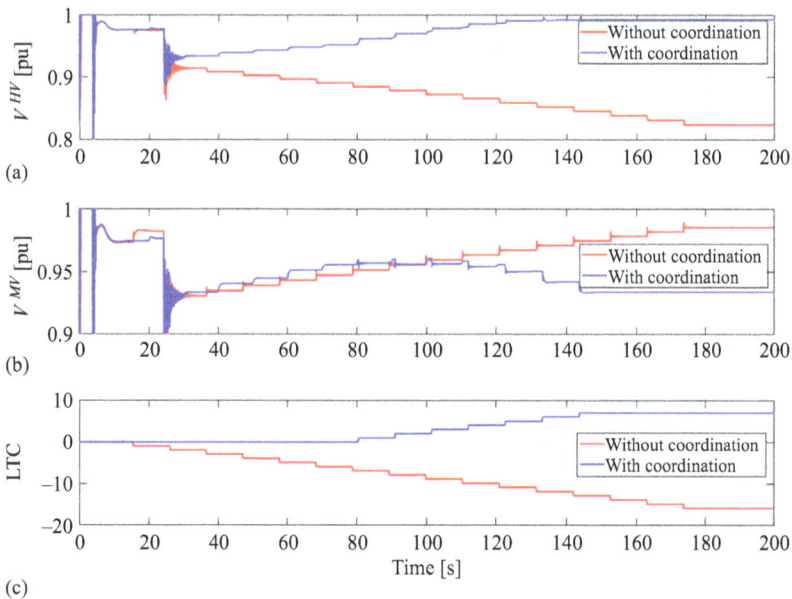

Figure 11.9   The voltage VHV and VMV in the case of with and without the coordination mechanism

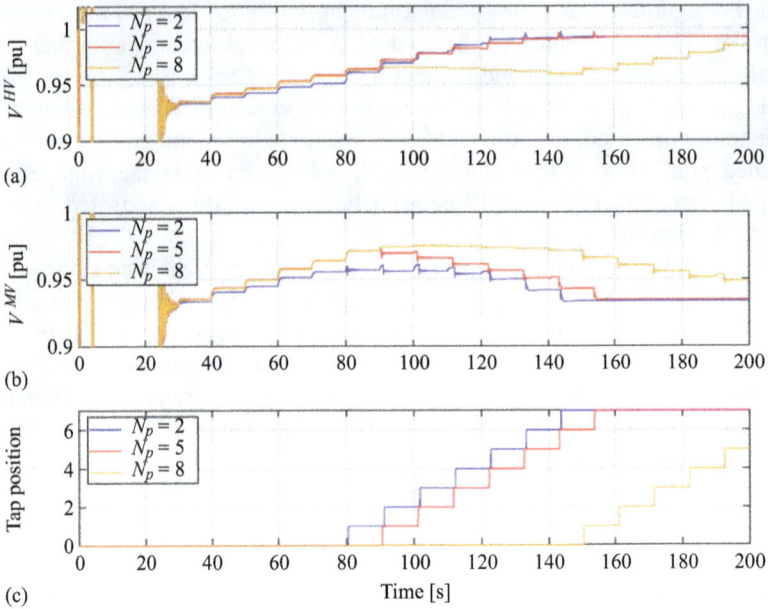

Figure 11.10   The voltage VHV and VMV in the case of different prediction horizons

Table 11.3   Performance comparison with different prediction horizons

Prediction step	LTC actions	Simulation time (min)
2	7	16.27
5	7	25.51
8	5	38.22

V^{HV} recovers faster when using least number of the prediction horizon with the cost of more LTC actions. On the contrary, the computational burden is increased with a longer prediction horizon. This trade-off can be reduced by using adaptive prediction horizon for MPC controller, which is out of the scope of this chapter. Table 11.3 shows the performance comparison with different prediction horizons.

11.5   Service provision

This section aims to another aspect of DERs in supporting the resilience enhancement of urban power grids, namely ancillary services provision. Different from the previous functionalities which are related to *short-term resilience* (i.e., during contingency conditions), the service provision refers to long-term (e.g., hours, or day ahead), market wise contribution of DERs. The increasing penetration of intermittent renewable energy resources in today urban power grid has required network operation to deploy more ancillary services, such as balancing reserves, frequency restoration, to increase the system reliability, resilience and continuity of supply.

Distributed ES systems with large capacity are considered as a promising technology to provide demand/generation balancing in term of frequency regulation. There are three difference types of frequency regulation procured by TSO [35,36]: the frequency containment reserve (FCR) can be considered as the fastest frequency control, which restores the balance between the generated power and load power demand within a few seconds using droop control, the automatic frequency restoration reserve (aFRR) restores the frequency within a few minutes in the secondary control layer, and the manual frequency restoration reserve (mFRR) optimizes the frequency in up to one hour.

Application of ESs in providing the balancing reserve at urban power grids level was investigated in many research in the literature [37,38]. The participation of urban power grids-located ESs in providing the ancillary service for TSO, by taking part in balancing reserve markets, was investigated in [39]. In [40], an ESs bidding mechanism and operation in the frequency regulation market is proposed to maximize the profit of prosumers when they participate in frequency reserve services. Some studies address the roles of distribution network aggregators to take part in a joint day-ahead and reserve markets to support the balancing reserve [41].

Figure 11.11 shows the overview on the proposed scheduling of the urban power grids-level ESs and their interactions with the balancing service provider (BSP) and TSO. In day-ahead stage, the network operator can estimate the flexibility bids based on the generation forecast of intermittent renewable sources (e.g., PV, wind generation), base load forecast, and the availability ESs by running a scheduling algorithm (e.g., stochastic optimization) to maximize the available flexibility that can be provided in the following day. The BSP is obliged to send aFRR energy bids to TSO based on the flexibility bids of each unit for each imbalance settlement period (ISP) during 15-min intervals. The submission time is no later than 14:45 UTC + 01:00, on the day prior the day of execution. TSO organizes capacity auctions via the action platform for ancillary services (APAS). In the executed day, if there is an imbalance happened in the system, the APAS sends delta-set points to the BSPs in order to activate their energy bids. The value of delta-set points is limited to the offered power values in the energy bids in day ahead stage. After receiving the allocation from BSP, based on the quarter-hour updated data inputs, the network operator runs optimization algorithm again to maximize the real-time profit. Then the decision to join the aFRR service will be issued.

11.5.1 aFRR service

If the network operator of urban power grids knows insightful value of the maximum profit, they can make decision to respond the aFRR service by running the optimization algorithm [42]. Equation (11.12) illustrates the objective function of urban power grids profit within 15 min. The objective function aims to optimize the income when they respond the flexibility requirements from TSO, considering

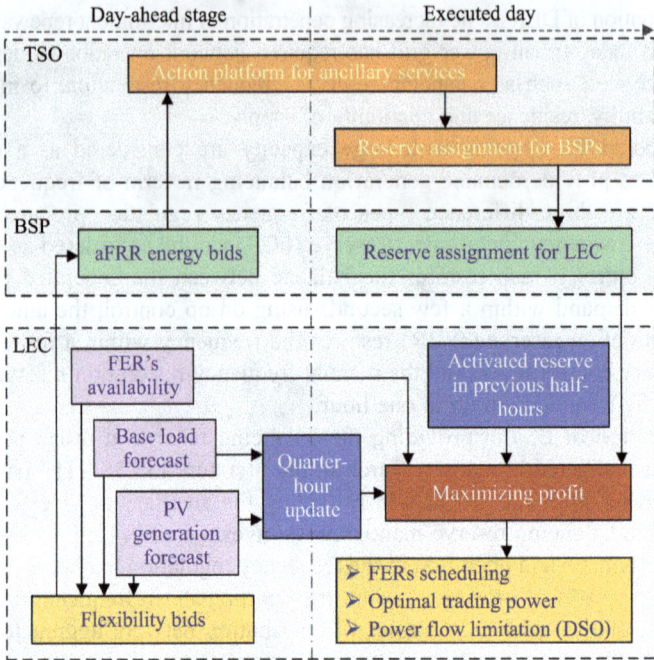

Figure 11.11   Overview on the proposed scheduling of the urban power grid-level ESs and their interactions with the balancing service providers and TSO

several general constrains (e.g., available power, charging/discharging rate of ESs):

$$\min \sum_{m=1}^{15} -Pf_m \tag{11.12}$$

Here, the profit Pf when responding to the aFRR service in 15-min period is expressed as follows:

$$Pf_m \underbrace{[fm + u^{fl}\lambda^{up}P^{fl,up} - (1 - u^{f1})\lambda^{dn}p^{fl,dn}]}_{A} + \underbrace{\varsigma^{B}[u^{B}P_m^{B,dis} + (1 - u^{B})P_m^{B,ch}]}_{B} + \underbrace{[\lambda^{exp}P_m^{exp} - \lambda^{imp}P_m^{imp}]}_{C}$$

$$\tag{11.13}$$

The A term is the income obtained from participating in aFRR service. fm is a fixed monetary amount which is paid for offering flexibility capacities (both upward and downward). μ^{fl} is a binary variable (1 when urban power grids offers the upward flexibility, otherwise 0). λ^{up} and λ^{dn} are the price when urban power grids offer the upward ($P_m^{fl,up}$) or downward ($P_m^{fl,dn}$) flexibility through the ES operation, respectively.

The B term is the cost when the ES discharges or charges to support the aFRR service. ζ^B is the cost rate for ES operation. μ^B is a binary value (1 when ES discharges, otherwise 0). $P_m^{B,dis}$ and $P_m^{B,ch}$ are the discharging or charging power of ES.

The C term is the income when urban power grids exchanges a surplus power P_m^{exp} or an imported power P_m^{imp} to/from the upstream grid with the selling price λ^{exp} or the buying price λ^{imp}, respectively, after fulfilling the aFRR service.

Equations (11.14) and (11.15) express the imported and the exported power to trade a surplus power from ES units in urban power grid with the upstream grid through a DSO or retailer after fulfilling the assigned flexibility value from TSO in minute time resolution during ISP:

$$P_m^{imp} = \left(P_m^{load} - P_m^{PV}\right) + P_m^{B,ch} - P^{fl,dn} \tag{11.14}$$

$$P_m^{exp} = -\left(P_m^{load} - P_m^{PV}\right) + P_m^{B,dis} - P^{fl,up} \tag{11.15}$$

where P_m^{load} and P_m^{PV} are base load forecast and renewable generation forecast in the previous 30 min. If $P_m^{imp} > 0$, there is a certain amount of power imported from upstream grid to the urban power grids after fulling the assigned downward flexibility. In contrast, if $P_m^{exp} > 0$, there is a surplus power exported from urban power grids to upstream grid after fulfilling the assigned upward flexibility.

11.5.2 Constraints

In the suggested model, it is assumed that TSO only assign the upward $P^{fl,up}$ or downward $P^{fl,dn}$ flexibility separately during ISP. On the other words, the status binary value of discharging/charging ES operation is μ^B or $1 - \mu^B$, respectively. As shown in (11.16), the SoC_m of the ES at the mth minute depends on its value at the $(m-1)$th minute (SoC_{m-1}), the discharging/charging status, and the capacity Cap_B. η^{dis} and η^{ch} are the efficiency of discharging or charging. The value of SoC_m is limited within the range $[SoC_{min}, SoC_{max}]$. The operation of BESS (discharging or charging power) cannot exceed their power limit rate ($P_{max}^{B,dis}$, $P_{max}^{B,ch}$), respectively:

$$SoC_m = \begin{cases} 50\%, & if \quad m = 1 \\ SoC_{m-1} + \left(\eta^{ch}P_m^{B,ch} - \dfrac{P_m^{B,dis}}{\eta^{dis}}\right)\dfrac{1}{Cap^B}, & else \end{cases} \quad \forall m \in [1, 15] \tag{11.16}$$

11.5.3 Numerical simulation

In the executed day, after receiving the upward/downward flexibility assignment from TSO, the BSP allocates the amount of flexibility from the urban power grids. The optimization problem introduced in (11.12)–(11.16) has been solved with a mixed-integer linear programming formulation using the Pyomo solver in executed day. The tested system is assumed to have PV generation and ES units as flexibility resources.

In the first case, at mid night from 00:00 to 00:14 am, the BSP assigns the upward flexibility while urban power grid consumes power from the grid without any PV generation in the left-sub-figure of Figure 11.12. Although urban power grid consumes the power from the grid (green line) without any PV generation (red line), the BSP assigns the 30 kW upward flexibility (blue line). That means ES units should discharge to support both the aFRR service and the load consumption. Figure 11.13 shows the power flow from urban power grid to

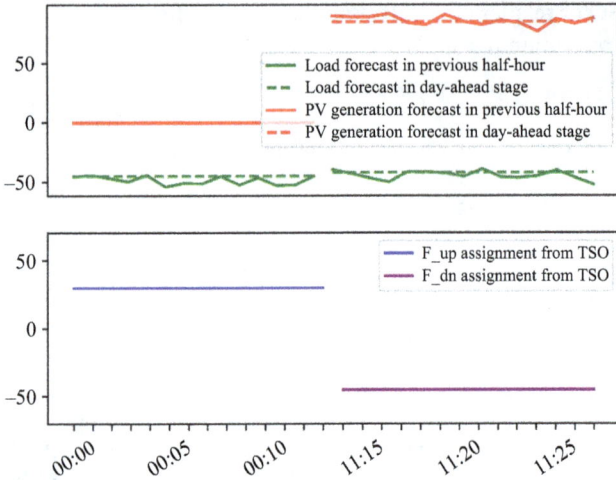

Figure 11.12   PV generation, load demand, flexibility of ES units in three case studies

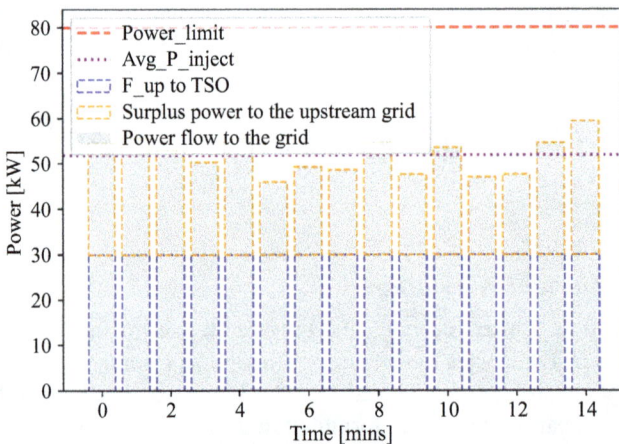

Figure 11.13   Urban power grid injects power to the grid at 00:00 to 00:14 am, case 1

the upstream grid (grey bar). The average value of the power delivered to the grid (purple dotted line) is smaller than the capacity limitation in the contract with DSO (red dashed line). In this case, ES units are discharged to cover the upward flexibility assignment from TSO (blue dashed line). To get the maximum profit during the aFRR activation period, after fulfilling the upward flexibility and the load consumption, ES units discharge as much as possible within the rate limitation and the surplus power are sold (orange dashed bar) to the upstream grid through DSO or retailer.

In the second case, at 11:15 to 11:29 am, the BSP assigns the downward flexibility while urban power grid has a surplus power due the very high PV generation. Urban power grid has a surplus power due to high PV generation while the BSP assigns the 50 kW downward flexibility (purple line). Figure 11.14 shows that urban power grid should charge BESS to support the aFRR service. After fulfilling the downward flexibility, urban power grid only absorbs the surplus power from PV generation to maximize the profit, even though BESS can afford to absorb more power from the grid. Figure 11.15 shows the power flow absorbed by Urban power grid from the upstream grid. There is an identical value between downward flexibility value (blue dashed bar) and the power flow value (grey bar). In other words, urban power grid does not buy power from retailers because they do not want to pay more from upstream grid after fulfilling the flexibility in order to maximize the profit.

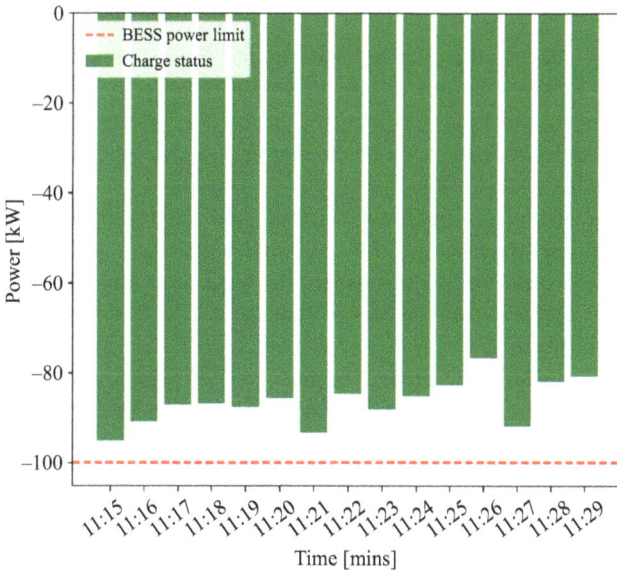

Figure 11.14   ES charges power from to cover load consumption and support the aFRR service at 11:15 to 11:29 am, case 2

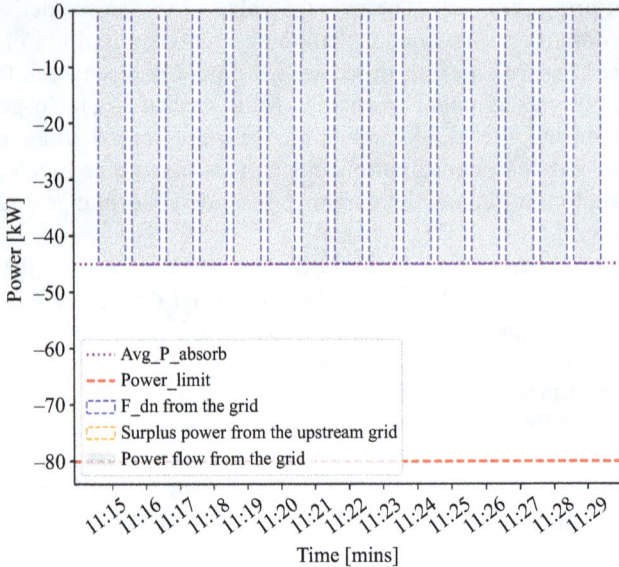

Figure 11.15   Urban power grid absorbs power from the grid at 11:15 to 11:29 am, case 2

11.6   Conclusion

In this chapter, the coordination mechanism for dispatchable sources, including intermittent renewable energies and distributed ESs, in today urban power grids is presented. The coordination mechanisms utilize the available generation from dispatchable sources to enhance both the restoration capability the critical areas in urban power grids under different disruptive events such as severe weathers, grid malfunctioning or human operation missteps, and maximize the benefit of these sources by optimal mobilizing the available generation to provide ancillary services. These coordinated mechanisms are based on a multiple time-scale hierarchical framework that can effectively manage a complex, multi-target requirement of the grid resilience.

A FII mechanism is discussed in this chapter which enables the network operator to detect the faults, isolate (a part) urban power grid from main distribution system and reconnect it back when the fault are clear. The proposed method also allows a seamless transition between grid-connected to off-grid mode. The hierarchical control structure consists of three control layers, each of which has its own control objective and different dynamic response. The FII detection based on active detection method is embedded into *Application Control Layer* to continuously observe the change in network impedance to generate control signals for DERs in the urban power grid to switch to proper operation modes. The simulation results show a good performance of the described methods in detecting the faults and switching between operation modes to maintain stable operation and continuity of supply of urban power grids.

A dynamic voltage stability control is presented to support the urban power grid to restore the voltage after emergency conditions. The method coordinates the operation of DERs and LTC controller of transformers to improve the dynamic voltage stability. The method uses simple voltage prediction technique to predict the evolution of the voltage, then served as a predictor for the MPC model to determine the optimal control action for LTC and DERs. The simulation results show how it is possible to mobilize the available generation from DERs to effectively restore the voltage at the grid edges back to the pre-fault conditions with a minimum number of LTC control actions.

References

[1] Pan J, Callavik M, Lundberg P, and Zhang L. A subtransmission metropolitan power grid: using high-voltage dc for enhancement and modernization. *IEEE Power and Energy Magazine*. 2019;17(3):94–102.

[2] Oboudi MH, Mohammadi M, Trakas DN, and Hatziargyriou ND. A systematic method for power system hardening to increase resilience against earthquakes. *IEEE Systems Journal*. 2021;15(4):4970–9.

[3] Forssén K and Mäki K (eds.). Resilience of Finnish electricity distribution networks against extreme weather conditions. *CIRED Workshop 2016*, June 2016, 2016 14–15.

[4] Deng Y, Jiang W, Hu F, Sun K, and Yu J. Resilience-oriented dynamic distribution network with considering recovery ability of distributed resources. *IEEE Journal on Emerging and Selected Topics in Circuits and Systems*. 2022;12(1):149–60.

[5] Eder-Neuhauser P, Zseby T, and Fabini J. Resilience and security: a qualitative survey of urban smart grid architectures. *IEEE Access*. 2016;4:839–48.

[6] Nguyen-Huu TA, Tran TT, Tran MQ, Nguyen PH, and Slootweg J (eds.). Operation orchestration of local energy communities through digital twin: a review on suitable modeling and simulation approaches. In *2022 IEEE 7th International Energy Conference (ENERGYCON)*, 9–12 May 2022.

[7] BSI. Certification Report BSI-CC-PP-0077-V2-2015: BSI Protection Profile. Tech Rep BSI-CC-PP-0077-V2-2015, 28 January 2015.

[8] ENISA. Guidelines for Smart Grid Cyber Security. National Institute of Standards and Technologies, 2010.

[9] Panteli M and Mancarella P. Modeling and evaluating the resilience of critical electrical power infrastructure to extreme weather events. *IEEE Systems Journal*. 2017;11(3):1733–42.

[10] Tran TT. Advanced Hierarchical Control Structure for Virtual Oscillator-Based Distributed Generation in Multi-Bus Microgrids Under Different Grid Dynamics and Disturbances, 1. *Auflage*, Aachen: RWTH Aachen University, 2020.

[11] Han Y, Shen P, Zhao X, and Guerrero JM. Control strategies for islanded microgrid using enhanced hierarchical control structure with multiple current-loop damping schemes. *IEEE Transactions on Smart Grid*. 2017;8(3):1139–53.

[12] Raisz D, Thai TT, and Monti A. Power control of virtual oscillator controlled inverters in grid-connected mode. *IEEE Transactions on Power Electronics*. 2019;34(6):5916–26.

[13] Zhong Q, Nguyen P, Ma Z, and Sheng W. Self-synchronized synchronverters: inverters without a dedicated synchronization unit. *IEEE Transactions on Power Electronics*. 2014;29(2):617–30.

[14] Tran TT, Sowa I, Raisz D, and Monti A. An average consensus-based power-sharing among VOC-based distributed generations in multi-bus islanded microgrids. *IET Generation, Transmission & Distribution*. 2021;15(4):792–807.

[15] Sowa I, Tran TT, Heins T, Raisz D, and Monti A. An average consensus algorithm for seamless synchronization of Andronov-Hopf oscillator based multi-bus microgrids. *IEEE Access*. 2021;9:90441–54.

[16] Tran TT, Tran MQ, Nguyen An, Nguyen HP, and Le AT. Virtual oscillator based hierarchical control strategy for multi-mode operation of microgrids. In *12th IEEE PES Innovative Smart Grid Technologies Conference Europe, ISGT Europe 2022*, Novi Sad, Serbia, 2022.

[17] Roos MH, Faizan MF, Nguyen PH, Morren J, and Slootweg JG (eds.). Probabilistic adequacy and transient stability analysis for planning of fault-initiated islanding distribution networks. In *2021 IEEE Madrid PowerTech*, 28 June–2 July 2021.

[18] Laaksonen H. Advanced islanding detection functionality for future electricity distribution networks. *IEEE Transactions on Power Delivery*. 2013;28(4):2056–64.

[19] Rintamaki O and Kauhaniemi K (eds.). Applying modern communication technology to loss-of-mains protection. In *CIRED 2009 – 20th International Conference and Exhibition on Electricity Distribution – Part 1*, 8–11 June 2009.

[20] Coffele F, Moore P, Booth C, *et al.* (eds.). Centralised loss of mains protection using IEC-61850. In *10th IET International Conference on Developments in Power System Protection (DPSP 2010) Managing the Change*, 29 March–1 April 2010.

[21] Ku Ahmad KNE, Selvaraj J, and Rahim NA. A review of the islanding detection methods in grid-connected PV inverters. *Renewable and Sustainable Energy Reviews*. 2013;21:756–66.

[22] Khamis A, Shareef H, Bizkevelci E, and Khatib T. A review of islanding detection techniques for renewable distributed generation systems. *Renewable and Sustainable Energy Reviews*. 2013;28:483–93.

[23] Panigrahi BK, Nandi R, Mahanta B, and Pal K (eds.). Islanding detection in distributed generation. In *2016 International Conference on Circuit, Power and Computing Technologies (ICCPCT)*,18–19 March 2016.

[24] Ten CF and Crossley PA (eds.). Evaluation of Rocof relay performances on networks with distributed generation. In *2008 IET 9th International Conference on Developments in Power System Protection (DPSP 2008)*, 17–20 March 2008.

[25] Menon V and Nehrir MH. A hybrid islanding detection technique using voltage unbalance and frequency set point. *IEEE Transactions on Power Systems*. 2007;22(1):442–8.

[26] Singh SK, Rawal M, Rawat MS, and Gupta TN (eds.). Hybrid islanding detection technique for inverter based microgrid. In *2021 2nd International Conference for Emerging Technology (INCET)*, 21–23 May 2021.

[27] Lu M, Dutta S, Purba V, Dhople S, and Johnson B (eds.). A grid-compatible virtual oscillator controller: analysis and design. In *2019 IEEE Energy Conversion Congress and Exposition (ECCE)*, 29 September–3 October 2019.

[28] CIGRE. Benmark Systems for Network Integration of Renewable Energy Resources. CIGRE Task Force C60402, 2011, 07.

[29] Capitanescu F, Bilibin I, and Ramos ER. A comprehensive centralized approach for voltage constraints management in active distribution grid. *IEEE Transactions on Power Systems*. 2014;29(2):933–42.

[30] Valverde G and Cutsem TV. Model predictive control of voltages in active distribution networks. *IEEE Transactions on Smart Grid*. 2013;4(4):2152–61.

[31] Moradzadeh M, Boel R, and Vandevelde L. Voltage coordination in multi-area power systems via distributed model predictive control. *IEEE Transactions on Power Systems*. 2013;28(1):513–21.

[32] Nguyen HM, Torres JLR, Lekić A, and Pham HV. MPC based centralized voltage and reactive power control for active distribution networks. *IEEE Transactions on Energy Conversion*. 2021;36(2):1537–47.

[33] Cai H, Ma H, and Hill DJ. A data-based learning and control method for long-term voltage stability. *IEEE Transactions on Power Systems*. 2020;35 (4):3203–12.

[34] Tran MQ, Tran TT, Nguyen PH, and Tuan LA (eds.). Coordination of load tap changer and distributed energy resources to improve long-term voltage stability. In *2022 IEEE 7th International Energy Conference (ENERGYCON)*, 9–12 May 2022.

[35] Nguyen-Huu T-A, Nguyen VT, Hur K, and Shim JW. Coordinated control of a hybrid energy storage system for improving the capability of frequency regulation and state-of-charge management. *Energies*. 2020;13(23):6304.

[36] Shim JW, Verbič G, Zhang N, and Hur K. Harmonious integration of faster-acting energy storage systems into frequency control reserves in power grid with high renewable generation. *IEEE Transactions on Power Systems*. 2018;33(6):6193–205.

[37] Merten M, Olk C, Schoeneberger I, and Sauer DU. Bidding strategy for battery storage systems in the secondary control reserve market. *Applied Energy*. 2020;268:114951.

[38]   Xu B, Shi Y, Kirschen DS, and Zhang B. Optimal battery participation in frequency regulation markets. *IEEE Transactions on Power Systems*. 2018;33(6):6715–25.

[39]   Lund PD, Lindgren J, Mikkola J, and Salpakari J. Review of energy system flexibility measures to enable high levels of variable renewable electricity. *Renewable and Sustainable Energy Reviews*. 2015;45:785–807.

[40]   Nasrolahpour E, Kazempour J, Zareipour H, and Rosehart WD. A bilevel model for participation of a storage system in energy and reserve markets. *IEEE Transactions on Sustainable Energy*. 2018;9(2):582–98.

[41]   Nezamabadi H and Nazar MS. Arbitrage strategy of virtual power plants in energy, spinning reserve and reactive power markets. *IET Generation, Transmission & Distribution*. 2016; 10(3):750–63. https://digital-library.theiet.org/content/journals/10.1049/iet-gtd.2015.0402.

[42]   Nguyen-Huu TA, Tran TT, Nguyen PH, and Slootweg J (eds.). Network-aware operational strategies to provide (flexibility) services from Local Energy Community. In *2022 57th International Universities Power Engineering Conference (UPEC)*, 30 August–2 September 2022.

Chapter 12

Energy storage impacts on intelligent electricity markets

Vinícius Braga Ferreira da Costa[1], Jorge Vleberton Bessa de Andrade[1], Mateus Gomes de Siqueira e Salles[1] and Benedito Donizeti Bonatto[1]

This chapter seeks to provide details of the impacts of distributed energy storage in intelligent electricity markets. Business models and ancillary services for energy storage systems are described, as well as the changes on electricity market agents due energy storage system deployment taking in consideration regulatory issues. The chapter also approaches the context of energy storage policies and programs like demand response in the context of high renewable resources penetration and energy storage. Lastly, policies and flexibility is briefly explored.

12.1 Initial considerations

As previously highlighted, the electric grid has undergone an intense transformation. Such transformation includes the increasing deployment of intermittent renewable generation, increased energy efficiency (e.g., efficient appliances), and the popularisation of ancillary services (e.g., advanced frequency and voltage control). In this context, energy storage systems (ESS) are being increasingly deployed, as they are essential to support (or even to allow) the grid edge transition. However, the popularisation of ESS requires technological improvements and a developed intelligent market and regulatory design. Therefore, the implemented business model regarding ESS must simultaneously ensure increasing deployment of ESS, efficient use of assets, and a fair market environment among all market agents (leading to a smart market establishment). Thus, technological, economic, environmental, market and regulatory design advances must occur concurrently to improve the efficiency of the electrical system.

A bibliographic search of 2,000 publications in the energy storage and electricity market topics (Web of Science database) and big data processing by the software R and the open-source tool Bibliometrix [1] resulted in Figure 12.1, which

[1]Institute of Electrical Systems and Energy, Federal University of Itajubá, Brazil

Most relevant words

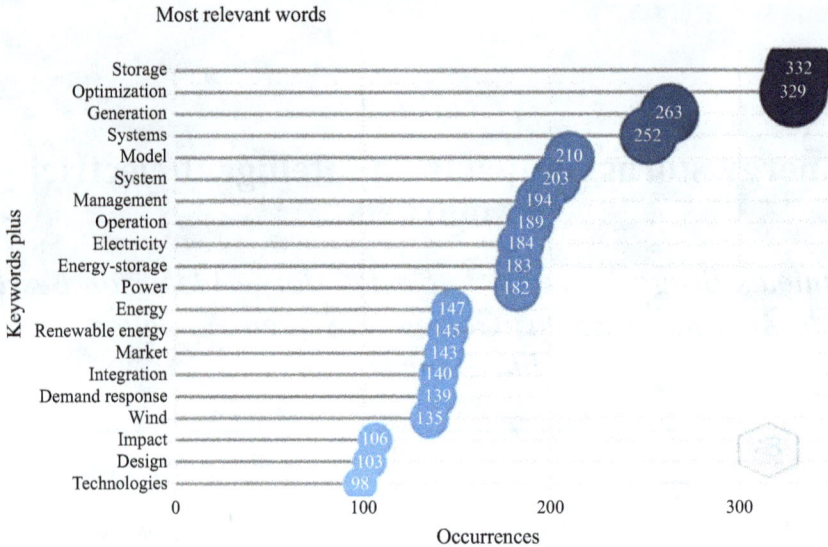

Figure 12.1 contains a chart with the vertical axis labeled "Keywords plus" and the horizontal axis labeled "Occurrences" (0, 100, 200, 300). The keywords and their occurrence values:

- Storage — 332
- Optimization — 329
- Generation — 263
- Systems — 252
- Model — 210
- System — 203
- Management — 194
- Operation — 189
- Electricity — 184
- Energy-storage — 183
- Power — 182
- Energy — 147
- Renewable energy — 145
- Market — 143
- Integration — 140
- Demand response — 139
- Wind — 135
- Impact — 106
- Design — 103
- Technologies — 98

Figure 12.1   Bibliographic search of most used keywords in the energy storage and electricity market

illustrates the most used keywords. Some of them are further analysed in this section.

12.2   Business models and ancillary services in the context of ESS

Several studies on developing an advanced market structure/regulatory/business model in the context of increasing ESS deployment have been found in the literature. However, Ref. [2] indicates that there are many opportunities for improvement since the profitability of ESS has remained ambiguous (opinions of researchers diverge considerably). The challenge of developing efficient business models is directly related to the vast range of applications, particularly regarding ancillary services. Each of these services requires notable differences in how energy storage is monetised. For instance, the business model must fairly remunerate backup energy providers and frequency regulation providers. Naturally, a business model based on ESS's usage time would be inefficient since backup energy providers barely use their assets (but when they do, it is essential). It is also challenging to develop efficient business models that equitably consider both the interests of the investors in ESS and the optimal operation of the electric grid. Moreover, Ref. [2] highlights that there are multiple ESS market roles (e.g., trading, production). Thus, an efficient business model for ESS must also consider all the ESS possible roles.

In the context of increasing ESS deployment, Ref. [3] focuses on an empirical collaborative business model among a network operator and initiatives on ESS. Ref. [4] assesses the techno-economic aspects that must be considered to develop efficient business models. Ref. [5] proposes a sharing economy scheme as a business model, i.e., the investors in ESS can offer their systems to multiple kinds of customers. In [6], a business model is proposed for an EV charging station.

In conclusion, the literature review demonstrated that continuous efforts had been made to develop efficient business models in the context of ESS. However, each author has their perspective on the topic and generally analyses a specific problem. Therefore, integrating these studies into an efficient, fair, replicable and flexible business model is the biggest challenge.

In addition, most studies conduct empirical assessments concerning business models. Given the importance of the theme, more analytical studies are required. According to [7], the effectiveness of business models is typically restricted to specific regions, and there is often more than one feasible model to implement; thus, the choice becomes generally empirical. It is also noteworthy that business models can be combined to enhance the deployment of ESS and develop electricity markets.

Another issue concerning business models that require attention is that schemes should be addressed as changeable. For example, the net metering policy in Brazil (DG business model) was recently ruled out and replaced by Ordinary Law 14300, which will soon decrease the compensation of the electricity injected into the grid by prosumers. In this context, studies should propose current business models and assess how they can evolve. Thus, market needs and regulatory issues yet can play an important role.

12.3 ESS and demand response programs in the context of high penetration of intermittent renewable sources

The increasing deployment of intermittent renewable generation leads to efficient strategies to ensure a reliable power supply and efficient power system operation. Among strategies commonly applied, stand out:

- Deployment of distributed ESS to balance demand and supply;
- Implementation of demand response (DR) methods, i.e., financial incentives (e.g., dynamic electricity pricing) to shift consumption strategically (usually to periods with low demand/high supply) [8].
- Implementation of capacity markets (CM) (e.g., implementation in the United Kingdom [9]), which ensures backup generators at times of grid stress.

While the three strategies mentioned above are practical to support the deployment of intermittent renewable generation, CM usually remunerates participants based on offered capacity (MW) [10]. This means that participants are remunerated even if backup generation is not necessary. Therefore, CM is generally viewed as a more straightforward strategy. However, as proven by [11], distributed energy storage combined with DR is expected to be a more efficient strategy since they achieve virtually the same supply reliability but at a lower cost. Therefore, with the continuous price decrease of ESS, it is safe to assume that solutions will naturally transition towards deploying distributed ESS.

Several studies focus on deploying ESS with DR to support high penetrations of intermittent renewable generation. For instance, Ref. [12] states that they are almost perfect solutions and proposes a method to calculate optimal energy storage investments. Results demonstrate that optimal investments highly depend on the applied DR method, i.e., the potential of energy shifting. Ref. [13] proposes a framework to evaluate the effectiveness of ESS and DR in supply reliability. The findings of [13] are more conservative as the authors affirm that ESS and DR can sometimes lead to more severe power interruptions. Ref. [14] analyses the specific case of the integration of photovoltaic (PV) generation and plug-in hybrid electric vehicles (PHEV) in the context of several DR programs [e.g., time-of-use (TOU) and real-time pricing (RTP)]. Results demonstrate that if carried out correctly, customers and utility companies can benefit from DR programs with PHEV integration. Finally, Ref. [15] analyses the operational value of ESS and DR in supply reliability, market and regulatory issues, including potential barriers to deployment.

12.4   New transitions in energy market infrastructure toward a hydrogen economy

Traditionally, it was assumed that electricity markets mainly focused on short-term profits, whereas investment policies rarely consider long-term consequences for society. The deregulation of electricity markets was expected to be suitably competitive, achieving a fair balance between demand and supply [16]. This paradigm has become even more challenging due to the establishment of distributed energy resources and environmental concerns. At the same time, throughout the last decades, significant changes in communication systems have offered greater possibilities for control and monitoring throughout the entire electrical system, especially in the future smart electricity market. Flexibility and efficient operation of the electricity grid at a lower cost are assumptions of the so-called smart grid, revolutionising the renowned electrical system. An intelligent network intelligently manages all connected entities, from generators to end-users and prosumers [17]. The smart grid aims to supply sustainable, economical, and safe electricity efficiently. Understanding it becomes essential to facilitate handling and increase resource and equipment efficiency, especially in the electrical system [18].

Smart grids are also being redesigned to support the hydrogen market. The current revolution in the hydrogen economy, resulting from the Paris Agreement, signed in 2016 by 196 nations to meet the goals of reducing carbon emissions (and hopefully the COP27 in 2022), is imposing changes in market infrastructure [19]. This, not to mention the Ukraine conflicts, which showed Europe's energy dependence on Russia, bringing to light the importance of Europe strengthening its transition to renewable sources, mainly hydrogen. In a press conference, Josep Borrell, Vice President of EU diplomacy, said, "Energy will not be out of this conflict" [20]. Europe has a dependence on Russian gas, and reducing this dependence will bring substantial turmoil in the markets, increasing prices, which consumers will feel.

Decarbonisation is experienced in several electricity markets, especially in integrating renewable energies, which fulfil the role of achieving net-zero emissions [21]. In the current context, decarbonising brings profound changes in production and consumption patterns and their significant synergistic effects decreases in atmospheric pollutants [22]. This challenging task will require increasing the current resources of electrical systems to face new challenges, especially in electricity-related sectors such as heating and transport, which have an urgent need for decarbonisation. Electrification and energy poverty reduction are more efficient ways to decarbonise these sectors.

In this way, the hydrogen economy becomes necessary for challenging sectors to decarbonise, with green hydrogen potentially being one of the most promising solutions. This sectoral integration will undoubtedly pose new challenges to the functioning of the electricity market and its infrastructure. In addition, the energy transition will require the support of energy storage and demand-side management, particularly with green hydrogen. Therefore, a multifaceted approach is needed to define how the infrastructures of the electricity markets will evolve, adapt to changes, and assess the techno-economic and social advantages of different decarbonisation means.

Hydrogen is valuable as an energy carrier! It allows for the efficient storage of large energy blocks over long periods and for mobility and low-carbon distributed energy generation [23,24]. In an energy storage context, hydrogen has a considerable capability for intermittent renewable generation [24]. Wind and solar represent an opportunity for hydrogen production by electrolysis (green hydrogen) when there is an excess supply of electricity of intermittent origin and, consequently, as an energy storage vector, enabling more outstanding entry of variable renewables in the electricity markets. Thus, hydrogen can interconnect the fuel, electricity, and industrial sectors, mainly green hydrogen generated from water electrolysis, with energy from variable renewable sources (primarily wind and solar).

The development of hydrogen storage technologies is at an emerging and increasingly punctual stage, especially when planning the infrastructure of smart grids and how these potential technologies can be integrated. In Table 12.1, technologies in hydrogen systems are listed (by applications) in the context of energy storage with characteristics that make hydrogen a vector of integration between energy systems and other networks. These technologies are in the Power-To-X (P2X) technology group, which concerns the different electricity conversion, storage, and reconversion paths that use hydrogen as a vector to produce gaseous, liquid, or solid fuels [25,26], as seen in Figure 12.2.

ESSs have become an essential solution for the energy transition from fossil fuel-based to renewable electricity, from the diversity of ways energy can be stored, transported, and consumed by society. Identifying which technology would be best for a given application is helpful. These systems can be classified according to Figure 12.3 and, in the current context, link directly to the subset of P2X technologies for energy storage as a crucial set of technologies to

Table 12.1 Power-to-X hydrogen systems in energy storage context

Technology	Concept	Scale	Application	Efficiency[a]	Source
Power-to-Power (P2P)	Power conversion to hydrogen via electrolysis stored and re-electrified	Large scale and small scale	Electrolysers [proton exchange membranes (PEM)]; fuel cells; hydrogen gas turbine	30–45%	[28–30]
				60–68% LHV	[31]
			Alkaline electrolysers; fuel cells; hydrogen gas turbine	29–40%	[28,32]
				63–71% LHV	[31]
Power-to-heat (P2H)	It involves using the electricity produced from renewable sources to provide heat for residential purposes	Large scale and small scale	Electrolysers; electric boilers; heat pumps	60–70%	[33–36]
Power-to-heat and hydrogen (P2HH)	Integrated utilisation solution of producing hydrogen while recycling heat	Small-scale and decentralised	Electrolysers (PEM); hydrogen storage boiler	90–100%	[33]
Power-to-gas (P2G)	Power conversion to hydrogen through electrolysis through the option of further combining it with CO_2 to produce methane.	Large scale and small scale	Hydrogen-enriched natural gas (HENG) excl. gas turbine	~73%	[28]
			Hydrogen-enriched natural gas (HENG) incl. gas turbine	~26%	[28,37]
			Methanation excl. gas turbine	~55%	[28]
			Methanation incl. gas turbine	~21%	[28,37]
Power-to-fuel (P2F)	Liquid fuel out of electricity by hydrogenating biomass or using CO_2 as a carbon source	Industrial scale	Sabatier process for methane	~72.1%	[38]
			Haber–Bosch process for ammonia	~52%	[38]
			Methanol synthesis	39–42%	[38]

[a]Efficiency based on HHV of hydrogen.
Note: HHV, higher heating value; LHV, lower heating value; excl., excluding; incl., including.

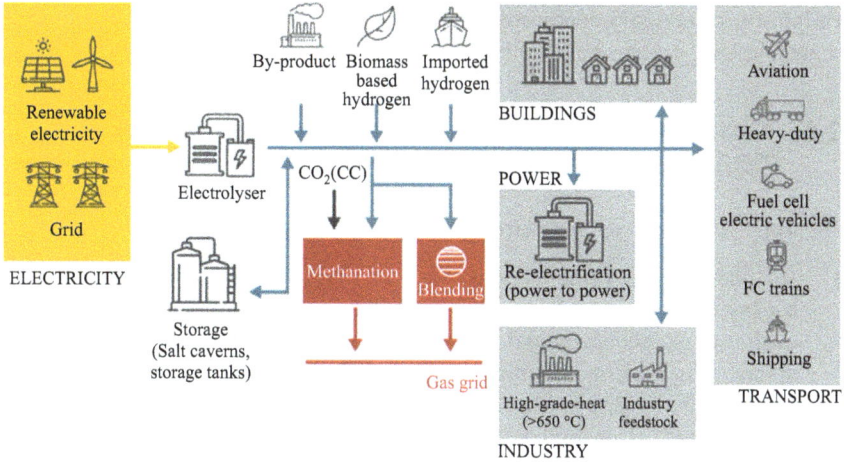

Figure 12.2   Integration of renewables into end uses using hydrogen technologies. Source: [25,26].

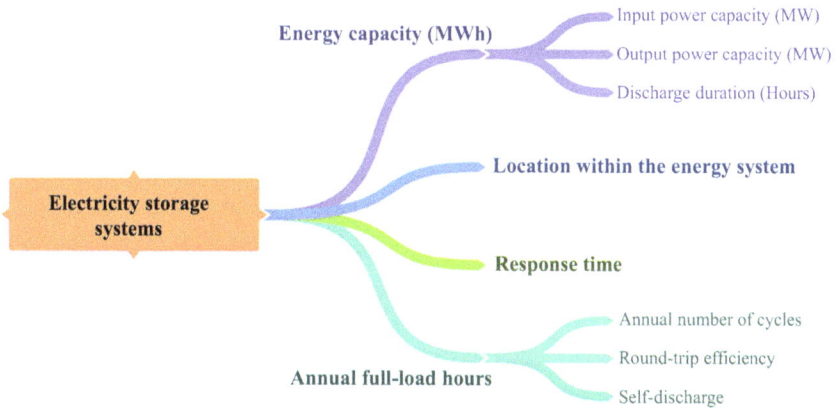

Figure 12.3   Classification of electricity storage systems. Source: Based on Ref. [28].

achieve net-zero emissions. In many cases, P2X encompasses the group of energy storage vectors closely resembling fossil fuels' energy and chemical properties and is well-positioned to smooth the energy transition [27]. It is expected that P2X technologies may be able to counterbalance daily fluctuations in renewable energy supply as long as prolonged lapses must be covered, for example, during low wind and solar radiation intensity. As these are flexible solutions for the different sectors, they can balance fluctuations over several months, making it possible to compensate for deficiencies in the seasonal availability of primary renewable energy, which in phases of excess supply of renewable energy, can be used by the P2X solutions.

12.5 Changes on electricity market agents due to ESS deployment and regulatory issues

Among the roles of ESS in electricity markets stand out [39]:

(i) Ancillary services applications, such as [39]: levelling the load, providing backup electricity, and ensuring grid safety and reliability; enhancing power quality (e.g., frequency and voltage regulation); deferring system expansion; diversifying portfolios; mitigating system fluctuations.

(ii) Transacting electricity in advantageous periods (e.g., selling electricity in peak periods). This application is more straightforward than advanced ancillary services and has been extensively studied [40].

It is essential to recognise that if regulation is precarious in the context of increasing distributed energy resources (DERs) deployment, conventional consumers might be harmed by technology integration. This is particularly true since technologies can affect the electricity tariff of regulated markets. Figure 12.4 illustrates the possible implications of an inefficient electricity market regulation (a process popularly known as the death spiral). Therefore, solutions must simultaneously focus on compensating ESS investors fairly, promoting more ESS investments, and economically safeguarding consumers who do not have access to technology (ensuring more social equality). In this context, business models must not only consider investors in ESS, their role, and the service they are providing, as there is a necessity to assess the whole picture (whole electricity market, including

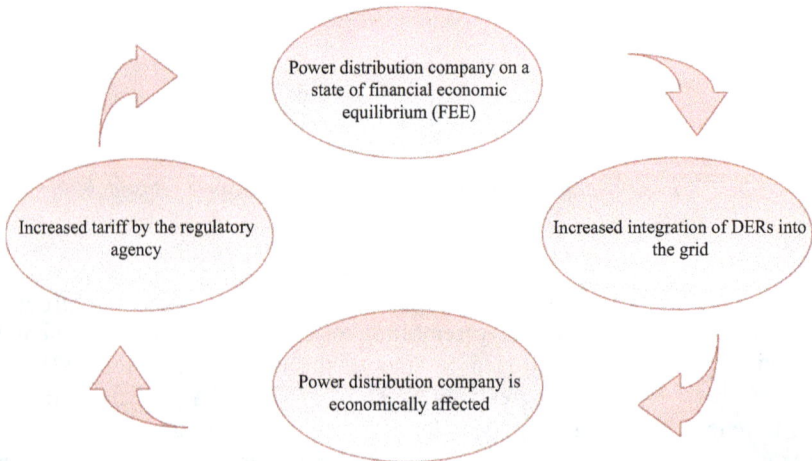

Figure 12.4 Possible implications of inefficient regulation in the context of increasing DERs deployment

the consumers, DERs owners, power distribution/transmission company, government, underrepresented market agents, etc.).

Each technology presents advantages and disadvantages (e.g., lifespan, efficiency, energy cost, etc.); thus, they contribute concurrently to fostering electricity markets. For instance, one particular technology might be more feasible for a particular ancillary service application. It is the case of supercapacitors in supporting the grid for short periods due to their high power density. It is also noteworthy that hybrid storage systems showcase great potential as they can combine the advantages of multiple technologies. Hybrid systems have been implemented in Aachen, Germany (hybrid Li–ion + lead–acid battery system) and in Braderup, Germany (hybrid Li–ion + redox flow battery system) to provide testing, demonstration, and balancing [39].

The development of the hydrogen economy enables the insertion of new agents in the electricity markets from the perspective of energy storage. The use of electrolysers provides greater flexibility in smart markets, whether regional or national and on a small or large scale, where the price of hydrogen is based on the sources used for its generation, for example, solar and wind [41]. In a competitive market, dispatchable plants can produce electricity when the price of electricity equals or exceeds its variable cost. On the other hand, electrolysers can start operating when electricity prices fall below the amount to be paid for the consumed electricity. Therefore, flexible electricity demand from green hydrogen electrolysers effectively and permanently halt renewables' market value decline. In other words, the reduction in wholesale electricity prices caused by renewables triggers commercial investment in electrolysers, which produce hydrogen whenever electricity prices are low, and for the reason that the additional demand for electricity to water electrolysis, in turn, stabilises market prices and, with them, the value of renewables [41]. Additionally, it is essential to emphasise that this new topology brings new challenges for ESSs in electricity markets, which must be supported by analytical frameworks that align the hydrogen economy with robust policies on energy storage in future markets.

12.6   Energy storage policies

As stated in [42], the absence of adequate regulations/policies is one of the most likely reasons why many electricity markets do not include energy storage. More specifically, due to a lack of satisfactory remuneration for ancillary services or proper business models (vide Section 12.2). This is particularly true in the initial stages of ESS deployment. Several countries have implemented policies to promote ESS deployment, which are expected to be the foundation for future advanced/ mature business models. Table 12.2 describes examples of implemented ESS policies worldwide.

12.7   Flexibility market

The electrical system is transitioning from centralised to distributed, with more DERs being integrated into the grid, such as DG, ESS, and EVs. This process

Table 12.2   Example of implemented ESS policies worldwide

Country	Summary of implemented ESS policies	Source	Notes
US	The system operator must review the tariff to recognise energy storage's physical and operational characteristics within 270 days	[43]	Description of general policies (policies might vary depending on the application). Policies might also vary depending on the state [44]
China	ESS promotion in two stages in the next ten years: transition from demonstration to commercialisation and then transition to large-scale commercialisation	[43]	Description of general policies (policies might vary depending on the application)
UK	• Removal of regulatory barriers • Subsidies • Funding of Research and Development (R&D) studies for further deployment of ESS	[44]	Description of battery-specific support instruments
Germany	Loan/subsidies for ESS in conjunction with PV systems	[45]	Description of battery-specific support instruments
Netherlands	• Grants for energy innovation/ demonstration projects; • Public–private sector collaboration toward new projects	[45]	Description of battery-specific support instruments
Italy	• Distinct tariffs (e.g., feed-in tariffs) • Funding of Research and Development (R&D) studies for further deployment of ESS	[45]	Description of battery-specific support instruments
Spain	• Priority dispatch • Premium tariffs • Grants for energy innovation/ demonstration projects	[45]	Description of battery-specific support instruments

creates the opportunity to implement local flexibility markets, which enable DERs to provide flexibility in demand or generation. More specifically, market signals can be implemented to promote supply when the demand is high and discourage supply when the demand is low. By doing so, several benefits can arise, such as better system efficiency, congestion relief, and investment deferral.

The interest in flexibility markets has been increasing due to the growing share of renewables, whose generation depends on the availability of the primary energy factor. Therefore, generation has become less dispatchable/controllable, requiring other means of balancing the electrical grid, such as by providing flexibility.

The concept of flexibility market call for a high degree of digitalisation, communication, and control (e.g., switching conventional meters for smart meters).

At the same time, it requires an advanced market design for proper remuneration of the market agents that provide flexibility. If the remuneration is unsatisfactory, the agents are unlikely to respond as intended. For such reasons, flexibility markets are challenging to implement, although they present significant potential.

Currently, some flexibility markets have been implemented in Europe (e.g., in the UK, Germany, and the Netherlands [45]). However, there is room for improvement, as most pilot platforms that include several services are still not ready for large-scale commercial deployment.

References

[1] Aria, M. and Cuccurullo, C. 'Bibliometrix: an R-tool for comprehensive science mapping analysis', *Journal of Informetrics*, 2017, 11(4), pp. 959–975.

[2] Baumgarte, F., Glenk, G., and Rieger, A. 'Business models and profitability of energy storage', *iScience*, 2020, 23(10), p. 101554.

[3] Proka, A., Hisschemöller, M., and Loorbach, D. 'When top-down meets bottom-up: is there a collaborative business model for local energy storage?', *Energy Research & Social Science*, 2020, 69, p. 101606.

[4] Ilieva, I. and Rajasekharan, J. 'Energy storage as a trigger for business model innovation in the energy sector', in *2018 IEEE International Energy Conference (ENERGYCON)* (IEEE, 2018), pp. 1–6.

[5] Lombardi, P. and Schwabe, F. 'Sharing economy as a new business model for energy storage systems', *Applied Energy*, 2017, 188, pp. 485–496.

[6] Martinsen, T. 'A business model for an EV charging station with battery energy storage', in *CIRED Workshop 2016* (Institution of Engineering and Technology, 2016), pp. 205, 4.

[7] da Costa, V.B.F., de Doile, G.N.D., Troiano, G., *et al.* 'Electricity markets in the context of distributed energy resources and demand response programs: main developments and challenges based on a systematic literature review', *Energies* 2022, 15(20), p. 7784.

[8] Office of Electricity: 'Demand Response' https://www.energy.gov/oe/demand-response, accessed April 2022.

[9] Engie: 'Understanding the Capacity Market' https://www.engie.co.uk/sites/default/files/inline-files/6.3%20-%20Industry%20Info%20-%20Capacity%20Market%20Guide21.pdf, accessed April 2022.

[10] Khan, A.S.M., Verzijlbergh, R.A., Sakinci, O.C., and De Vries, L.J. 'How do demand response and electrical energy storage affect (the need for) a capacity market?', *Applied Energy*, 2018, 214, pp. 39–62.

[11] Dvorkin, Y. 'Can merchant demand response affect investments in merchant energy storage?', *IEEE Transactions on Power Systems*, 2018, 33(3), pp. 2671–2683.

[12] Zhou, Y., Mancarella, P., and Mutale, J. 'Modelling and assessment of the contribution of demand response and electrical energy storage to adequacy of supply', *Sustainable Energy, Grids and Networks*, 2015, 3, pp. 12–23.

[13]   Zhao, J., Kucuksari, S., Mazhari, E., and Son, Y.-J. 'Integrated analysis of high-penetration PV and PHEV with energy storage and demand response', *Applied Energy*, 2013, 112, pp. 35–51.

[14]   Ma, O. and Cheung, K. *Demand Response and Energy Storage Integration Study*. Office of Energy Efficiency and Renewable Energy, Office of Electricity Delivery and Energy Reliability, 2016.

[15]   Ramsebner, J., Haas, R., Auer, H., *et al.* 'From single to multi-energy and hybrid grids: historic growth and future vision', *Renewable and Sustainable Energy Reviews*, 2021, 151, p. 111520.

[16]   De Castro, L., and Dutra, J. 'Paying for the smart grid', *Energy Economics*, 2013, 40, pp. S74–S84.

[17]   Hu, Z., Li, C., Cao, Y., Fang, B., He, L., and Zhang, M.. 'How smart grid contributes to energy sustainability', *Energy Procedia*, 2014, 61, p. 858861.

[18]   United Nations Framework Convention on Climate Change: 'The Paris Agreement', https://unfccc.int/process-and-meetings/the-paris-agreement/the-paris-agreement, accessed March 2022.

[19]   European External Action Service (EEAS): 'Informal video conference of Defence Ministers: remarks by High Representative/Vice-President Josep Borrell at the Press Conference', www.eeas.europa.eu/eeas/informal-video-conference-defence-ministers-remarks-high-representativevice-president-josep-0_en, accessed March 2022.

[20]   Welton, S. 'Electricity markets and the social project of decarbonization', *Columbia Law Review*, 2018, 118 (4), pp. 1067–1138.

[21]   Zhang, S. and Chen, W. 'Assessing the energy transition in China towards carbon neutrality with a probabilistic framework', *Nature Communications*, 2022, 13(1), p. 87.

[22]   Blanco, H. and Faaij, A. 'A review at the role of storage in energy systems with a focus on Power to Gas and long-term storage', *Renewable and Sustainable Energy Reviews*, 2018, 81, pp. 1049–1086.

[23]   Carmo, M. and Stolten, D. 'Energy storage using hydrogen produced from excess renewable electricity', in *Science and Engineering of Hydrogen-Based Energy Technologies* (Elsevier, 2019), pp. 165–199.

[24]   Böhm, H., Moser, S., Puschnigg, S., and Zauner, A. 'Power-to-hydrogen & district heating: technology-based and infrastructure-oriented analysis of (future) sector coupling potentials', *International Journal of Hydrogen Energy*, 2021, 46(63), pp. 31938–31951.

[25]   Nastasi, B. and Lo Basso, G. 'Power-to-gas integration in the transition towards future urban energy systems', *International Journal of Hydrogen Energy*, 2017, 42(38), pp. 23933–23951.

[26]   Cesaro, Z., Nayak-Luke, R.M., and Bañares-Alcántara, R. 'Energy storage technologies: power-to-X', in *Techno-Economic Challenges of Green Ammonia as an Energy Vector* (Elsevier, 2021), pp. 15–26.

[27]   Bowen, T., Chernyakhovskiy, I., and Denholm, P.L. 'Grid-scale battery storage: frequently asked questions', National Renewable Energy Lab (NREL), 2019.

[28] Körner, A., Tam, C., Bennett, S., and Gagné, J. *Technology Roadmap Hydrogen and Fuel Cells*. (International Energy Agency (IEA)): Paris, France, 2015).

[29] Rozzi, E., Minuto, F.D., and Lanzini, A. 'Dynamic modeling and thermal management of a power-to-power system with hydrogen storage in microporous adsorbent materials', *Journal of Energy Storage*, 2021, 41, p. 102953.

[30] Nastasi, B. and Lo Basso, G. 'Hydrogen to link heat and electricity in the transition towards future Smart Energy Systems', *Energy*, 2016, 110, pp. 5–22.

[31] Buttler, A. and Spliethoff, H. 'Current status of water electrolysis for energy storage, grid balancing and sector coupling via power-to-gas and power-to-liquids: a review', *Renewable and Sustainable Energy Reviews*, 2018, 82, pp. 2440–2454.

[32] Nikolic, V.M., Tasic, G.S., Maksic, A.D., Saponjic, D.P., Miulovic, S.M., and Marceta Kaninski, M.P. 'Raising efficiency of hydrogen generation from alkaline water electrolysis – energy saving', *International Journal of Hydrogen Energy*, 2010, 35(22), pp. 12369–12373.

[33] Fu, C., Lin, J., Song, Y., Zhou, Y., and Mu, S. 'Model predictive control of an integrated energy microgrid combining power to heat and hydrogen', in *2017 IEEE Conference on Energy Internet and Energy System Integration (EI2)* (IEEE, 2017), pp. 1–6.

[34] Saxe, M. and Alvfors, P. 'Advantages of integration with industry for electrolytic hydrogen production', *Energy*, 2007, 32(1), pp. 42–50.

[35] Thess, A. 'Thermodynamic efficiency of pumped heat electricity storage', *Physical Review Letters*, 2013, 111(11), p. 110602.

[36] Datas, A., Ramos, A., and del Cañizo, C. 'Techno-economic analysis of solar PV power-to-heat-to-power storage and trigeneration in the residential sector', *Applied Energy*, 2019, 256, p. 113935.

[37] Parra, D., Valverde, L., Pino, F.J., and Patel, M.K. 'A review on the role, cost and value of hydrogen energy systems for deep decarbonisation', *Renewable and Sustainable Energy Reviews*, 2019, 101, pp. 279–294.

[38] Bargiacchi, E., Antonelli, M., and Desideri, U. 'A comparative assessment of power-to-fuel production pathways', *Energy*, 2019, 183, pp. 1253–1265.

[39] Hu, X., Zou, C., Zhang, C., and Li, Y. 'Technological developments in batteries: a survey of principal roles, types, and management needs', *IEEE Power and Energy Magazine*, 2017, 15(5), pp. 20–31.

[40] Costa, V., de Souza, A.C.Z., and Ribeiro, P.F. 'Economic analysis of energy storage systems in the context of time-of-use rate in Brazil', in *2019 IEEE Power and Energy Society General Meeting (PESGM)* (IEEE, 2019), pp. 1–5.

[41] Ruhnau, O. 'How flexible electricity demand stabilizes wind and solar market values: the case of hydrogen electrolyzers', *Applied Energy*, 2022, 307, p. 118194.

[42]   Xiao, Y., Gao, Y., Kuang, S., *et al.* 'Comparative analysis on energy storage policies at home and abroad and its enlightenment', *IOP Conference Series: Earth and Environmental Science*, 2019, 267, p. 032019.

[43]   US Department of Energy: 'Energy Storage Policy Database'.

[44]   European Commission: 'Battery Promoting Policies in Selected Member States'.

[45]   Valarezo, O., Gómez, T., Chaves-Avila, J., *et al.* 'Analysis of new flexibility market models in Europe', *Energies* 2021, 14(12), p. 3521.

Chapter 13

Integrated operation of energy storage in urban grids

A.C. Zambroni de Souza[1], Gabriel F. Alvarenga[1] and Pedro N. Vasconcelos[1]

Energy storage devices are already an important asset for power system planners to deal with uncertainty and changes promoted by the development of smart grid technologies and modernizing public policies. In the urban landscape, every electricity customer archetype may pursue a different goal with respect to energy and benefit from both local and grid-scale investments in terms of energy storage. This chapter aims to stress the value added by energy storage applications for residential, commercial, and industrial customers, as well as the seamless integration of electric vehicles (EVs) as mobile sources of energy both in the forms of privately owned resources and in public transportation. Finally, as different market models arise thanks to the proliferation of such resources, some approaches and practical outcomes related to the community and (locally or virtually) aggregated use of energy storage assets are presented with the intent to integrate all aspects previously discussed and further evidence the significance of energy storage for the present and future of the power sector.

13.1 Introduction

Energy storage devices will soon become common assets in modernizing electrical power systems, especially in urban areas. As we pursue net zero carbon emissions, batteries and plug-in EVs may help small and large-scale energy projects operating as an electrical load or source, depending on the operational needs. This will enable the procurement of green energy and increase system functionalities as these devices may sum up in a way that the system may guarantee electricity supply for critical loads in emergency conditions using the stored amounts of energy.

The evolution of battery design and the increasing number of applications tend to create a challenging environment for academic and industrial professionals and a holistic perspective of their impacts cannot be neglected. Coordination with the

[1]Institute of Electrical Systems and Energy, Federal University of Itajubá, Brazil

current electricity utilities' practices is paramount and incorporating the interests of all stakeholders (i.e., system planners, operators, prosumers, and so on) in a cooperative environment seems to be a good strategy. However, the greatest changes promoted by energy storage systems may lie in the transportation sector, potentiated by the development of hydrogen-based energy generation and storage.

In this context, this chapter presents examples of the potential use of energy storage by different types of electricity customers, mainly the residential, commercial, and industrial archetypes. Once the employment of such devices may change the current interactions between these agents, this chapter also provides an overview of their impacts on the power grid, the transportation sector, and the current electricity market platform.

It is important to mention that this chapter is not an exercise of futuristic scenarios. Rather than that it focuses on technological transformations undergoing, and their applications are already a reality. At the same time, it is expected that the emerging power systems will be shaped by these technologies, changing habits, and placing new paradigms. It is expected that such a scenario improves the system reliability and helps including people in developing countries, paving a road of transformation for all societies.

13.2   Use of energy storage of different customer archetypes

13.2.1   Residential microgrids

There are now a large number of technologies for residential energy storage, with different performance and cost parameters, making them interesting for different situations. The battery type that stands out the most today is the lithium-ion battery, with more than 90% of the total storage installed. The shares of sodium–sulfur batteries and lead batteries are also notorious.

A residential Battery Energy Storage System (BESS) system can be directly connected between the household load and the distributed generation system, as depicted in Figure 13.1, or it can be connected in an aggregated way, joining the electrical loads from a group of houses, community energy storage (CES) systems, Figure 13.2.

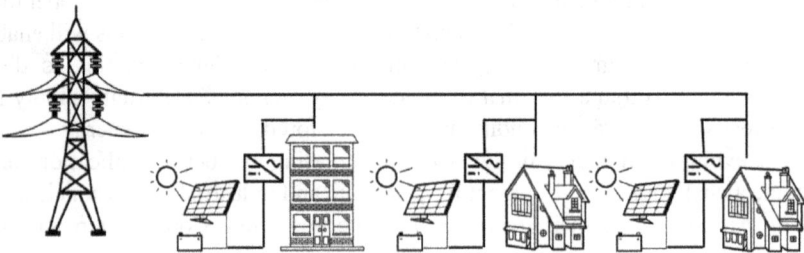

Figure 13.1   Directed connected BESS to the houses

Figure 13.2   Aggregated BESS connected to the houses

The use of a shared battery is a good alternative to overcome problems such as limited space, high initial investment cost, safety, and better use of stored energy. The most used technology for these applications is lithium-ion batteries and flow batteries.

Lithium-ion batteries have the greatest advantage in residential use due to their high charge density, and being able to store large amounts of energy in a compact storage volume, although the constant advances related to this technology, still present some risks [1], such as fire risk and ability to hold a charge fades over time. With incentives and appropriate sizing, photovoltaic with battery systems compete with grid prices [2].

The residential storage battery market is led by two South Korean players: LG-Chem Ltd. and Samsung SDI Co. Ltd. Beyond these, there is still important participation of the companies Panasonic Corporation, Sonnen GmbH, Saft Groupe SA, Hitachi Chemical Co. Ltd., Sunverge Energy LLC, Tesla Inc., and Deutsche Energieversorgung GmbH (SENEC).

The flow batteries, unlike lithium-ion, store liquid electrolytes in an external reservoir, reducing the fire hazard and presenting fewer issues with capacity fade. An example of flow batteries is the zinc bromide battery, which has been developed and used as BESS, each of its modules delivers 25 kW of electricity, enough to power 5–7 homes for 5 h during peak energy demand and 12–15 h during off-peak.

Another type of flow battery used today, the iron flow battery, comes in a shipping container, and they can deliver power from 100 kW of power for 4 h to 33 kW for 12 h, using an electrolyte entirely of iron, salt, and water. They are applied to urban residential loads and even to public transport.

A factor that is crucial for the development and consolidation of batteries in an urban context is their scalability. Mainly due to their modular characteristic, it is possible to increase the capacity of lithium batteries by placing more batteries in parallel, or by increasing the reservoir in the case of flow batteries. This capability allows customization according to the needs of each consumer.

As a way of reducing costs and making battery technology more accessible, the residential use of batteries in their second-life cycle is currently being studied, that is, after the maximum storage capacity of a lithium battery drops to 80% of its initial value, it will be sent to residential use. In its first cycle of use, it can be used by EVs and public transport, where storage density is more relevant. When it can no longer be used in the building, the battery is sent for recycling and new batteries are made using its materials. This cycle is maintained in order to use fewer natural resources and make technology cheaper. Figure 13.3 shows us this cycle. Second life battery is economically beneficial and environmentally sustainable [3].

13.2.2   Commercial/industrial microgrids

In addition to the objective in common with residential batteries, reducing the cost of electricity, the battery can be used by commercial and industrial loads in order to have an alternative energy source in case there is a problem in the main network, playing a role of source secondary power supply. This configuration can supply the energy of an elevator in a smart building, breathing equipment in a hospital, as seen in Figure 13.4, or other equipment that requires uninterrupted power.

The Energy Storage System (ESS) is connected to the load using an AC–DC converter. It can also be connected in parallel with a second-generation source, such as a diesel generator, or solar panels. A controller is needed in these cases to define when it will be necessary to charge the batteries, managing the energy that comes from the grid and the second generation source.

ESSs have many applications to support commercial and industrial loads, unlike residential applications, the power and quantity of storage in these cases

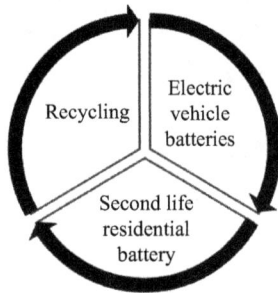

Figure 13.3   EV battery cycle

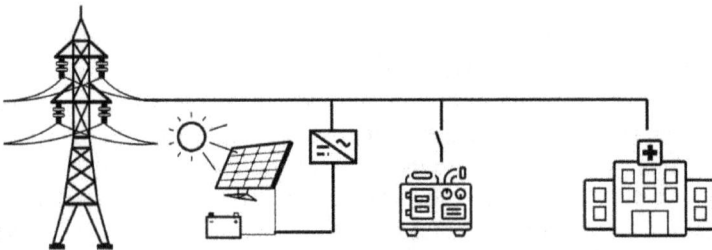

Figure 13.4   Uninterruptible power supply configuration

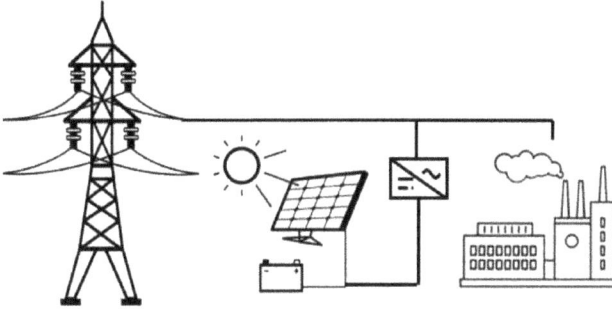

Figure 13.5 Power conditioning configuration

have larger dimensions, requiring the use of technologies adapted to the specific needs of the load, such as high demand, life cycle, and cost.

Another use that ESS has great advantages is in high-precision industries, such as the manufacture of microprocessors, where power quality problems can cause a loss in the final product quality, leading to losses. The batteries, in this case, act as a signal conditioner, connected in series with the load using power electronics equipment and filters, as seen in Figure 13.5. The batteries mitigate power fluctuations occurring in PV power generation due to weather conditions, thus enabling smooth power generation. This leads to increased stability and quality of the overall system [4].

Since batteries can absorb and deliver real energy, BESS can be used to help regulate grid frequency, just like in a microgrid. PCS uses special inverters to convert DC battery power into three-phase AC voltage. The AC voltage is smoothed by filter elements and then increased to mains voltage by a transformer. The PCS can also be used as a Statcom device for the network. Statcom is an electronic system capable of supplying and absorbing reactive energy (VAR) to the grid. Thus, when the PCS is not doing its main function of charging and discharging the battery, the PCS can be used to improve the power factor or regulate the voltage on the grid.

13.3   Transportation and batteries

Technology has always been attached to the way mankind evolved. The invention of wheels, along with the domestication of animals played a crucial role in agriculture, changing habits and enabling people to connect with other regions, creating the environment to sow a culture and language. People took advantage of wheels, initially, for moving heavy stuff. However, it was for agricultural purposes that its usage spread around, helping people to create roots and establish in places that became their land. Using wheels for agriculture posed a lighter burden than moving heavy stuff since the land was limited and could be pre-worked to ease the movement. Later, people used chariots for centuries for transportation and that did not change even during the important revolution of the great navigations.

The first industrial revolution made available the steam machines, enabling the steam-driven train. Around 1,830 trains were already popular, shortening distances and allowing a massive interchange of goods and services. Curiously, public transportation, was, therefore, available, whereas short distances were still overcome by chariots. Finally, in 1886, the German Karl Benz designed the first automobile, whose production became massive by Ford from 1913. The United States, a continental country, is a good example of the change of paradigm imposed by automobiles. By the time the T ford was made available, the country had around 30,000 km of paved paths (not exactly highways in the sense known nowadays). Thus, it was a product that needed a huge change of infrastructure to be viable and could be neglected by many investors.

But it prevailed and changed the way people move, igniting an energy sector that would change the economic profile of the world: the petroleum. Its discovery brought new opportunities and shaped the geopolitics of the twentieth century. The advent of automobiles also placed a new habit in public transportation, since they became available to the medium-class people, creating urban traffic problems and pollution. It is worth mentioning that modern cars do not pollute much as a few decades ago, but petroleum is still a point of concern among environmental researchers around the world. This section is proposed to analyze how emerging technologies may change the public and private transportation modes. The section focuses on some advantages that trams and EVs may present to their users.

13.3.1 EVs

When it comes to modern power systems, some concepts have the appeal of being new and revolutionary, which, indeed is true in many times. However, the Dutch windmills, for example, remind the modern wind-powered generation, even though the former exist since the eleventh century. EVs also have a similar story, since the first electric cars were proposed in the years 1830s. It is also worth mentioning that by 1900, almost one-third of the cars produced in the USA were electric. At that time, already, driving an electric car was easier than fuel-driven cars, since the driver did not need to change gears, and igniting the electric car was trivial. However, the known technology of batteries and the charging time of batteries made fuel-based cars more viable. The automobile industry, thus, crossed the whole twentieth century ignoring the vehicle cars as a commercial possibility, which changed dramatically over the last years. Initially, the main problems of the batteries of EVs were their weight, low autonomy, and the charging time. Currently, the advances in technology enabled the EV as a commercial product. The main advantages may be listed as:

- **Cost of electricity:** In comparison with conventional fuel, it is cheaper and poses an interesting appeal.
- **Maintenance:** Some components of fuel-driven vehicles are not present in EVs, making their maintenance cheaper and simpler.
- **Gas emission:** An obvious characteristic of these vehicles, they do not pollute. Note, however, that considering EVs along with renewable-based sources

plays a crucial role. It is not wise, for example, to deploy coal-based plants to feed EVs.
- **Noise pollution:** This is a great characteristic of these vehicles. Indeed, it is considered, sometimes, as a problem, since pedestrians need to be aware in public places as drivers do.

Nowadays, the cost of the vehicle is still a barrier, even though the price has been consistently reduced. Two other problems may be pointed out, since the infrastructure is still incomplete, demanding a larger number of charging stations; and discarding the batteries place an ecological problem.

13.3.2 V2G and G2V

The grid-to-vehicle (G2V) concept is a trivial idea and is part of the plug-in class of vehicles. In this case, the owner may charge their car at home, or any place prepared for that. Since most owners tend to charge their cars overnight, an intelligent monitoring of the network is required, in order to avoid excessive losses and undervoltage [5]. Another problem about the charging regards the long time required to charge an EV. For this sake, a Danish company is working on the development of a 350 kW buffer-high power charger, which may charge an EV at the same time as conventional vehicles are fueled nowadays [6].

There are many PHEVs with different configurations. The main differences refer to motor/engine coupling to the powertrain, which could be a series or parallel hybrid [7] and battery energy storage system models. Regarding the electric motors, it is more common to use synchronous units with permanent magnets and rated power varying from 50 kW to 160 kW. As electric motor development continues, some of the key areas are identified, such as multiple motor design concepts including variable-voltage traction motors and sintered or bonded magnets for permanent magnet motors. Besides increasing performance, research efforts are being made to reduce materials, parts, and manufacturing costs [8]. Table 13.1 shows a summary of the batteries available to EVs. Then, G2V allows owners to charge their vehicles at the best time for them, initially, overnight. However, vehicle-to-grid (V2G) technology explained next may open new possibilities, such that the owner may work cooperatively with the utility, injecting energy into the system.

13.3.3 G2V and V2G

A plug-in hybrid EV (PHEV) has at least a 4 kWh battery energy storage system; may perform a 16.1 km length in electric-drive mode only, and may be recharged from an external electricity source [9].

The advent of EVs unfolds several concerns and possibilities to grid operators and researchers since they tend to provide sustainable alternatives of oil-free transportation. Even the classical concept of grid-to-vehicle (G2V) brings some important concerns, since these vehicles may cause problems such as limits violations, power losses increasing, harmonics, and thermal limits violations on distribution transformers and conductors. Also, taking into account the whole energetic matrix, the primary sources feeding these vehicles must be clean.

Table 13.1   Battery technology and autonomy capacity of current EV models

Model	Technology	Rated capacity	Autonomy[a]	Price
Kandi K27 (BEV)	Li-Ion	17.7 kWh	95 km	$12,900
Hyundai Ioniq (PHEV)	Li-Po	8.9 kWh	47 km	$29,850
Nissan Leaf (BEV)	Li-Ion	62 kWh	364 km	$29,900
Chevrolet Bolt (BEV)	Li-Ion	65 kWh	416 km	$31,900
Toyota Prius Prime (PHEV)	Li-Ion	8.8 kWh	40 km	$34,000
Hyundai Kona (BEV)	Li-Po	39.2 kWh	415 km	$34,200
Ford Fusion Energi (PHEV)	Li-Ion	7.6 kWh	35 km	$37,490
Kia Niro (BEV)	Li-Po	64 kWh	385 km	$39,900
Polestar 2 Single (BEV)	Li-Ion	78 kWh	434 km	$45,900
Tesla Model 3 (BEV)	Li-Ion	50 kWh	437 km	$46,990
Kia EV6 (BEV)	Li-Po	77.4 kWh	499 km	$46,990
BMW i4 (BEV)	Li-Ion	80.7 kWh	484 km	$55,400
Tesla Model Y (BEV)	Li-Ion	75 kWh	488 km	$62,990
Ford Mustang Mach-E (BEV)	Li-Ion	88 kWh	505 km	$67,900
Tesla Model S (BEV)	Li-Ion	100 kWh	652 km	$99,990
Mercedes Benz EQS (BEV)	Li-Ion	107.8 kWh	547 km	$102,300
Tesla Model X (BEV)	Li-Ion	100 kWh	523 km	$114,990
Lucid Air Touring (BEV)	Li-Ion	118 kWh	830 km	$139,900
Mini Cooper SE (PHEV)	Li-Ion	32.6 kWh	177 km	$249,900
BMW i3 (BEV)	Li-Ion	37.9 kWh	322 km	$304,950

[a]Estimated by manufacturers based on the US Environmental Protection Agency (EPA) Multi-Cycle test procedure, with 45% highway and 55% city driving. Autonomy refers to a full depletion of the battery capacity, although most manufacturers limit the depth of discharge (DOD) to around 70%.

Reference [10] addressed the problem of minimizing power losses during the charging period. The results obtained, along with the conclusions discussed in [11,12], flag losses monitoring as a good strategy for charging purposes. Reference [13] employs an artificial immune systems-based approach to minimize the losses and keep the voltage level within its limits. The proposed approach yields a charging process in such a way that at dawn all the vehicles are charged following the voltage and loss monitoring.

Note, also, that V2G may offer possibilities in markets and demand a change of infrastructure. Quoting [14], the advent of EVs may provide possibilities, such as:

- Vehicle identification and billing, allowing payment for charging at public charging stations, but also individual billing of used energy to the user's account when the vehicle is charged at any outlets connected to a smart meter.
- Charge cost optimization by choosing the most appropriate time window where electricity rates are the lowest.
- Grid load optimization by controlling charger ampacity in function of grid demand.
- Peak-shaving functionality by using EVs connected to the grid as a spinning reserve (vehicle-to-grid).
- Appropriate billing and user compensation functions for vehicle-to-grid operation.

An EV, however, may play an important role as a storage device in emergency conditions. In this case, it may provide energy to the system, enabling the vehicle to grid (V2G) concept. Such a concept takes advantage of the bidirectional characteristic of these vehicles. Thus, if required they may provide energy to the system. The main possibilities are described next:

• Market-driven condition. In this case, the owner may decide to charge their car in cheap periods and supply its own energy when it is economically viable. This option, however, is dependent on market orientation. Reference [15] develops a model for EV participation under the zonal and modal model discussion. Reference [16] proposes an optimal bidding for ancillary services by considering electricity market uncertainties. Reference [17] discusses the economical impact that V2G may have in the Chinese market by 2030. For this purpose, a general framework is proposed to test different scenarios.
• Island condition. In this case, market conditions may still drive the decision of the owner. The emergency welfare is the focus in this case. Thus, the owner has a contract with the utility which foresees this emergency condition. Since the disconnection time may be unknown, the problem demands an energy based solution. Thus, defining the most priority loads to be supplied during the disconnection time is the core of this study. This enables a holistic approach to the problem, as explored in [18], where a strategy of plug-in EVs for unbalanced smart microgrid environments is proposed.

13.3.4 Electric trams and regenerative break

The advent of EVs and their subsequent growing penetration into the market brings benefits to society as a whole. Thus, one may cite gas emission and noise reduction as some of the advantages of these vehicles. As for public transportation, trams may use an interesting approach to reduce the electric consumption and increase efficiency. This is done by employing the regenerative breaks, which allow to recover part of the energy usually lost as heat when breaking the car [19].

Regenerative systems are based on the fact that a generator may work as a motor. A tram has three stages of operation. First, it requires energy to accelerate from a stopping condition; second, it requires to work in steady-state condition when the cruise speed is reached; and finally, the breaking process is activated when stopping is required. Normally, this energy spent during the breaking was lost as heat. With the help of the regenerative breaks, part of the energy is stored by the tram facilities, and another part is sent to the electric system, enabling another tram to use it for its accelerating process.

The advances in power electronic converters and energy storage technologies enable researchers and companies to enhance their projects so that energy recovery is maximized. In this sense, they are adherent to the topic of this chapter. Note also that the absence of storage devices would not inhibit the classification of regenerative breaks as storage units, since the energy recovered would be sent back to the system. Using local storage, however, makes the system more sophisticated and efficient. Two ways of storing may be employed: the first is stationary, so that the storage devices are located somewhere in the grid, collecting energy from the trams and liberating when required, as shown in Figure 13.6. The second, also known as onboard, has the batteries located on top of the tram, enabling immediate use when required, as shown in Figure 13.7.

The storage system adopted may vary according to a range of batteries available. Three of them are briefly described below:

- **Lead-acid batteries:** Shown here because of their long-time applications industry, this battery may be placed on top of trams, operating in charging and discharging modes according to the need of the tram. This kind of battery presents poor energy densities, short lifetime, high cost of maintenance, and

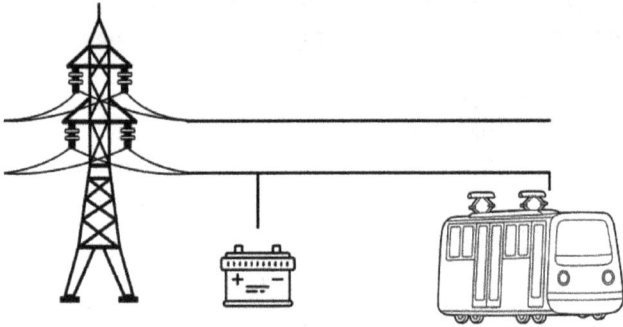

Figure 13.6   Tram with battery connection across the grid

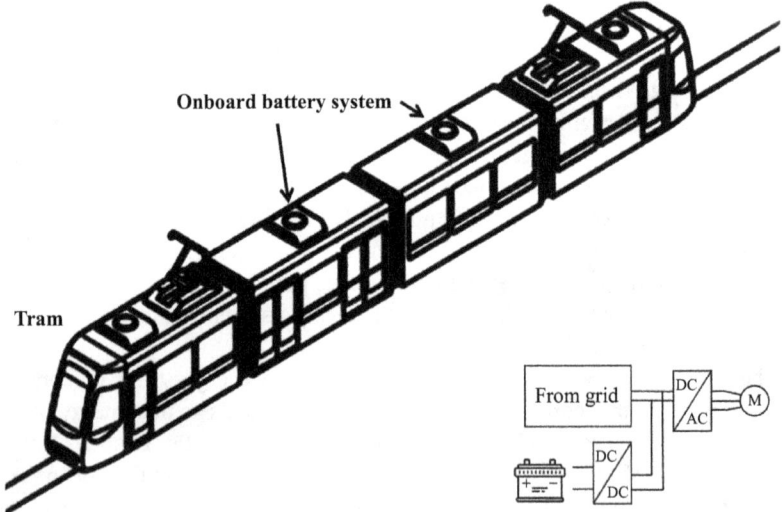

Figure 13.7   Onboard battery tram configuration

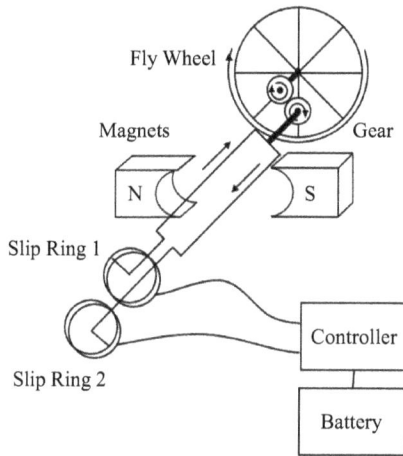

Figure 13.8   Flywheel and battery diagram

does not work well in low temperatures. Besides, its discard poses an environmental problem that does not meet the clean energy agenda of EVs.

- **Lithium-based batteries:** This kind of battery presents the best trade-off among all the batteries available. Its main advantages are the energy density and the wide range of temperature operation. Its high cost of protection is a limitation, as well as the discard process.
- **Flywheel:** A flywheel is a mechanical device (a disk) used to store energy through its kinetic energy. A carbon-based material is currently used for this purpose, enabling the storage of high values of energy, since it is proportional to the square to the rotational speed. Building it in a vacuum avoids overheating while avoiding the deformation of the disk due to its high speed.

The application of flywheels has gained increasing attention lastly, especially because of hybrid vehicles. However, its use has been reported since the decade of 1940 in Switzerland. Nowadays, Formula 1 cars adopted it under the name of Kinetic energy recovery system (KERS). Figure 13.8 shows the schematic diagram of a flywheel incorporated into a tram. Its use, in the proposed structure, is triggered for the acceleration stage of the tram, saving energy from the electric grid.

Flywheels, thus, may provide important operational characteristics, since they may store and provide energy as required and causes no discarding problem, once it has a long lifespan.

13.4   Integration of energy storage and aggregation alternatives

The increasing number of energy storage assets scattered throughout the electricity grid – be it in the form of standalone battery units or EVs – demands a change in the way these resources are managed by utility operators at the distribution level, as well as in how current electricity markets are organized once the community and aggregated-use of such resources enable interactions between emergent agents, especially in urban environments.

In summary, the increasing adoption of ES has been driven by a combination of technological improvements, favorable legislation, and lowering costs. Much of the ES integration has been in the form of utility-scale efforts, while recently particularly smaller capacity applications have been introduced at the consumer level, i.e. behind the meter, in the form of distributed energy resources. The collective management of such dynamic energy reserves may benefit not only their owners but also provide additional grid stability, while sustainably lowering costs for all other stakeholders [20].

For customers connected to the main distribution system at any voltage level, the use of ES introduces much-welcomed flexibility to deal with the variability of local renewable resources, dynamic tariff models at electricity retail, and unintended disturbances that may affect the supply coming from the main bulk transmission system. The presence of dozens of ES units scattered across buildings or neighborhoods creates an energy network that can be used to deliver immediate value for the electric system and further improve the existing functionalities and benefits of ES for different archetypes of urban grid customers, as previously discussed in this chapter.

For large electricity market players, such as transmission and distribution utilities, aggregators, and local system operators, the control of multiple ES assets concurrently – acting as a single entity – enables new sorts of benefits. Among these, lie the ability to aggregate and control the existing distributed resources as in a virtual power plant (VPP) or, more specifically, a virtual energy storage system (VESS), with standard attributes such as minimum/maximum capacity, ramp-up, ramp-down, and so on. In this sense, the aggregated management of distributed ES allows small energy sources to participate in wholesale and retail markets to sell electricity or provide ancillary services [21].

The increase in the participation of smaller suppliers in electricity markets can further provide additional benefits for the system stakeholders in terms of economic efficiency. With a decline in market concentration and increased competition, the largest expected benefits from this change can be summarized as follows [22]:

• Lower prices due to a downward pressure on the profit margins of generators and suppliers forced by competition.
• Lower prices due to an increased energy trade, which facilitates inter-system deals and a better allocation of existing resources, thus reducing costs for the electricity supply and infrastructure investments.
• Higher labor productivity and development of new energy services.

In this sense, since the introduction of statewide policies in California in 2010, facilitating investor-owned utilities (IOUs) to manage aggregated distributed energy resources and deliver electricity to their customers, electricity rates currently are lower than the California Independent System Operator (CAISO), ranging from −0.1% to −2.1% lower [23]. Similar effects and a gradual decline in market concentration for the residential, commercial, and industrial sectors in the United Kingdom can also be observed considering the same period [24].

Finally, besides the technical and economic advantages raised at the system-wide level, the collective use of battery ES at the local level potentially is a game changing strategy for the energy transition and decarbonization of the sector, while accommodating the needs and expectations of citizens and local communities. With the increasing prevalence of prosumers and the number of individual DER units, it makes sense the need for the engagement of all actors, even the community engagement, to benefit from this technology trend. The community energy storage (CES), in this context, involves the accomplishment of either individual households all hooked into one large battery deployed at the neighborhood level, or with smaller household batteries linked together so that neighboring customer units could borrow and lend power as needed.

It is expected that this kind of shared battery scheme would be much more cost-effective for households, as well as more efficient in terms of materials required for battery production – therefore, potentially more environmentally friendly in the long run [25]. The aggregation of energy storage, CES, as well as other strategies and pilot projects that have the potential to shape future society and societal needs in terms of energy are described in the next sections.

13.4.1 CES

The concept of CES embodies many facets to deliver a more equitable distribution of benefits and control of the local grid. There are multiple definitions of CES in practice among organizations and scholars that can lead to different interpretations. For example, the California Public Utility Commission (CPUC) defines community storage as storage connected at the distribution feeder level, associated with a cluster of customer load [23]. Reference [26] refers to it as energy storage located at consumption level with multiple applications and positive impacts on end-users and network operators. Both definitions are limited to the location of energy storage and do not provide attention to the community aspect, ownership, and benefits.

Complimentarily, Ref. [27] defines CES as energy storage introduced for a community that can be shared between members who are typically, but not exclusively located in the local community. This definition is interesting as it opens the possibility for virtual CES, which in turn enables different business models, other than direct utility ownership and control. The cases and business models including utility-owned, virtual, and energy storage systems shared among local customers are discussed later in this section. For now, a comprehensive definition that embodies previous ones and stresses the implications of the community ownership of electricity assets considers CES as:

> ... an energy storage system with community ownership and governance for generating collective socio-economic benefits such as higher penetration and self-consumption of renewables, reduced dependence on fossil fuels, reduced energy bills, revenue generation through multiple energy services as well as higher social cohesion and local economy [25].

As the authors later stated, this definition excludes defining CES as a purely residential or utility-scale application since theoretically, it should lie between these two applications. Therefore, different dynamics of CES in the energy systems as well as within the local community such as coordination and interaction among actors and components of CES should be explored.

In practice, the main tasks of these systems in distribution networks can be summarized as peak-shaving, smoothing DER's intermittencies (output shifting/leveling), power quality improvements (voltage and reactive power supports), islanding during outages, frequency regulation, maximizing self-consumption, and so on [28]. Typically, a CES consists of a battery (mostly lithium-ion), a four quadrant inverter, and a measurement/control system that includes a battery management system (BMS), an inverter, and a monitoring/control (IC) unit. The main difference between CES and DGs lies in the fact that CES has a fully dispatchable four-quadrant inverter that enables it to bi-directionally exchange active and reactive powers. In other words, CES can regulate voltage and frequency and inject/absorb reactive power to/from the grid on demand. Figure 13.9 presents the main structure of CES.

Finally, community storage can theoretically encompass a wide range of storage technologies, including batteries and EVs, as well as thermal storage such as ice storage, electric space heaters, and water heaters. Some existing projects, pilots,

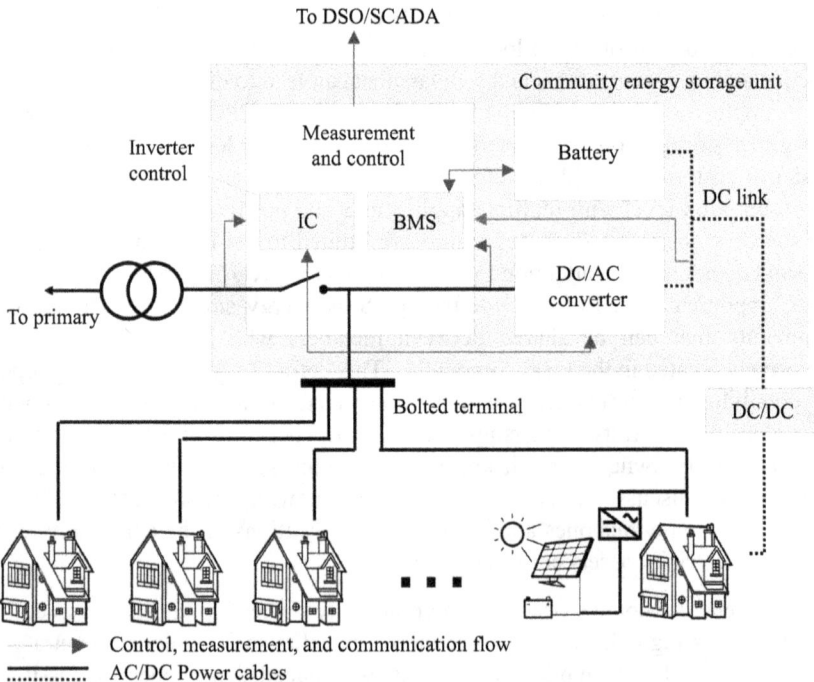

Figure 13.9   The main structure of a CES system

as well as business models, and policy discussions are summarized in the next few sections.

13.4.1.1   Shared residential energy storage

There are few consolidated cases of customer-owned and shared residential energy storage in practice and several pilot projects are being implemented worldwide to demonstrate the value of the CES. To date, the most common step that state and system regulators have taken is to open community solar programs to solar and storage projects simultaneously. For example, the states of New York and Oregon in the United States recently expanded their rebate and technical assistance and predevelopment programs to assist affordable housing providers in installing community solar plus storage [29,30]. This model is also popular in Australia, where a drop in compensation for exported solar power has led many PV owners to retrofit their systems with batteries [31].

Among several pilots being implemented in the United States, the one being employed by the Sacramento Municipal Utility District (SMUD) in California, namely Anatolia III solar Smart Homes Community, has been successful since its deployment [32]. The project consists of a network with fifteen 7.7 kWh residential Li-ion battery systems installed behind the customers' meters and connected to 2 kW of PV arrays. Results show success in managing high penetrations of PV by shifting the load that mitigates ramping in a system-wide power supply and smoothing the PV capacity by balancing out fluctuations that may occur. However, this pilot is utility-owned and controlled and local communities have no roles and responsibilities except being able to monitor their daily electricity consumption as well as export.

13.4.1.2   Shared local energy storage

Projects with aggregated storage from residential, behind-the-meter batteries in tandem with rooftop PV or other renewables represent a small share of current CES systems. Many solutions consist of a larger-scale communal battery sited on the local feeder and each household may purchase fractional shares and collectively benefit from services such as peak shaving, higher self-consumption of local generation and lower energy cost [33].

This centralized configuration can also be employed on campuses or multi-family buildings and shared among multiple tenants. In a pilot in Stockholm, Sweden, a conventional solar power inverter was replaced by intelligent battery management and inverter control system to include a single 32.4 kWh storage unit to transfer energy to different apartments in a multi-tenant building [34]. In Detroit, Michigan, the Monroe County Community College received funds from the government to install a 500 kW solar system and a 500 kWh lithium-ion battery to support the main campus, two EV charging stations, plus 20 other near residential homes [32].

A prominent project takes place in Feldheim, Germany. This 37-household village – whose energy entrepreneurship dates back to 1995 with the installation of the first wind turbine – achieved energy self-sufficiency with a 10 MWh CES, 81.1

MWp wind, 2.25 MWp Solar PV, and 500 kW biomass plant [35]. The energy community, funded by EU subsidies, constructed its own electricity and heating network and meets all its energy demands locally, selling surplus generation to the national grid. Currently, Feldheim also started to provide primary frequency control services to the local transmission network.

13.4.1.3   Shared virtual energy storage

There are a couple of virtual energy storage networks being developed in areas with the support of net metering, independent of the established utilities and location [25,35]. For example, the SonnenCommunity® project, born in Germany, is a growing network of above 10,000 end-users who produce, store, use, and share energy. It functions as a utility and constitutes a community of producers, consumers, and energy storage owners who can supply each other with self-generated as well as stored electricity. Recently, the company announced its participation in a NY state research program to constitute a VPP with more than 200 residents and a total of 80 MWh of storage capacity by 2024 [36].

In late 2016, the largest VPP involving residential behind-the-meter batteries was announced internationally in Australia, with the sale and installation of 1,000 batteries in metropolitan Adelaide. The project, up to 2.765 GWh of total capacity, ran until 2019 when the installation of the last battery was conducted. The interested reader is referred to [37] for a rich description of the main outcomes of this CES configuration and a thorough analysis of its functionality and performance.

13.4.2   *Community choice aggregation*

In regions with deregulated energy markets, the community choice aggregation (CCA) programs enable local leaders to aggregate purchasing power and enroll their communities under a single energy provider. By representing a large group of individual customers, the CCAs can participate in wholesale markets and negotiate with generators to pursue common goals such as lowering tariffs, providing an alternative energy supply contract, or allowing consumers greater control of their energy mix, mainly by offering greener generation portfolios than local utilities.

In practice, CCAs are local, not-for-profit, public agencies that aggregate a customer base – usually in a specific city or region – and take on the decision making role about sources of energy for electricity generation. Despite the introduction of a new agent in the process, in a CCA service territory, the incumbent utility continues to own and maintain the transmission and distribution infrastructure, as well as metering and billing. The main interactions between agents in a CCA contract are represented in Figure 13.10.

Once the local officials vote to constitute a CCA, all the electric customers in their jurisdiction are notified and are automatically signed up for an opt-out program. Three different energy programs are often offered to customers: a default program, a program for solar, and a more expensive program advertising use of 100% renewable sources. In this sense, several CCA programs across the United States have reported the interesting combination of significantly higher renewable

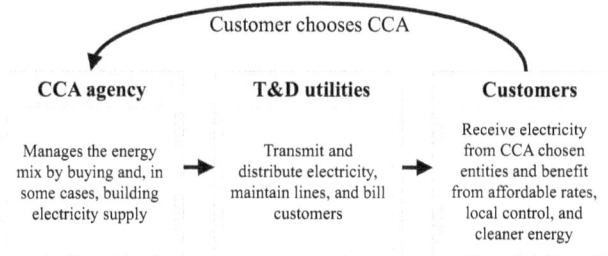

Figure 13.10   Interactions between the local CCA agency, transmission & distribution utilities, and customers

energy portfolios while maintaining rates that are competitive with conventional fossil and nuclear-based utility power.

As of 2022, the state of California – which currently leads the growing momentum of CCAs – holds 23 active programs serving more than 11 million customers, which correspond to more than 51 TWh of load consumed per year and almost 20% of the local market share. Across the state, CCA programs offer electricity from renewable sources ranging from 40% to 100% with a state average of 52%.

For example, consider the second largest CCA in California, i.e. the East Bay Community Energy (EBCE) program, which manages 2.3% of the total yearly state load. The regular EBCE option, with at least 50% electricity from carbon-free generation sources, offers rates 1% lower than PG&E rates. Moreover, the 100% renewable plan option offers electricity sourced exclusively from EBCE-contracted wind and solar plants located in the state with a total capacity of 157.5 MW of wind, 669 MW of solar, and 884 MWh of storage [38]. In total, the 23 CCAs in California account for contracting the capacity of 6.1 GW in solar panels, 1.0 GW in wind turbines, 2.6 GWh in energy storage, and 21 MW in geothermal and bio-mass plants.

13.4.2.1   Grid impact and storage significance

The push for a greater customer choice and greener electricity will certainly affect the renewable energy share of the energy mix in regions experiencing the pro-liferation of CCAs. Some effects are expected even to reverberate to local IOUs as they may lose customers due to the decentralization of the energy market. However, an increased amount of renewables being procured also impose new challenges to the grid, such as dealing with high (variable) renewable penetration, transmission congestion, and resource adequacy. In all cases, energy storage pro-jects may be an essential part of the solution.

Most of the time, power plants are constructed at locations that ensure the highest capacity factor and not necessarily where the load demand is located. Therefore, with the installation and procurement of new large-scale plants to increase the renewable share of the energy portfolio, it could lead to an intensified use of long-distance transmission lines, which in turn increases the prevalence of

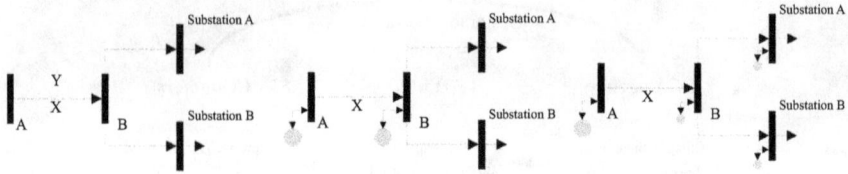

Figure 13.11   Transmission congestion and energy storage

energy losses and transmission congestion. This problem can also be aggravated by the fact that CCAs, despite focusing on regional development, may purchase power from far (even out-of-state) distances and there is evidence that most current CCAs contracts are currently established with new generators [23].

Besides energy storage units providing a myriad of benefits as discussed earlier in this chapter, these resources can also contribute to the alleviation of transmission congestion, even if located at distribution levels. Consider the section of a power system represented in Figure 13.11, where line X between buses A and B is congested. Common transmission alternatives are to construct a new line segment Y or reconduct line X (see Figure 13.11(a)). However, a combination of storage devices at buses A (charging) and B (discharging) would have an identical effect and would have the same impact on the network as the addition of line Y. This is shown in Figure 13.11(b). However, it is important to note that this effect can also be produced by substituting the storage device at bus B operating in a centralized fashion for several existing other storage devices connected to the grid. In theory, this could be implemented using assets (even EVs) at the distribution level – properly paid for the services provided – constituting a VESS. This is represented in Figure 13.11(c).

CCAs, as well as any other load-serving electricity utility, are subject to resource adequacy requirements to ensure a safe and reliable operation by the local system operator. Some of these requirements include the obligation to procure capacity to meet the expected peak demand, to procure a minimum capacity in areas with transmission limitations, and to provide flexible resources needed to manage grid stability over the largest ramps experienced by the system. As storage gets increasingly cheaper, CCAs can invest in this resource to meet the grid requirements and help reduce ramping needs and intermittency resulting from renewable energy generation. Once more, these efforts may be achieved by both grid-scale applications and the constitution of a shared network of energy storage units.

13.5   Final remarks

Energy storage is an essential tool for greater renewable energy penetration, both on the transmission grid and the distribution network. Within the distribution level, its presence has been steadily increasing and a variety of applications have emerged, including the enhancement of EV technologies, and local, shared, and aggregated use by residential, commercial, and industrial electricity customers. Additionally, storage provides a number of important benefits and greatly improves the value of distributed generation, as solar peak production does not always coincide with peak load demand. These benefits are wide-ranging and include frequency regulation, voltage support, congestion relief, and transmission and distribution grid upgrade deferral.

With planning, EVs could be used to solve several grid issues. The right policies and rate designs could prevent EVs from amplifying demand ramp curves and even alleviate the problem of oversupply during off-peak hours. Smart EV chargers can also be viewed as a resource to form virtual power plants that can decide when to charge a car depending on market conditions – similar to how demand response works.

In conclusion, the modernization of the current legislature in areas with deregulated electricity markets is paramount to favor the existence of different market models, which can further improve the benefits for customers at the distribution level. The recent launch of CCAs and the wide diversity among them makes it hard to draw general conclusions and predict future trends, and this may become increasingly true as more CCAs are being planned across different regions of the world. There is therefore a need for future research to examine the impacts of these shifting trends, especially related to the significance of energy storage in these models and its contributions to the present and future of the power sector.

References

[1] Kong L, Li C, Jiang J, *et al.* Li-ion battery fire hazards and safety strategies. *Energies.* 2018;11(9):2191.
[2] Tervo E, Agbim K, DeAngelis F, *et al.* An economic analysis of residential photovoltaic systems with lithium ion battery storage in the United States. *Renewable and Sustainable Energy Reviews.* 2018;94:1057–1066.
[3] Shahjalal M, Roy P, Shams T, *et al.* A review on second-life of Li-ion batteries: prospects, challenges, and issues. *Energy.* 2022;241:122881.
[4] Devassy S and Singh B. Performance analysis of solar PV array and battery integrated unified power quality conditioner for microgrid systems. *IEEE Transactions on Industrial Electronics.* 2021;68(5):4027–4035.
[5] Oliveira D, Zambroni de Souza A, and Ribeiro P. Overview of plug-in electric vehicles technologies. In: Rajakaruna S, Shahnia F, Ghosh A, editors. *Plug In Electric Vehicles in Smart Grids.* Singapore: Springer; 2015. p. 1–24.
[6] High Power Charger For Electric Vehicles. In: Commission E, editor, 2019. https://cordis.europa.eu/project/rcn/211472/factsheet/en.
[7] Emadi A, Lee Y, and Rajashekara K. Power electronics and motor drives in electric, hybrid electric, and plug-in hybrid electric vehicles. *IEEE Transactions on Industrial Electronics.* 2008;55(6):2237–2245.
[8] Wirasingha S, Schofield N, and Emadi A. Plug-in hybrid electric vehicle developments in the US: trends, barriers, and economic feasibility. In: *2008 IEEE Vehicle Power and Propulsion Conference*, 2008. p. 1–8.

[9]   El-Refaie, A. M. Motors/generators for traction/propulsion applications: a review. *IEEE Vehicular Technology Magazine.* 2013;8(1):90–99.

[10]   Sovacool B and Hirsh R. Beyond batteries: an examination of the benefits and barriers to plug-in hybrid electric vehicles (PHEVs) and a vehicle-to-grid (V2G) transition. *Energy Policy.* 2009;37(3):1095–1103.

[11]   Clement-Nyns K, Haesen E, and Driesen J. The impact of vehicle-to-grid on the distribution grid. *Electric Power Systems Research.* 2011;81(1):185–192.

[12]   Liu M, MacNamara P, Shorten R, *et al.* Residential electrical vehicle charging strategies: the good, the bad and the ugly. *Journal of Modern Power Systems and Clean Energy.* 2015;3(1):190–202.

[13]   Oliveira D, Zambroni de Souza A, and Delboni L. Optimal plug-in hybrid electric vehicles recharge in distribution power systems. *Electric Power Systems Research.* 2013;98:77–85.

[14]   Van den Bossche P. Electric vehicle charging infrastructure. In: Pistoia G, editor, *Electric and Hybrid Vehicles.* Amsterdam: Elsevier, 2010. p. 517–543.

[15]   Santos P, Zambroni de Souza A, Bonatto B, *et al.* Analysis of solar and wind energy installations at electric vehicle charging stations in a region in Brazil and their impact on pricing using an optimized sale price model. *International Journal of Energy Research.* 2021;45(5):6745–6764.

[16]   Ansari M, Al-Awami A, Sortomme E, *et al.* Coordinated bidding of ancillary services for vehicle-to-grid using fuzzy optimization. *IEEE Transactions on Smart Grid.* 2015;6(1):261–270.

[17]   Ji Z, Huang X, Zhang Z, *et al.* Evaluating the vehicle-to-grid potentials by electric vehicles: a quantitative study in China by 2030. In: *2020 IEEE Power and Energy Society General Meeting*, 2020, vol. 1, no. 1, pp. 1–5.

[18]   Rodrigues Y, Zambroni de Souza A, and Ribeiro P. An inclusive methodology for plug-in electrical vehicle operation with G2V and V2G in smart microgrid environments. *International Journal of Electrical Power and Energy Systems.* 2018;102(1):312–323.

[19]   Sadiq D. Review of Energy Storage Systems in Regenerative Braking Energy Recovery in DC Electrified Urban Railway Systems: Converter Topologies, Control Methods and Future Prospects.

[20]   Abgottspon H, Schumann R, Epiney L, *et al.* Scaling: managing a large number of distributed battery energy storage systems. *Energy Informatics.* 2018;1(20):55–71.

[21]   Yi Z, Xu Y, Gu W, *et al.* Aggregate operation model for numerous small-capacity distributed energy resources considering uncertainty. *IEEE Transactions on Smart Grid.* 2021;12(5):4208–4224.

[22]   Priddle R. *Competition in Electricity Markets.* Paris: International Energy Agency (IEA), 2001. https://iea.org/reports/competition-in-electricity-markets. Accessed Feb. 18, 2022.

[23]   DeShazo J, Gattaciecca J, and Trumbull K. *The Growth in Community Aggregation Choice: Impacts on the California's Grid.* University of

California, Los Angeles, Luskin School of Public Affairs, 2018. https://next10.org/publications. Accessed Feb. 18, 2022.

[24] Mettrick A. *Competition in UK Electricity Markets*. Department for Business, Energy and Industrial Strategy, 2021. https://gov.uk/government/collections/energy-trends. Accessed Feb. 18, 2022.

[25] Koirala B, van Oost E, and van der Windt H. Community energy storage: a responsible innovation towards a sustainable energy system? *Applied Energy*. 2018;231:570–585.

[26] Mair J, Suomalainen K, Eyers D, *et al.* Sizing domestic batteries for load smoothing and peak shaving based on real-world demand data. *Energy and Buildings*. 2021;247:111109.

[27] Barbour E, Parra D, Awwad Z, *et al.* Community energy storage: a smart choice for the smart grid? *Applied Energy*. 2018;212:489–497.

[28] Manbachi M. Impact of distributed energy resource penetrations on smart grid adaptive energy conservation and optimization solutions. In: Zare K and Nojavan S, editors. *Operation of Distributed Energy Resources in Smart Distribution Networks*. London: Academic Press, 2018. p. 101–138.

[29] New York State Energy Research and Development Authority. Affordable Solar Predevelopment and Technical Assistance, 2020. https://nyserda.ny.gov/All-Programs/NY-Sun/Communities-and-Local-Governments/Predevelopment-and-Technical-Assistance. Accessed Mar. 04, 2022.

[30] Oregon Department of Energy. Oregon Solar Plus Storage Rebate Program, 2022. https://oregon.gov/energy/Incentives/Pages/Solar-Storage-Rebate-Program.aspx. Accessed Mar. 04, 2022.

[31] Kurmelovs R. Community batteries: what are they, and how could they help Australian energy consumers? 2021. https://theguardian.com/environment/2021/apr/05/community-batteries-what-are-they-and-how-could-they-help-australian-energy-consumers. Accessed Mar. 04, 2022.

[32] Takata E. Analysis of solar community energy storage for supporting Hawaii's 100% renewable energy goals. *Master's Projects and Capstones*, University of San Francisco 2017;544:1–64.

[33] Fleischhacker A, Auer H, Lettner G, *et al.* Sharing solar PV and energy storage in apartment buildings: resource allocation and pricing. *IEEE Transactions on Smart Grid*. 2019;10(4):3963–3973.

[34] Stockholmshem Housing Association. Safe Energy Storage for Multi-Tenant Apartments, 2019. https://nilar.com/wp-content/uploads/2019/05/Customer-case-Stockholmshem-EN.pdf. Accessed Mar. 04, 2022.

[35] Koirala B, Koliou E, Friege J, *et al.* Energetic communities for community energy: a review of key issues and trends shaping integrated community energy systems. *Renewable and Sustainable Energy Reviews*. 2016;56:722–744.

[36] Archer R. Sonnen Launches SonnenCommunity® Solar Battery Program for N.Y. Residents, 2021. https://cepro.com/control/energy-power/sonnen-launches-sonnencommunity-solar-battery-program-for-n-y-residents/. Accessed Mar. 04, 2022.

[37]   AGL Energy. Virtual Power Plant in South Australia. Australian Renewable Energy Agency, 2020. https://arena.gov.au/assets/2020/10/virtual-power-plant-in-south-australia.pdf. Accessed Mar. 04, 2022.

[38]   Newbold A. EBCE Launches First-of-its-kind Home Battery Backup Program, 2020. https://ebce.org/news-and-events/ebce-launches-first-of-its-kind-home-battery-backup-program/. Accessed Mar. 24, 2022.

Index

fault detection algorithm 254
feasible region 224–6, 245
Federal Energy Regulatory
    Commission (FERC) 76, 154
feedback linearization 111–12
feed-in tariff (FIT) 75
fixed power thresholds 208
fixed time operation mode 209
flexibility market 285–7
flywheel 301
forecast model 100–1
fossil fuels 18
frequency containment reserve (FCR)
    267
frequency control service (FCAS) 74
frequency regulation 54–5, 105–6, 160
frequency/voltage regulation 46
front-of-the-meter (FTM) 41

Gauss–Seidel method 239
genetic algorithms 21
geopolitical conflicts 66
green energy technology 82
greenhouse gas emissions 28, 87
green tariff 211
grid-connected electric vehicles 168
grid connection model 110
    feedback linearization 111–12
    grid forming mode 113–16
    PLL synchronization 112–13
grid-feeding converters 110
grid forming mode 113
    frequency and inertial support
        115–16
    voltage and current control loop
        113–15
grid operators 162
grid regulation 29
grid resilience 249

grid-scale ES 45
    planning objectives and
        constraints 46
grid-to-vehicle (G2V) 297–9
Gurobi solver 231

harmonic phasor (HP) 178
harmonic phasor estimation (HPE)
    168, 188–9
harmonics 137–8
Heuristic convex hull algorithm 232–4
Heuristic moment machine (HMM) 48
high impact low-probability
    (HILP) 44
high-voltage transmission network 15
home energy management system
    (HEMS) 49
HomePlug Green Phy Standard 140
hosting capacity 158, 223–4
hosting capacity region 224
    assessment 227
        distribution grid model 227–32
        linearized DistFlow model 232–8
        relaxed DistFlow model 238–42
    coordinate shifting of 244
    with ESS 226–7
    grid power delivery potential
        exploitation 225–6
    grid problem format transformation
        226
    reshaping with storage 242
        case study 244–5
        general optimal storage capacity
            design 244
        minimal energy storage capacity
            quantification 242–4
household ES 45
hybrid energy storage system (HESS)
    94, 97–9, 116
hydrogen 281

www.ingramcontent.com/pod-product-compliance
Lightning Source LLC
Chambersburg PA
CBHW050508190326
41458CB00005B/1477